GEOGRAPHIES OF DEVELOPMENT

Geographies of Development

ROBERT B. POTTER, TONY BINNS,
JENNIFER A. ELLIOTT AND DAVID SMITH

LONGMAN

Addison Wesley Longman Limited
Edinburgh Gate
Harlow
Essex CM20 2JE
England
and Associated Companies throughout the world

Published in the United States of America
by Addison Wesley Longman Inc., New York

Visit Addison Wesley Longman on the world wide web at:
http://www.awl-he.com

First published 1999

ISBN 0 582 29825 3

British Library Cataloguing-in-Publication Data
A catalogue record for this book is available from the British Library

Library of Congress Cataloging-in-Publication Data
A catalog record for this book is available from the Library of Congress

Typeset by 35 in 9.5/11 pt Times
Produced by Addison Wesley Longman Singapore (Pte) Ltd.,
Printed in Singapore

Contents

List of boxes

Preface

We came together as a team of authors convinced of the pressing need for a new approach to teaching and learning in the fields of development geography and development studies. This perceived need is fully articulated in the introduction. The team was assembled after the publisher approached the first author, with an invitation to prepare an up-to-date textbook for university students reading for courses on geography and development.

The authors, therefore, would like to thank Sally Wilkinson, who was initially involved with the commissioning of the volume, and in particular, Matthew Smith, who subsequently took on the task of overseeing the project as it evolved. Throughout the period of preparation and writing, the publishers were enthusiastic, facilitative and extremely positive; we are delighted to have had the chance to work with them in such a purposeful and constructive manner. We hope that, like us, they will consider all their efforts to have been worthwhile. We are very grateful to Shuet-Kei Cheung for her hard work throughout the production process.

The point is made at the end of the introduction that teaching about places, cultures and environments other than those in which we were born, or have lived for considerable periods of time, carries special responsibilities and challenges, albeit ones that must be met. All four authors have considerable experience of undertaking first-hand research in developing areas, particularly those of north, west and southern Africa, Asia and the Pacific, and the Caribbean and parts of South America. Further, in our different ways, we have also been involved in the production of textbooks for university students and sixth-formers dealing with a variety of development-oriented topics. But it has been a particularly exciting task to work together in producing a comprehensive overview text such as *Geographies of Development*.

Thanks are due to Kathy Roberts of the Department of Geography at Royal Holloway, University of London, for so ably assisting with the preparation of the preliminaries and the like, along with the management and formatting of the computer disks. She also helped to ensure that the authors maintained good communications during a very hectic period.

Rob Potter, Tony Binns,
Jenny Elliott and David Smith
March 1998

Acknowledgements

We are grateful to the following for permission to reproduce copyright material;

Blackwell Publishers for figures 1.1, 2.1 and 3.6 from P.W. Preston's *Development Theory: An Introduction* (Blackwell: Oxford, 1996); Routledge for figure 1.2 from K. Wellard and J. Copestake *Nongovernmental Organizations and the State in Africa* (Routledge: London, 1993); Addison Wesley Longman for figure 2.2 from Peter Taylor's *Political Geography* (Addison Wesley Longman: Harlow, 1985); Routledge for the figure in box 2.3 from ILL Griffiths *The Atlas of African Affairs* (Routledge: London, 1994); Penguin Books for figure in box 3.1 from P. Gould and R. White *Mental Maps* (Penguin: Harmondsworth, 1974); Private Eye for figure 3.8b; Oxford University Press for figures 3.9, 3.13, 9.5 and 9.6 from R.B. Potter's *Urbanization in the Third World* (OUP: Oxford, 1992); Mr. Paul Fitzgerald (Polyp) for figure 3.12; Cambridge University Press for figure 4.3 from D.J. Keeling in P.L. Knox and P.J. Taylor *World Cities in a World System* (CUP: Cambridge, 1995); Paul Chapman for figures 4.4, 4.5 and 4.6 from P. Dicken's *Global Shift (3e)* (Paul Chapman: London, 1998); Yaffa Advertising and King Features for figure 4.8; Oxford University Press for figure 4.10 from J. Allen and C. Hamnett *A Shrinking World* (OUP: Oxford, 1995); *The Independent* for figure 5.3; Pluto Press for figure 6.5 and table 2 in box 6.1 from N. Middleton, P. O'Keefe and S. Moyo *Tears of the Crocodile* (Pluto Press: London, 1993); Mr. David Hughes for figure 6.9; Blackwell Publishers for figure 6.11 from A.J. Reading *et al. Humid Tropical Environments* (Blackwell: Oxford, 1995); Oxfam for the figure in box 8.1 from *Behind the Weather: Lessons to be Learned. Drought and Famine in Ethiopia* (Oxfam: Oxford, 1984); So Chin-Hung for the two maps in box 8.2 from So Chin-Hung *Economic Development, State Control and Labour Migration of Women in China* (unpublished PhD thesis: University of Sussex, Brighton, 1997); Oxford University Press for figure 9.2 from J. Gugler's *The Urban Transformation of the Developing World* (OUP: Oxford, 1996); Methuen for figure 9.13 from A Gupta's *Ecology and Development in the Third World* (Methuen: London, 1988); Routledge for figure 10.1 from J. Dickenson *et al. A Geography of the Third World (2e)* (Routledge: London, 1996); John Wiley & Sons for figure 10.2 from T. Binns' *People and Environment in Africa* (John Wiley & Sons: Chichester, 1995).

Still Pictures for plates 4.1, 4.3, 4.4, 4.6, 6.3, 7.1, 8.5, 10.4 and 10.6; Panos Pictures for plates 3.2, 3.3, 3.6, 4.7, 6.1, 6.2, 7.2, 8.2, 10.1, 10.2

and 10.7; The Surveyor's Office of the government of Zimbabwe for plate 10.3.

We were unable to trace the copyright holder for figure 3.14 (the J.F. Batellier cartoon) and plate 6.4.

Whilst every effort was made to trace copyright holders, in a few cases this has proved impossible and we would like to take this opportunity to apologise to any copyright holders whose rights we may have unwittingly infringed.

Introduction

The intention of *Geographies of Development* is to provide a comprehensive introductory textbook for students, primarily those taking courses in the fields of development geography and the interdisciplinary area of development studies. Although the text is mainly aimed at the undergraduate market, specifically second-year students who are taking whole course units on development, given the global importance of the subject-matter dealt with, the book should be just as appropriate for first years taking broader courses, along with those reading for more specialist options in the third year of their degree programmes, or indeed those at the taught masters level.

Given this overall remit, the distinctive aim of *Geographies of Development* is to move away from what has become the traditional structure of development and geography texts, which all too frequently start with definitions of the Third World and colonialism, and then proceed to consider, step by step, topics such as population and demography, agriculture and rural landscapes, mining, manufacturing, transport, urbanisation, development planning, and so on. Having provided detailed accounts on such topics, many texts terminate at this juncture, but those which have provided a broader picture generally go on to present a selection of country-based case studies.

Geographies of Development endeavours to break this mould of development-oriented textbooks in a manner which reflects the rapidly changing concerns of development itself. In this sense, its *raison d'être* is to provide a text for learning and teaching about development that will take students into the twenty-first century and beyond. Reflecting this, the book is divided into three broad and relatively equal parts (see figure overleaf), dealing respectively with theories of development (Part I), development in practice (Part II) and the spaces of development (Part III).

Part I provides a detailed overview of the ideas, concepts and ideologies which have underpinned writings about the nature of development as well as pragmatic attempts to promote development in the global arena. It gives detailed consideration to important topics such as histories, meanings and strategies of development, the emergence of the Third World, the nature of imperialism and colonialism and its various stages of mercantile, industrial and late colonialism, together with key concepts such as the new international division of labour and the new international economic order. Part I also provides thorough reviews of relevant and related topics such as modernity, enlightenment thinking, the possible relevance of post-modernity to the Third World, antidevelopmentalism, the socialist

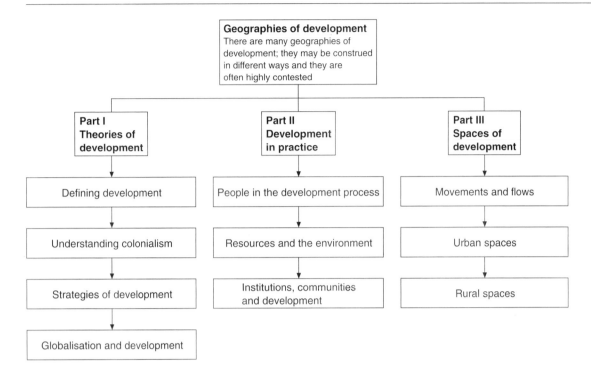

```
┌─────────────────────────────┐
│ Geographies of development  │
│ There are many geographies of│
│ development; they may be     │
│ construed in different ways  │
│ and they are often highly    │
│ contested                    │
└─────────────────────────────┘
```

Part I Theories of development	Part II Development in practice	Part III Spaces of development
Defining development	People in the development process	Movements and flows
Understanding colonialism	Resources and the environment	Urban spaces
Strategies of development	Institutions, communities and development	Rural spaces
Globalisation and development		

Third World, responsibility to distant others, globalisation, global shifts and time–space convergence.

Just like the other parts of the book, these early chapters exemplify the title and the overarching theme of the volume. Part I makes it clear that ideas concerning development have been many and varied, and have been highly contested through time. Thus, definitions of, and approaches to, development have varied from place to place, from time to time, from country to country, region to region and group to group within the general populace. It is essentially this plural nature of development that *Geographies of Development* seeks both to exemplify and to illustrate. Furthermore, this part of the book demonstrates that current global processes are not leading to the homogenisation of the world's regions. Far from it, the evidence shows all too clearly that contemporary global processes are leading to increasing differences between places and regions, and thus to the generation of unequal patterns of development and change. Hence the emphasis is on multiple geograph*ies* of development.

Part II covers what may be regarded as the basic components of the development equation – people, environments, resources, institutions and communities – together with the complex and multifaceted interconnections which exist between them. The inclusion of a chapter specifically dealing with institutions and communities as the primary decision makers involved in the development process, serves to exemplify the utility of the overall approach adopted in *Geographies of Development*. The decision makers considered in this account extend from the agents of global governance – the United Nations, World Bank, International Monetary Fund and World Trade Organisation – via the country level, involving the role of the state, down to community participation and the empowerment of the individual, embracing non-governmental and community-based organisations. This account serves to stress the plurality of decision makers affecting geographies of development, just as the detailed expositions on population, resources, environment and development exemplify the diversity of views on salient topics of the moment. These include the character of sustainable development, people–resource relations, the concept of limits to growth, the environmental impacts of development, biodiversity loss, land degradation, pollution and global warming, health, education and human rights.

Part III focuses on what development means in relation to particular places and people. This is achieved by consideration of the flows and movements that occur between geographically separate locales, and in terms of the distinctive issues raised by development and change in both urban and rural areas. Once again, notwithstanding the difference in focus, the theme is the diversity and complexities of the movements and flows of people, capital and innovations, along with the diverse realities of transport and communications. Pressing topics of current import, such as international tourism, the realities of world trade and the debt crisis, receive detailed attention in this part of *Geographies of Development*. The nature and scale of urbanisation in the contemporary Third World, evolving urban systems and the incidence of unequal development, the need for urban and regional planning, the salience of basic needs and human rights and the quest for sustainable cities in relation to the brown agenda are the prominent topics reviewed in relation to urban spaces and development imperatives. Rural spaces are analysed with particular reference to diverse rural livelihood systems, plus the examination of the multiple meanings and outcomes of approaches to rural development, such as land reform, the green revolution, irrigation and the promotion of non-farm activities. Forming the last major part of the book, these chapters draw heavily on earlier accounts presented in Parts I and II, and they make frequent reference to the realities of globalisation, convergence, divergence, urban bias, industrialisation and sustainable development, and other topics.

The thematic structure and orientation of *Geographies of Development* means that important contemporary development issues such as postmodernity, globalisation, gender and development, structural adjustment, environmental degradation, human rights, basic needs, empowerment are not dealt with in distinct chapters, but are treated as appropriate at various points in the text, and from a variety of different perspectives. This approach reflects the complexity of these issues in the context of multiple geographies of development. A case in point is the relationships between tourism and development. This is first covered in detail in Part I in considering processes of globalisation. The issue of tourism is returned to in Part II, where Chapter 6 deals with resources in relation to the environmental and social costs of tourism as an explicit development policy. Finally, international tourism reappears when global movements and flows are considered in Part III.

Geographies of Development focuses on the processes that are leading to change, whether for better or worse. In this sense, the book follows Brookfield's (1975) simple and straightforward definition of development as change, whether positive or negative. Thus, although the primary remit of the volume is the so-called Third World, the focus of the book is very much on development as change, regardless of where or how it is occurring. We can take the case of tourism once more, and the first major example of the use of mass tourism as a strategy of development is provided by Spain, a European colonial power. This is presented as an example of the early stage of internationalisation in relation to development in Part I.

Every effort has been made in *Geographies of Development* to provide clear and cogent examples of the issues under discussion, in the form of diagrams, maps, tables and photographs. In addition, boxed case studies and examples are presented throughout the chapters. These seek either to extend definitions of basic concepts or to provide detailed illustrations of the generic topics under consideration. These are as diverse as Aboriginal town camps in Alice Springs, Australia, globalisation and the production of athletic footwear, golf-course development in the Far East, the environmental impacts of tourism, and China's one-child population policy.

In Part I of *Geographies of Development*, the nature and definition of the Third World is the subject of detailed discussion. It can generally be argued that the term *Third World* still serves to link those countries which, all bar a very few, are characterised by a colonial past and relative poverty in the current global context. It is for this reason that the term *Third World* is employed as a shorthand collective noun for what must be appreciated as a diverse set of nations. When referring to nations and areas which make up this broadly defined category, the expressions 'developing countries' and 'developing areas' are used interchangeably, as befits the overarching title *Geographies of Development*. But the title of the book implicitly recognises that these specific terms can be used just as readily in relation to the former socialist states of Eastern Europe, to southern Spain, or indeed, Aboriginal Australia.

Teaching, learning and researching about territories other than the ones in which we live, and of which we have direct experience, are demanding, but vitally important tasks (Potter and Unwin, 1988; 1992; Unwin and Potter, 1992). Without them, there would not be an international view in what is clearly becoming an increasingly interdependent, yet highly unequal world. It is the ultimate aim of *Geographies of Development* to assist students and teachers alike in structuring their observations and discussions of the multiple meanings of development in the increasingly complex and interdependent contemporary world.

PART I   Theories of development

CHAPTER 1

Defining development

Introduction

Development is a word that is currently ubiquitous within the English language. Many firms have research and development divisions in which the evolution of new products, from running shoes to car exhausts, is given high priority. At the level of the state, national development plans are formulated to guide the development process. Indeed, one of the main divisions of the globe is between developed countries and developing countries; the developed countries assist the developing countries by means of development aid. But what do we mean by *development*? Do individuals, firms, states and global institutions all have the same interpretation (Table 1.1)? Most development processes are influenced by development planning, and most plans are in turn shaped by development theories which ultimately should reflect the way in which development is perceived; in other words, the ideology of development. But the development process is affected by many factors other than ideologies (Tordoff, 1992), although ideologies often condition state and institutional reactions to these.

This initial chapter is about the way various actors in the development process think about development;

how they seek to define it, determine its components and conceptualise the purpose of development. For the most part, development here is discussed in relation to developing countries, but development relates to all parts of the world at every level, from the individual to the global. However, development has become most often linked with the Third World, itself a value-laden term, the emergence of which has been closely associated with the rapid evolution of the concept of development in the second half of the twentieth century. The second part of this chapter will therefore examine the emergence, use and persistence of what appears to be an outmoded terminology, and will associate this with thinking about development itself. The chapter will conclude with a brief discussion of the changing relationship between geography and development.

Thinking about development

Histories of development

Most development and antidevelopment theorists (P.W. Preston, 1996; Sachs, 1992; Escobar, 1995)

Table 1.1 Alternative interpretations of development.

Good	Bad
Development brings economic growth	Development is a dependent and subordinate process
Development brings overall national progress	Development is a process creating and widening spatial inequalities
Development brings modernisation along Western lines	Development undermines local cultures and values
Development improves the provision of basic needs	Development perpetuates poverty and poor working and living conditions
Development can help create sustainable growth	Development is often environmentally unsustainable
Development brings improved governance	Development infringes human rights and undermines democracy

Source: Adapted from Rigg (1996).

situate the origins of the modern development process in the late 1940s, more precisely in a speech by President Truman in 1949 in which he employed the term *underdeveloped areas* to describe what was soon to be known as the Third World. Truman also set out what he saw as the duty of the west to bring 'development' to these countries. Effectively he was establishing a new colonial, or neocolonial, role for the United States within the newly independent countries that were emerging from the decolonisation process. Truman was encouraging the underdeveloped nations to recognise their condition and to turn to the United States for assistance. It is undoubtedly true that the genesis of much modern(ist) development theory and practice lay in the decade between 1945 and 1955. Modernism is the belief that development is all about transforming 'traditional' countries into modern, Westernised nations. For many Western governments, particularly former colonial powers, such views represented a continuation of the late colonial mission to develop colonial peoples within the concept of trusteeship (M. Cowan and Shenton, 1995; Chapter 2). There was little recognition that many traditional societies might have been content with their ways of life. Indeed, development strategists often had to persuade them otherwise. Rigg (1996: 33) cites the American advisers to the Thai government of the 1950s as trying to prevent the monks from preaching the virtues of contentedness, which was seen as retarding modernisation.

Many other writers, however, recognise that the teleological origins of development lay in the rationalism and humanism of the eighteenth and nineteenth centuries respectively. During this period, the simple definition of development as change became transformed into a more directed and logical form of evolution. During the Enlightenment it was believed that by applying rational, scientific thought to the world, change could become more ordered, predictable and valuable. Those who could not adapt to such views became thought of as 'traditional' and 'backward'. Thus, the Australian Aborigines were denied any rights to the land they occupied by the invading British in 1788, because they did not organise and farm it in a systematic, rational manner.

In the nineteenth century, Darwinism began to associate development with evolution, a change towards something more appropriate for survival (Esteva, 1992). When combined with the rationality of the Enlightenment thinking, the result became a more narrow but 'correct' way of development, one based on Western social theory. During the Industrial Revolution, this became heavily economic in its nature, but by the late

nineteenth century, a clear distinction seems to have emerged between the notion of 'progress', which was held to be typified by the unregulated chaos of pure capitalist industrialisation, and the 'development' which was representative of Christian order, modernisation and responsibility (M.P. Cowan and Shenton, 1996; P.W. Preston, 1996).

It is this latter notion of development which, as Chapter 2 discusses, began to permeate the colonial mission from the 1920s onwards, firmly equating development in these lands with an ordered progress towards a set of standards laid down by the West; or as Esteva views it, 'robbing people of different cultures of the opportunity to define the terms of their social life' (Esteva, 1992: 9). Little recognition was given to the fact that 'traditional' societies had always been responsive to new and more productive types of development; had they not done so, they would not have survived. Furthermore, the continued economic exploitation of the colonies made it virtually impossible for such development towards Western standards and values to be achieved. In this sense underdevelopment was the creation of development, an argument that is considered in several of the following chapters.

Conventional development

Chapter 3 discusses in detail the theories and strategies by which development is portrayed as a materialist process of change. This section overviews the process and its impact on the revision of 'development' as a set of ideas about change. President Truman, in his speech of 1949, notes how the underdeveloped world's poverty is 'a handicap and threat both to them and more prosperous areas . . . greater production is the key to prosperity and peace. And the key to greater production is a wider and more vigorous application of modern scientific and technical knowledge' (Porter, 1995). Enlightenment values were thus combined with nineteenth-century humanism to justify the new trusteeship of the neocolonial mission, a mission that was to be accomplished by 'authoritative intervention', primarily through advice and aid programmes (P.W. Preston, 1996); see Figure 1.1. Clearly the 'modern notion of development' (Corbridge, 1995: 1) had long and well-established antecedents.

It is therefore perhaps not too surprising that, in its earliest manifestation in the 1950s, development became synonymous with economic growth. One of the principal 'gurus' of this approach, Arthur Lewis, was uncompromising in his interpretation of the

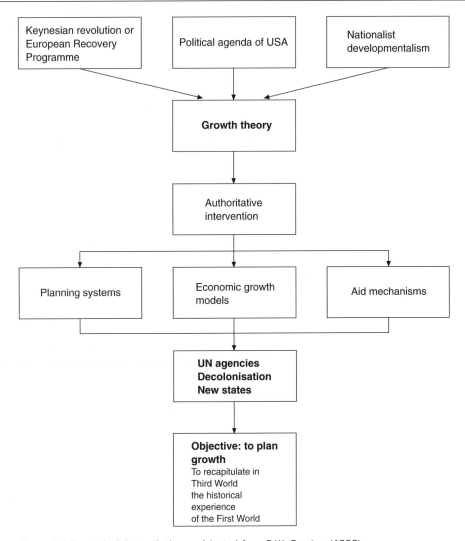

Figure 1.1 Post-colonial growth theory. Adapted from P.W. Preston (1996).

modernising mission, 'it should be noted that our subject matter is growth, and not distribution' (Esteva, 1992: 12). During the second half of the twentieth century, therefore, the development debate has been dominated by economists. This is not to say that other aspects of development have not contributed, often crucially, to this debate, particularly sociologists and geographers in the field of the social and spatial unevenness of development, but the dominant influence in both theory and practice has been economics.

The prominence and influence of development economics in the 1950s and 1960s have clear repercussions on other terminologies related to development, most notably the way in which underdeveloped countries were identified and described (see below). The earliest, and for many, still the most convenient way of quantifying underdevelopment has been through the level of gross national product (GNP) per capita. As Michael Watts (1996) notes, this is still a principal way in which the poverty of the Third World and the failure of development are blandly laid out in the statistical sections of World Bank and United Nations development reports. Some preferred a classification linked to resource potential, the bases of which were equally shaky. For example, in the 1950s, the eminent economist Myint (1964) argued that small, overpopulated states – a category in which he included Hong Kong, Singapore, South Korea and

Taiwan – faced the direst prospect for the future; little did he know.

To be sure, the notion of development as growth has broadened over the years to incorporate social indicators and political freedoms (Box 1.1), but the most recent of the composite indicators which enables international comparisons to be made, the Human Development Index of the United Nations, is still underpinned by economic growth and is expressed as a single statistical measure. In similar vein, the model for economic development is still seen to be that of the capitalist West, as this quote from Nobel prizewinner Douglas North (Mehmet, 1995: 12) seems to confirm:

> The modern western world provides abundant evidence of markets that work and even approximate the neo-classical ideal. . . . Third World countries are poor because the institutional constraints define a set of payrolls to political/economic activity that do not encourage productive activity.

Critiques of development

Criticism of development as conceptualised and practised in the ways described above has been continuous since the 1960s, and has clearly influenced theory and strategy (Chapter 3). Even the narrow focus on economic growth failed to produce a convergence of income indicators. Indeed, there is evidence that in-

equality between and within countries has increased substantially (Griffin, 1980). Trickle-down economics had not worked and the call came for a more diversified and broader interpretation of development (Dwyer, 1977). Explanations were sought and offered for the failure of the modernisation project (Brookfield, 1978) and new strategies were devised; but in most of this discussion, development as a linear and universal process was seldom questioned or addressed. What was debated was the variable and erratic nature of development, and explanations were sought in relation to both its chronological and spatial unevenness.

For some critics the answer was straightforward; it was, and still is, the Eurocentricity (European orientation) of economic development theories which had distorted development, especially through their pseudo-scientific rationale (Table 1.2). Mehmet (1995), in particular, has been virulent in his criticism:

> As a logical system, Western economics is a closed system . . . in which assumptions are substituted for reality, and gender, environment and the Third World are all equally dismissed as irrelevant. . . . [However], mainstream economics is neither value-free nor tolerant of non-western cultures.

Of course, Eurocentrism is a criticism that can be levied at more than mainstream economics and its associated modernisation strategy (T.G. McGee, 1995; Hettne, 1990). Indeed, as Chapter 3 indicates, almost

---

BOX 1.1   Measuring development

By the end of the first UN Development Decade, not only was concern arising over the interpretation of development as economic growth, there was also considerable criticism of GNP per capita (total domestic and foreign value added divided by total population) as *the* indicator of such growth, particularly since it gives no indication of the distribution of national wealth. Nevertheless, as Seers (1972: 34) points out, to argue that GNP per capita is an inappropriate measure of a nation's development is to weaken the significance of the growing GNP per capita gap between rich and poor nations. In other words, the serious criticisms that one can make of development statistics do not deny them some use in the analysis of the development process, particularly its unevenness.

Seers himself, with his egalitarian leaning, suggested the use of three criteria to measure comparative development: poverty, unemployment and in-

equality. He accepted that statistical difficulties were considerable but argued that they produced data that were no less reliable than GNP per capita, and were a far better reflection of the distribution of the benefits of growth. Although Seers considered them to be economic criteria, they clearly contain social dimensions; indeed, Seers suggested social surrogates for their measurement. The 1970s and 1980s were conspicuous for the appearance of a whole series of social indicators of development, such as those related to health, education or nutrition, which were produced either as tables attached to major annual reviews, such as the World Bank's annual Development Report, or less frequently as maps which accompanied attempts to identify the developing world *per se*. Eventually these social indicators were broadened further still to incorporate measures of gender inequality, environmental quality and political and human rights.

Box 1.1 continued

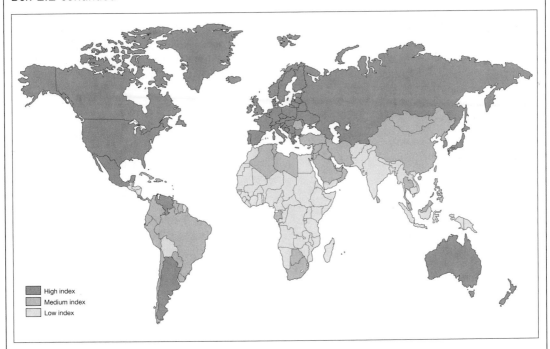

The 1990 UNDP Human Development Index.

As with all statistical measures, these data are open to a variety of criticisms, some technical, some interpretational. For example, how does one measure human rights when cultural interpretations are not consistent, as the recent East–West dispute has indicated (Drakakis-Smith, 1997b). Moreover, by the late 1980s a plethora of economic, social and other indicators were being produced on an annual basis that were not always consistent with one another and could be manipulated to show that some 'development' had occurred almost anywhere. The consequence was, not surprisingly, that as indicators multiplied, so there emerged a renewed enthusiasm for the single composite measure. Such measures did not always produce results which matched the GNP-based categories of development that have graced the pages of the *World Bank Development Report* for so long. In Richard Estes' (1984) Index of Social Progress, the United States was ranked well below countries such as Cuba, Colombia and Romania. As usual, one can always prove a point with statistics. Other measures were even more complex in an effort to be all-embracing. Tata and Schultz (1988) constructed a human welfare index

from ten variables using factor analysis. The final scores, however, were more or less arbitrarily divided into three sets, producing a table and map little different from those of the three worlds currently in vogue.

Single measures, usually in conjunction with multiple tables of individual indicators, are still popular as an easily digestible summary of world development trends. One of the most widely used is the United Nations Human Development Index (see figure), currently calculated from data on life expectancy, literacy, income, environmental quality and political freedom. As Esteva (1992) notes, human development is thus translated into a linear process indicated by measuring levels of deprivation, or how far countries depart from the Western ideal. Moreover, if one chooses other similar variables in the same categories, quite different overall indices can be obtained. And yet we cannot dispose of development indicators too readily, for above all, they indicate trends over time and even the antidevelopment critics use collated statistics of this nature in order to consolidate their starting point that 'development' has been a myth. Indeed, Ronald Horvath (1988) has conceded that he was 'measuring a metaphor'.

Table 1.2 Eurocentricity: some principal points of criticism.

Denigration of other people and places
Ideological biases
Lack of sensitivity to cultural variation
Setting of ethical norms
Stereotyping of other people and places
Tendency towards deterministic formulations
Tendency towards empiricism in analysis
Tendency towards male-orientation (sexism)
Tendency towards reductionism
Tendency towards the building of grand theories
Underlying tones of racial superiority
Unilinearity
Universalism

all of the major strategies for development have been Eurocentric in origin and in bias, from modernisation through neo-Marxist to the neoliberal 'counter-revolution' of the 1980s. Moreover, it can be argued that all equate 'development' with capitalism (J. Harriss and Harriss, 1979). Certainly all were universalist in their assumptions that development is a big issue that needs to be understood through grand theory or, so-called metanarratives. It is not surprising that such arrogant approaches began to be criticised, approaches in which the developed nations devised the parameters of development, set the objectives and shaped the strategies. Not only was this 'Westernised' development not working for most Third World countries, but the West itself was continuing to be the beneficiary of the distorted development that it produced. Since 1960, the start of the first United Nations Development Decade, disparities of global wealth distribution have doubled, so that by the mid 1990s, the wealthiest quintile of the world's population controlled 83 per cent of global income, compared to less than 2 per cent for the lowest quintile (UNDP, 1996). However, despite the extensive criticism that began to appear, we must also recognise the fact that some societies were able to absorb selectively from this imposed development to their own advantage; the Asian industrialising societies provided ample evidence of this.

Two principal sets of voices began to be heard in the widespread criticism of the general situation. The first was characterised by a stance which recommended greater input into defining development and its problems from those most affected by it – 'development from below', as Stöhr and Taylor (1981) expressed it, or 'putting the last first' as Robert Chambers (1983)

memorably termed it. The second set of views exhibited similar values but its supporters were not prepared to work within what they regarded as an unfair and heavily manipulated dialogue of development in which the West, through the medium of international development agencies and 'national governments', arrogates to itself the ability to speak and write with authority about development (Corbridge, 1995: 9). This group has become known, therefore, for its 'antidevelopment' stance; perhaps a somewhat misleading description as we will see later. Some of the values of this group cut across the opinions of what might be termed postmodern development, with its repudiation of metatheory and its embracing of meso- or microapproaches to development problems, which would include gender and environmental issues. Stated simply, post-modern development is development which rejects the tenets of modernity and the Enlightenment, a theme considered in this chapter as well as Chapters 3 and 4.

Alternative or 'other' developments, which have been much discussed in the 1990s, are not necessarily recent phenomena. Even in the 1960s there were reactions against the idea that development could be narrowly defined and superimposed upon a variety of situations across the Third World. More locally oriented views began to emerge; for example, in the Dag Hammarskjöld Foundation's concept of 'another development', one which was more human-centred. These approaches were soon co-opted into official development policies, underpinning the 'basic needs' strategies of the 1970s, which fragmented the monolithic targets for development into what were seen as more locally and socially oriented goals. Unfortunately, these worthy objectives relating to shelter, education or health, not only competed with one another for funding, but were also tackled with the same universalist solutions that had bedevilled earlier development strategies.

However, the concept of locally oriented, endogenous development was by now firmly established and was given a considerable boost by Robert Chambers (1983) with his 'development from below' philosophy. Although initially discussed with reference to peasants and rural development in general, the philosophy of community participation has been widely adopted as an interpretation of development that is people-oriented (Plate 1.1) (J. Friedmann, 1992). However, as Munslow and Ekoko (1995) have suggested, empowerment of the poor has been stronger on rhetoric than in reality; in particular, the widening of political participation has been very slow compared to the improvement in social and economic rights that

Plate 1.1 Rural hawker and child in Guyana: ultimately development is about improving the life chances of people (photo: Rob Potter)

has occurred. Nevertheless, facilitating 'people's participation' now has a place on the agenda of the major development institutions; see the extensive review of participation and democracy in UNDP (1993).

A major facilitator in this process of empowering the poor through participation in the development process have been non-governmental organisations (NGOs); see Chapter 7. This blanket term covers a wide variety of community-based organisations (CBOs); the largest have operating budgets greater than those of some developing countries, whereas the smallest struggle on with little official encouragement or funding, blending almost imperceptibly with social movements (Figure 1.2, page 10). The role of NGOs has been scrutinised intensively over recent years, with some seeking to promote linkages away from the purely local, community-level projects and become involved with more comprehensive larger-scale planning, building stronger bridges with the state (Korten, 1990).

Others, however, already see NGOs, particularly the larger, Western NGOs, as extensions of the state, helping to maintain existing power relations and legitimising the political system (Botes, 1996). Indeed, within the changes that have accompanied structural adjustment, NGOs may be seen as facilitators in the process of the privatisation of welfare functions, freeing the state from its social obligations within development.

Many other criticisms have been levelled at NGOs and the role of outsiders in community participation; most of them have been lucidly summarised by Botes (1996). These cover the paternalistic actions of development experts who see their role as transferring knowledge to those who know less, disempowering them in the process; selective participation of local partners, often bypassing the less articulate or visible groups; favouring 'hard' issues, such as technological matters, over the more difficult and time-consuming 'soft' issues, such as decision-making procedures or community involvement; promoting 'gatekeeping' by local elites; and, particularly important, accentuating product at the expense of process (Figure 1.3 overleaf). Development from below *can* be qualitatively different from conventional development as envisaged by modernisationists but this process must be realistic rather than romantic in its praxis. Societies, even at the local scale, can be heterogeneous, divided and fractious; and grassroots development, keen to encourage participatory development, must take this into account. Of course, neoliberal development strategists would argue that their recommendations encourage empowerment of individuals through greater freedom of choice within an open market economy. This approach is criticised by Munslow and Ekoko (1995: 175) as the 'fallacy of empowerment' and 'mirage of power to the people'. In reality, they argue, 'participatory democracy is really about a transfer of power and resources, [if] not to people directly, [then] to NGOs and other representatives at grassroots level'. This is a theme taken up by another group of developmental thinkers, the antidevelopmentalists.

There is considerable overlap between populist interpretations of development and the antidevelopmentalists who have emerged in recent years to challenge the notion of development as a whole, although as Corbridge (1995) argues, there are long antecedents to anti (Western) developmentalism stretching back to the nineteenth century. It is also claimed that the failures of neo-Marxism 'to provide practical assistance to those on the front lines of development' (Watts and McCarthy, 1997: 79) have turned disillusioned radicals towards the antidevelopmentalists (D. Booth, 1993).

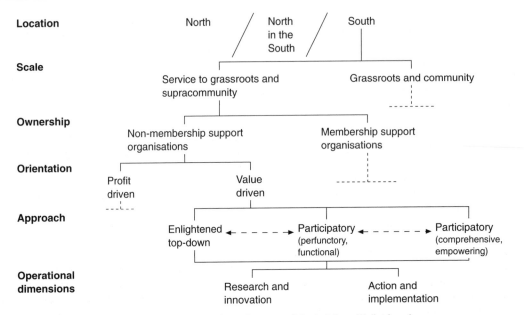

Figure 1.2 Non-governmental organisations: how they vary. Adapted from Wellard and Copestake (1993).

Process versus product	Decision-making dynamics	Underlying assumptions	Emphasis
Process less important than product	Developer-centred approach: characterised by top-down decisions taken by development elite	Rely on formal know-how and expertise to resolve development problems in the shortest possible time	Time and product
Process more important than product	People-centred approach: characterised by bottom-up decisions taken by community members or their legitimate leaders	The immediate resolution of a development problem is less important than the way in which the process of problem solving is taking place, even if it requires a longer time. Build on the saying: It is the approach rather than the outcome of the message that spells success	Participation, consultation and process

Figure 1.3 Community participation: process versus product. Adapted from Botes (1996).

In essence the theses of antidevelopmentalism are not new since they are essentially based on the failures of modernisation, and the criticism that development is a Western construction in which the economic, social and political parameters of development are set by the West and are imposed on other countries in a neocolonial mission to normalise and develop them in the image of the West, thus ignoring the local values and potentialities of 'traditional' communities. There is much of the 'globalisation steamroller' about the work of the antidevelopmentalists, particularly in their assumption that the universalism of contemporary development discourse is obliterating the local. The central thread holding these ideas together is that the discourse or language of development has been constructed by the West and promotes a specific kind of intervention 'that links forms of knowledge about the Third World with the deployment of terms of power and intervention, resulting in the mapping and production of Third World societies' (Escobar, 1995: 212). Thus, Escobar argues, development has 'created abnormalities' such as poverty, underdevelopment, backwardness, landlessness and has proceeded to address them through a normalisation programme that denies value or initiative to local cultures.

There are, in these arguments, many similarities to Said's orientalism, similarities that are both implicit and explicit in the views of the antidevelopmentalists who see both the 'problems' of the Third World and their 'solutions' as the creations of Western development discourse and practice. Of course, there is a recognition that they are not static but change according to contemporary power structures. However, a consistent factor within the antidevelopmental discourse is the role of the Third World state in facilitating the 'Westernisation' of the development mission. It follows, therefore, that the restructuring of development must come from below. Here the antidevelopmentalist in general, and Escobar in particular, place enormous emphasis not just on grassroots participation but more specifically on new social movements as the medium of change (Box 1.2).

The nature of these new social movements is allegedly quite different not only from the class-based group of the nineteenth century (P.W. Preston, 1996) but also from those which Castells (1978, 1983) has written about in the 1970s and 1980s. Escobar dismisses these as 'pursuing goals that look like conventional development objectives (chiefly, the satisfaction of basic needs)' (Escobar, 1995: 219). In contrast, the new social movements upon which Escobar pins so many of his hopes are antidevelopmental, promoting egalitarian, democratic and participatory politics within which they seek autonomy through the use and pursuit of everyday knowledge. Indeed, some observers have gone even further, and claim that the new social movements 'transcend any narrow materialist concerns' (P.W. Preston, 1996: 305–6). Escobar warns that such movements must be wary of being subverted into the developmentalist mission through compromised projects such as 'women and development' or 'grassroots development'.

Not surprisingly, antidevelopmentalism in these terms has been subject to some stinging criticism, particularly by Watts and McCarthy (1997), who point out

that Escobar is guilty of considerable reductionism in his critique of development, painting a picture very much resembling the dependency theories of the 1970s, in which a monolithic capitalism, particularly in the guise of the World Bank, monopolises development within a largely complacent Third World. As Rigg (1996) observes, Escobar is very selective in his evidence, with little discussion of those Asia-Pacific countries that might contradict his polemic. Corbridge (1995) also argues that Escobar ignores the many positive changes that Western-shaped development has brought about in terms of improved health, education and the like, no matter how uneven this has been. Escobar also attributes to Third World people his own mistrust of development, a view they may not share. Indeed, some would argue that the intellectual tradition in many Asian universities is to support the state and its policies rather than to criticise them (Rigg, 1996). Assumptions of widespread antidevelopmentalism are therefore as arrogant as assumptions of widespread approval of the modernist project. Indeed, as many observers have noted, the poor of the Third World simply get on with the business of survival; holding views on development is a luxury of the privileged. Indeed, many poor people in the Third World are quite conservative and resent imposed or introduced change of any kind, despite the fact they are often very innovative and adaptive in their own coping mechanisms and survival strategies.

Corbridge (1995) suggests that antidevelopmentalism romanticises and universalises the lifestyles of indigenous peoples. Do the actions of the poor and vulnerable really constitute a resistance to development or are they simply seeking to manipulate development to improve their access to basic resources and to justice? Certainly, the antidevelopmentalists have reinforced our sense of the local in the face of what appears to be an overwhelming process of globalisation. Indeed, the alleged retreat of the state and return to democracy which have occurred in some parts of the world, have opened up new spaces in which social movements can seize the initiative. But, as Watts and McCarthy (1997: 84) note, 'a central weakness of the social movements-as-alternative approach is precisely that greater claims are made for the movements than the movements themselves seem to offer.' Moreover, what is wrong with social movements having modest, self-centred aims which focus on basic needs, if it results in improvements in the quality of life for a group that subsequently disbands? Who are we to castigate this as mere self-seeking satisfaction of conventional development appetites?

BOX 1.2   Urban social movements in Australia

Over the years there has been considerable debate as to what constitutes an urban social movement and Drakakis-Smith (1989) considers it in the Australian context of this case study. We can define it here as a collective, territorially based action, operating outside the formal political system, with the objective of defending or challenging the provision of urban service against the interests and values of the dominant groups in society. Although such urban social movements (USMs) are essentially local and non-political in origin, their effectiveness in improving the quality of their life is strongly influenced by the broader social, political and economic contexts in which they are situated, not only at the urban level but also at the regional, national and international levels.

This case study will illustrate how the success of USMs can vary, even in apparently similar situations, in relation to the way in which local circumstances can be manipulated by more broadly based processes. It relates to the attempts by Aboriginal-based USMs in the towns of Darwin and Alice Springs to obtain improved housing conditions. Both towns are located in the Northern Territory of Australia but the outcome of the respective USMs was quite different. To understand this situation, it is necessary to discuss briefly both the general history of Aborigines in the development of Australia and the specific circumstances in each town.

When the British arrived in Australia in 1788 they simply took possession of the land in the name of the crown. There were no negotiations with the Aborigines and no treaty was signed. Exploitation of Aboriginal land for agriculture, pastoral industries and mining was accompanied by exploitation of Aboriginal labour as stockmen and domestic servants. In the present century Aboriginal culture has been exploited by the tourist industry. Indigenous Australians were effectively ignored by the Australian government throughout this period, being largely confined to missions/reserves where their labour was reproduced. The breaking of the link between the Aboriginal people and their land devastated their culture and spiritual basis, and by the mid twentieth century, those who had drifted into towns tended to live in small shanty towns in communities ravaged by alcoholism (Rowley, 1978). Even the incorporation of the Aboriginal population into the state welfare system in the 1960s failed to halt this

situation, but it did create thousands of jobs for white Australians 'servicing' the Aboriginal community.

One of the areas in which the Aboriginal community was poorly served was housing. Excluded from the private sector by their poverty, they were also excluded from state housing by virtue of their inability to meet rental payments and by their 'inability to cope' with a state house. This was certainly the situation in Alice Springs, where by the 1970s around 80 per cent of the housing stock was occupied by white families, whereas most Aboriginals lived in some thirty camps outside the town (see Plate 1.2).

This unequal situation was the end product of a set of ideological forces (Aboriginal people deserved no rights), ethnic prejudice and class antagonism. It was changed in Alice Springs not just by new economic circumstances, but by the translation of broader, global civil rights movements into national and local situations. First, the federal state, under external and internal pressure, made it possible for Aborigines to claim lease rights to the land they occupied around Alice Springs. Despite opposition from entrenched White interest groups, Aborigines were successful in the bids for leaseholds largely because they formed a collective of camp leaders called the Tangatjira Association. This took on the role of adviser to various camp groups and later became a facilitator for the introduction of appropriate housing technologies and building maintenance, all of which also created employment for local Aborigines.

Although the federal state, through the Department of Aboriginal Affairs, certainly facilitated this process, the success in obtaining tenure security and improved shelter undoubtedly came from the activities of a very focused urban social movement. The situation in Darwin, however, in similar global and national circumstances was quite different. There Aboriginal groups failed to mobilise into a social movement to challenge the system, despite the fact they were even more marginalised than in Alice Springs. In many ways this is the consequence of local political and economic circumstances.

As Darwin is the territory capital, it has a large percentage of both territorial and federal employees, most of whom received large incentives, usually in the form of housing subsidies. The housing rental market, both private and public, is therefore marked by high rents and those not favoured with subsidies

Box 1.2 continued

Plate 1.2 Aboriginal town camp on the outskirts of Alice Springs (photo: David Smith)

have to share in order to obtain adequate shelter. The low-income groups in Darwin who cannot afford rents, live in a variety of accommodation from camps to caravan parks or boats. However, in contrast to Alice Springs, there is no ethnic unity as many of the poor comprise White males who have 'dropped out' of conventional Australian society. Compared to Alice Springs, Aboriginal groups therefore comprise a smaller proportion not only of Darwin as a whole, but of the low-income group. No urban grouping has

therefore coalesced around a social issue such as housing. Moreover, as Darwin is the state capital, individuals with the potential for group leadership have tended to be co-opted into the political system, thus reducing the danger of social mobilisation.

These two examples illustrate the fragile distinction between success and failure in urban social movements, even in similar situations, and they emphasise the fact that local circumstances can impede as well as facilitate mobilisation.

Much of the problem with antidevelopmentalism is that it 'overestimates the importance of discourse as power . . . when the political economy of changing material relations between capital and native remain central to the process of capital accumulation' (Watts and McCarthy, 1997: 89). In short, the intention to develop is confused with the process itself. Although they can and do overlap, there is much more to the unsatisfactory nature of development than its intellectual discourse. However, the antidevelopment movement *has* brought about a re-emphasis of the importance of the local in the development process, as well as the important skills and values that exist at this level; it also reminds us what can be achieved at the local level in the face of the 'global steamroller', although few

such successes are free of modernist goals or external influences (Box 1.2).

Of all the recent changes within developmental thinking, perhaps the most successful and the least heralded has been the shift away from large-scale theory to meso-conceptualisations which focus on specific issues or dimensions of development in an attempt, not merely to separate out a slice of development for scrutiny, but to see how it relates to the development process as a whole and to local situations. A good illustration of this might be the recent fusion of gender and shelter debates which has made a strong impact on theory and policy in the 1990s (Chant, 1996). Some might claim that this illustrates the influence of post-modernism in development studies (Corbridge, 1992), involving a liberation of thought, a recognition of a local 'otherness' and support for small-scale development. However, one could argue that such approaches have been part of development geography for some time, reflecting its empirical traditions. As T.G. McGee (1997) points out, the accumulated experiences (histories) of empirical studies are invaluable in bringing out a sense of the local within the development process. On the other hand, post-modernism has also been interpreted as merely 'the cultural logic of late capitalism, effectively representing the new conservatism . . . preoccupied with commodification, commercialisation and cheap commercial developments' (Potter and Dann, 1994: 99). In this context, most of the new meso- and micronarratives of development thinking have little in common with post-modernism (Chapter 3 discusses post-modernism and development).

Reviewing development

So, what are we left with after all this discussion of development? For the antidevelopmentalists, development has become 'an amoeba-like concept, shapeless but ineradicable [which] spreads everywhere because it connotes the best of intentions [creating] a common ground in which right and left, elites and grass roots fight their battles' (Sachs, 1992: 4). But in its naivety, the antidevelopment alternative of 'cosmopolitan localism' based on regeneration, unilateral self-restraint and a dialogue of civilisations unfortunately seems no more than a Utopia for New Age travellers.

Despite its eighteenth- and nineteenth-century origins, 'development has never been a scientific concept, it has always been ideology' (J. Friedmann, 1992: v). Development can mean all things to all people; poor squatters may have a completely different view of what constitutes change for the better in their lifestyle than a senior politician or national planner. This is clearly evident in recent discussions of the brown agenda, in which Satterthwaite (1997) and others have pointed out that many of the concerns of the international agencies and national planners with global warming and ozone layers reflect a 'Northern' agenda that is far away from the clean water needed by most squatter households. As Hettne (1990: 2) notes, 'there can be no fixed and final definition of development, only suggestions of what development should imply in particular contexts'.

The debates over the definition of what development is and how people think about it are not simply academic, although for some this is the limit of their interest. Thoughts and views about the development condition underlie policy formulation and subsequent implementation. A common example of conflict is that which often occurs between the national government with its economic impetus and external linkages, and NGOs or CBOs which tend to emphasise democratisation, political involvement and the local, immediate needs of these disadvantaged groups (D.H.L. Thomas, 1992). Development is an historical process of change which occurs over a very long period but it can be, and usually is, manipulated by human agency. It is often forgotten that culture (particularly religion) can play an important role in characterising national and local development strategies. Many of the industrialising states in Asia claim to have followed an 'Asian way' to development, although this in itself has been criticised for its reductionism and selectivity (Rigg, 1996). It is the nature of these manipulations and their goals that will be discussed in Chapter 3. However, what we have now established is that development is not unidirectional. Improvement in the human condition has many different dimensions and the speed of change may vary enormously for any individual or community. And although a fair and balanced development may be a desirable goal, for most of the world's population it is far from being a realistic one (J. Friedmann, 1992). It is to the definition of this proportion of the world's population that this chapter now turns.

Spatialising development

This section will examine one of the terms most commonly employed to refer to spatial contrasts in types of development, different levels of development and different patterns of development. Specifically it will trace the evolution of the term *Third World*, its proponents, its critics and the alternatives they have posed, such as *the South*. Almost inevitably there will be a degree of overlap with the first half of the chapter, since in many ways we are examining the public lexicon of the more theoretical issues discussed previously. Indeed, the wider currency that the terms *development* and *underdevelopment* allegedly experienced as a result of President Truman's inaugural address of 1949 (Sachs, 1992; Esteva, 1992) was to a certain extent clarified by the new 'three world' terminology that was also emerging at the same time. Thus, the First World was promoting development, the Second World was opposing it, and the Third World was the object of the exercise. In the rigidities and absurdities of Cold War politics in the 1950s this

Plate 1.3 People making a living by a variety of means, Old Delhi (photo: Rob Potter)

seemed to make good sense to some. However, over the years, the association between the notion of three worlds and the development process has changed considerably. It will be useful to examine this association within a loose chronological framework.

The 1950s and the 1960s: emergence of the Third World

As with *development*, the antecedents of the term *Third World* go back beyond 1949, although not much further. Moreover, in contrast to the current largely economic interpretation of the Third World (Milner-Smith and Potter, 1995), particularly in terms of poverty, the origins of the term were political, largely centring around the search for a 'third force' or 'third way' as an alternative to the Communist–Fascist extremes that dominated Europe in the 1930s. In the Cold War politics of the immediate postwar years, this notion of a third way was revived initially by the French Left who were seeking a non-aligned path between Moscow and Washington (Wolfe-Phillips, 1987; Worsley, 1979; Pletsch, 1981). It is this concept of non-alignment that was seized upon by newly independent states in the 1950s, led in particular by India, Yugoslavia and Egypt, and culminating in the first major conference of non-aligned nations at Bandung in Indonesia in 1955. Indeed, at one point, 'Bandungia' appeared to be a possibility for their col-

lective description. J. Friedmann (1992: iii) claims that through this meeting 'the Third World was an invention of the non-western world', in spirit if not etymologically.

The sociologist Peter Worsley (1964) played a major role in the popularisation of the term *Third World* in his book of the same name. For Worsley the term was essentially political, labelling a group of nations with a colonial heritage from which they had recently escaped and to which they had no desire to return within the ambit of neocolonialism (or new forms of colonialism). Nation building was therefore at the heart of the project and it is no coincidence that the loudest voices came from those states with the most charismatic leaders. For India, Yugoslavia and Egypt, therefore, read Nehru, Tito and Nasser. But the Third World then was not quite the same as it is today; many countries had still to gain their independence and Latin American countries were not present in Bandung. Moreover, both the Bandung group and Worsley's Third World 'excluded the communist countries' (Worsley, 1964: ix). Nevertheless, for a while in the 1950s and 1960s, this Afro-Asian bloc did attempt to pursue a middle way in international relations. In economic terms, however, it was a different story.

Almost all newly independent states lacked the capital to sustain their colonial economies, let alone expand or diversify. Most remained trapped in the production of one or two primary commodities, the prices

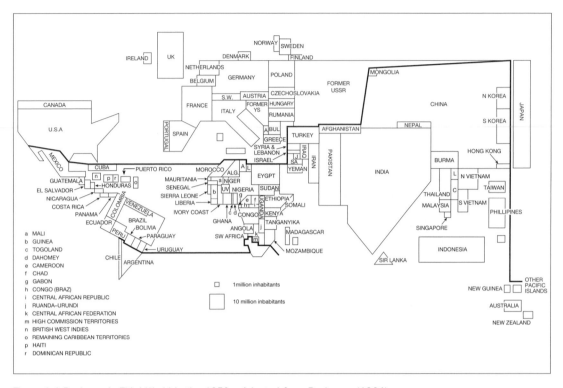

a MALI
b GUINEA
c TOGOLAND
d DAHOMEY
e CAMEROON
f CHAD
g GABON
h CONGO (BRAZ)
i CENTRAL AFRICAN REPUBLIC
j RUANDA–URUNDI
k CENTRAL AFRICAN FEDERATION
m HIGH COMMISSION TERRITORIES
n BRITISH WEST INDIES
o REMAINING CARIBBEAN TERRITORIES
p HAITI
r DOMINICAN REPUBLIC

1 million inhabitants

10 million inbabitants

Figure 1.4 Buchanan's Third World in the 1950s. Adapted from Buchanan (1964).

of which were steadily falling in real terms, unable to expand or improve infrastructure and their human resources. Once Worsley had identified the common political origins of the Third World (anticolonialism and non-alignment), he cemented this collectively through the assertion that its current bond was poverty (Plate 1.3). This feature had also been noted by Keith Buchanan (1964) in the first substantial geographical contribution to the debate. Buchanan's Third World (Figure 1.4) bears a close resemblance to that of the Brandt Commission twenty years later, but makes somewhat more geographical sense.

The 1960s witnessed a major shift in interest on the part of several social science disciplines towards the nature of development and underdevelopment, prompted largely by the failure of modernisation strategies to bring predicted growth to what was increasingly becoming called the Third World. It is important to note, however, that much of this economic debate was predicated on deeper political concerns: the fear that widespread and persistent poverty would lead to insurrection and a further round of communist coups. In Asia the puppet regimes of South Vietnam and South Korea were looking somewhat shaky,

whereas the continuing strength of Castro's revolution in Cuba raised fears of a Caribbean domino effect. The principal concern, particularly of development economists, was to find out what had gone wrong and where the problems were located. In geography too this was the era of the quantitative revolution, and from both disciplines there arose a series of measurements designed to rank Third World nations in terms of needs (Box 1.1) with the usual signifier being gross national product (GNP) per capita. Within some of the individual states, 'modernisation surfaces' were produced which indicated spatially uneven development by means of multiple indices of development 'attributes', most of which closely mirrored the spatial variations of colonialism (Soja, 1968; P.R. Gould, 1970). This approach is considered in detail in Chapter 3.

Despite the largely uninformative nature of these academic developments, the term *Third World* was now in widespread use, even by its constituent states in forums such as the United Nations (Wolfe-Phillips, 1987). Conceptually, therefore, the world was firmly divided into three clusters, namely the West, the Communist bloc and the Third World, but these terms were etymologically inconsistent. The first is an abstract geographical term (west of where?), the second is a political epithet and the last is numerical; hardly an example of consistent logic, but one which had obtained a significant measure of popular acceptance by the 1970s.

The 1970s: critiques of the Third World

By the early 1970s, the rather loose combination of political and economic features that constituted the Third World had already come in for criticism. The French Socialist Debray (1974: 35) argued that is was a term imposed from without rather than within, although more developing nations were beginning to use the term. Antidevelopmentalists consider this to be a critical point in the development process, a time when the Third World was beginning to recognise its own underdevelopment, adopting Western evaluations of its condition. Many other critics, however, also felt the term was derogatory since it implied that developing countries occupied third place in the hier-

archy of the three worlds (Merriam, 1988). An even more valid criticism was that users of the term had still failed to situate the socialist developing states in the three-world terminology (Box 1.3).

The main cause of the doubts that emerged during the 1970s was related to the growing political and economic fragmentation of the Third World as it 'post-modernised' from a 'metaregion' into a plethora of subgroupings. Ironically, perhaps the biggest impetus to the break-up of the Group of 77 non-aligned nations came from within when the OPEC nations raised the price of their oil massively in 1973/4, with a second wave in 1979 following the fundamentalist revolution in Iran. Initially conceived as a political weapon against the West for its support of Israel, the price rise had a much greater effect on non-oil-producing countries of the developing world, many of which were following an oil-led industrial and transport development programme. The result was a widening income gap between developing countries.

This was further reinforced by the new international division of labour in the 1970s, in which

---

BOX 1.3   The socialist Third World

Within the evolution of the term *Third World*, the place of the socialist developing countries always seemed to pose difficult conceptual problems. Although most writers seemed to have no problem in distinguishing between developed and developing capitalist states, the same did not hold true for the centrally planned economies. Peter Worsley solved this problem by not discussing the issue; others have added on various developing states to the main Socialist bloc, effectively shifting them from the Third World to the Second World at the stroke of a pen.

Part of the problem has emanated from the difficulty of defining exactly what a socialist state is, particularly as socialists and Marxists often disagree fundamentally about this. In the mid 1980s, before the collapse of the Second World, Thrift and Forbes (1986) listed the attributes of a socialist government as follows:

- one-party rule
- egalitarian goals
- high or increasing degree of state ownership of industry and agriculture
- collectivisation of agriculture
- centralised economic control

Using these criteria in a non-doctrinaire way, Drakakis-Smith, Doherty and Thrift (1987) identified a surprisingly extensive map of socialist Third World states (see figure overleaf). Of course, such a map cannot be static as governments come and go and, moreover, there are immense political, economic and social differences between the socialist states. In some, such as Tanzania, socialism has clear rural origins, often from an anticolonial struggle, leading to persistent suspicion of urbanites and reflected in anti-urban policies. However, few if any Third World socialist states reflect the nature and roots of European socialism in the class contradictions of industrial societies.

Perhaps the most important distinction between the socialist Third World states is the nature of their political structure – put crudely, the distinction between authoritarian and democratic socialism. The massive transformations in Eastern Europe were bound to have an impact on the socialist Third World, but it is difficult to separate this from other major events that have affected the Third World in general since the mid 1980s, particularly structural adjustment. It is too easy and arrogant to read into the

Box 1.3 continued

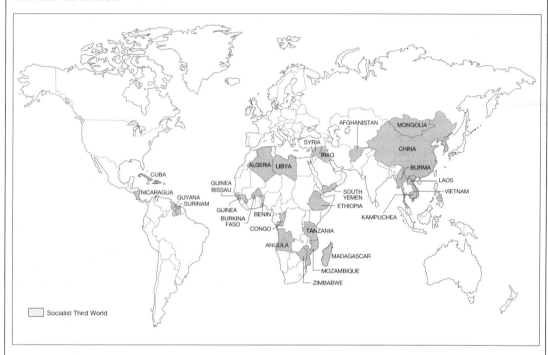

The socialist Third World in the 1980s. Adapted from Drakakis-Smith, Doherty and Thrift (1987).

events in Eastern Europe and the Soviet Union the 'end of history' in the rest of the world.

Perhaps surprisingly, socialist states have persisted throughout the Third World in quite considerable numbers, but often with substantially modified forms of socialism. The region that has seen most change is sub-Saharan Africa, where change of government has been as much the result of structural adjustment policies, with their insistence on 'good' (meaning Western-style democratic) governance as one of the conditions for renewed loans, as the breakdown of support from the Soviet Union. In several states, such as Mozambique, there has been a shift towards a more democratic form of socialism, at least as indicated by multiparty elections. In some states, however, this has released forces which the socialist government has struggled to suppress. This has been very evident in some Muslim countries, where Islamic fundamentalism has challenged the socialist state – successfully in Afghanistan and substantially in Algeria.

Elsewhere it is the *rapprochement* with capitalism that has been more noticeable, particularly in Pacific Asia, where economies have been opened to investment in the 1990s, with immediate and substantial impact on both economic and social life (Drakakis-Smith and Dixon, 1997). However, despite such changes, all these states are still nominally socialist, particularly in terms of their being one-party states. In these countries, such as China, the struggle for democracy within socialism is still ongoing and is linked more with the individual economic freedoms associated with capitalism rather than fundamental changes in belief at government level. This has resulted in a heated debate on the nature of democracy between the West and many Asia-Pacific states, with both socialist and capitalist states in the region arguing for democracy through collective responsibilities rather than individual freedoms.

On paper, therefore, the socialist Third World looks almost as extensive as ever, but fundamental changes have occurred in the nature of its internal structures and organisation. Above all, there has been a steady, but often contested, shift towards a more democratic form of socialism, although this shift is under threat from indigenous reactionary forces and also from the inherent inequalities of the free market economy.

capital investment via multinational corporations and financial institutions poured out of Europe and North America in search of industrial investment opportunities in developing countries. Most of this investment was highly selective and cheap labour alone was not sufficient to attract investment: good infrastructure, an educated and adaptable workforce, local investment funds, docile trade unions and the like were also important. The result was that investment focused on a handful of developing countries (the four Asian tigers, Mexico and Brazil) where GNP per capita began to rise rapidly, further stretching the economic and social contrasts in the Third World.

The widening differences began to exercise academic minds and journals such as *Area* and *Third World Quarterly* were filled with articles about the merits and demerits of the term as a descriptive concept (O'Connor, 1976; Auty, 1979; Worsley, 1979; Mountjoy, 1976). The debate soon spread to some of the 'serious' journals of the popular press, where various ways of regrouping the developed and developing countries were suggested. *Newsweek* identified four worlds; the Third World comprised those developing countries with significant economic potential and the Fourth World consisted of the 'hardship cases'. Not to be outdone, *Time* magazine subsequently put forward a five-world classification in which the Third World contained those states with important natural resources, the Fourth World equalled the newly industrialising countries (NICs) and the Fifth World comprised the 'basket cases'. Many academics joined in this semantic debate. Goldthorpe (in Worsley, 1979) produced a list of nine worlds; at the lower end came the better-off poor, the middling poor, the poor and the poorest – indefinable refinements of poverty that were of little conceptual value and even less comfort to those under such scrutiny. To cause even greater confusion, the term *Fourth World* was also coming into general use to describe underdeveloped regions within developed nations, particularly where this referred to the exploitation of indigenous peoples, such as Canadian Inuit or Australian Aborigines (see the special edition of *Antipode* **13**(1) in 1981).

The changes were reflected to a certain extent in the classification system employed by the World Bank in its annual development reports. In the early 1980s, the developed countries were classified by their dominant mode of production (industrial), the socialist states of Eastern Europe and the Soviet Union were politically identified as 'centrally planned'. The developing countries were, if not 'oil exporters', divided on a wealth basis into low- and middle-income states.

Subsequently, following the (apparently worldwide) demise of socialism, the classification has regressed to an entirely income-based classification. After thirty years of constant criticism, GNP per capita still rules as a development indicator with the World Bank.

The 1980s: the 'lost decade' for development in the Third World

Despite the regression at the World Bank to an economically based stratification of the Third World, the 1980s in general saw considerable widening of the range of indicators used for classifying the various nations of the developing world and soon they were being amalgamated into composite indices of well-being or quality of life (Box 1.1). However, such indices did little to address the debate on the concept of the Third World *per se*. Nevertheless, during the 1980s a growing critique of the term began to emerge from the new right-wing development strategists who argued that the Third World is merely the result of Western guilt about colonialism, a guilt which is exploited by the developing countries through the politics of aid. Economist Lord Bauer (1975: 87), one of the leading exponents of this view, has expressed it like this: 'The Third World (is) the collection of countries whose governments, with the odd exception, demand and receive foreign aid from the West . . . the Third World is the creation of foreign aid; without foreign aid there is no Third World.' In the eyes of the New Right, virtually all developing countries are tainted with socialism and their groupings have invariably been anti-Western and therefore anticapitalist, a view which has effectively been taken to task by John Toye (1987). Ironically, many Marxists too found it difficult to accept the term *Third World* because they regarded the majority of its constituent countries as underdeveloped capitalist states linked to advanced capitalism. Thus, in their eyes there were only two worlds, capitalism and Marxian socialism, with Marxian socialism subordinate to capitalism. Unfortunately, there was little agreement among Marxists as to what constituted the socialist Third World (Box 1.3).

The notion of two worlds perhaps represented the most concerted challenge to the three-world viewpoint and, indeed, most of the semantic alternatives which we currently use are structured around this dichotomy, namely rich and poor, developed and underdeveloped (or less developed), North and South. The last pair in particular received an enormous boost in popularity with the publication of the Brandt Report (1981). As many critics have noted, the Brandt Report set out a rather naive and impractical set of recommendations for overcoming the problems of underdevelopment, relying as it did on the governments of the South to pass on the recommended financial support from the governments of the 'North' (Singer, 1980). Moreover, much of the impetus behind the new 'concern' for development was fuelled by the economic crises of the developed countries and their search for new markets in the rest of the world (Frank, 1980). The heads of state assembled at Cancun in the early 1980s to discuss the report and the plight of the world and, having been publicly seen to be concerned, duly dispersed affirming their faith in the market rather than Willy Brandt.

From a developmental perspective, one of the Brandt Report's major defects was its simplistic subdivision of the world into two parts based on an inadequate conceptualisation of rich and poor (Figure 1.5, opposite). Some critics have claimed that this is spatial reductionism of the worst kind, apparently undertaken specifically to divide the world into a wealthy, developed top half and a poor, underdeveloped bottom half – North and South, them and us – although the terms did no more than rename pre-existing spatial concepts. However, the labels *North* and *South* do seem to be used with disturbing geographical looseness since the South includes many states in the northern hemisphere, such as China and Mongolia, whereas Australasia comprises part of the North. No definitions were discussed in the report, and the contorted dividing line which separates the two halves of the world stretches credulity more than a little as it is bent around Australia and New Zealand, totally ignoring the many small island states of the Pacific, generously, but erroneously, giving them developed status.

One problem with the North–South division of the Brandt Report was that it lacked explanatory power and compares unfavourably with another dichotomous model that also became popular in the 1980s. This is the core and periphery model (Wallerstein, 1979). Such an interpretation does not allow for, or explain, the immense variety that exists both in the core and periphery, nor does it incorporate change over time, whether growth or decline. To accommodate this, a semiperiphery was introduced; this is a category of countries allegedly incorporating features of both the periphery and the core (Figure 1.5). Effectively this gives us another division of the world in three sections, but although apparently based on very different principles from those identified earlier, the various components are still bound together by the overarching operations of capitalism. A map of core, semiperiphery and periphery in the 1990s, following the break-up of the Soviet Union would clearly look quite different over much of Central Asia and Eastern Europe.

As the 1980s wore on, however, the old Truman goal of development toward the Western capitalist model began to fade. The finishing line had in any case been moving away from most Third World countries faster than they were moving towards it. The unified social and economic objectives of the second United Nations Development Decade began to look rather limp in the face of worsening world recession and a harder attitude towards a set of nations that were now being looked upon as a drag on world development through their incessant demand for aid and their growing debt defaulting. The dualism of North and South thus took on a much harsher complexion as the World Bank, the IMF and the regional banks began to impose their structural adjustment programmes on the Third World.

The growing confusion over whether the world should be divided into two or three components, both in conceptual and policy terms, was further accentuated in the 1980s by a feeling among some commentators that the original universalism of the United Nations had somehow been lost and that we should return to thinking of the world as a single entity. Allan Merriam (1988) notes that views on the unity of humanity are long established and cites the seventeenth-century Czech educator Comenius, who stated that 'we are all citizens of one world, we are all of the same blood . . . let us have but one end in view, the welfare of humanity'. Such sentiments have featured frequently in the speeches of some Third World leaders, such as Indira Ghandi and Julius Nyerere, although often as rhetoric rather than reality.

Much of this growth in one-worldism was sustained by a belief that development in the Third World is characterised by a convergence along those paths experienced by the West towards the current lifestyles and political–economic structure of developed countries (Armstrong and McGee, 1988). For many the thought of such convergence is alarming since pursuit of the same economic ends by the same means will only lead to a faster use of the earth's finite resources and will only exacerbate its environmental problems. However, many of these concerns spring from self-interest, in that ultimately it is our own way of life which may be threatened. The people of the Third World are therefore being asked to

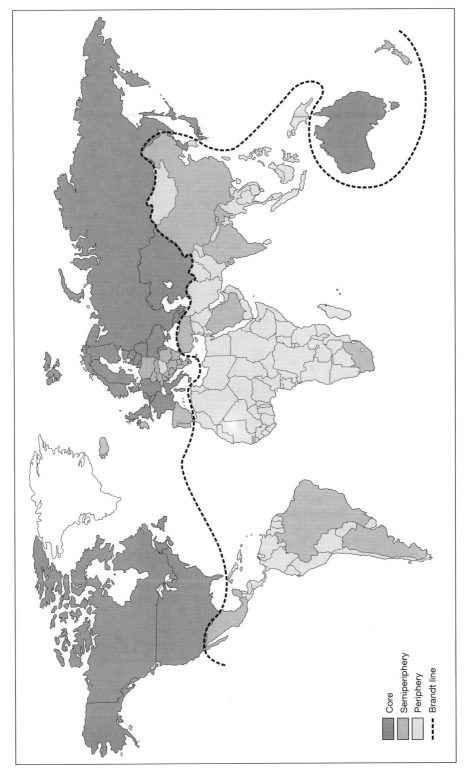

Figure 1.5 Models in the 1980s: North and South; Core, Periphery and Semiperiphery. Adapted from Brandt 1980, 1983.

make sacrifices 'for the greater good of human-kind,' sacrifices that we in the West have never made.

Sachs (1992) sees the convergence theories of the ecologists as yet another example of universalism, perpetuating the single goal, single strategy of the Truman doctrine and denying any role or opportunity for diversity. He claims that the 'one world or no world' warnings of environmental scientists suggest that preservation of our fragile global ecosystem demands that everyone has a responsible and particular role to play. 'Can one imagine a more powerful motive for forcing the world into line than that of saving the planet?' he asks (Sachs, 1992: 103). As the Third World poor have been conveniently found to be the worst offenders in resource destruction, so their re-education could usefully be combined with scaled-down poverty reduction programmes through 'sustainable development'. The West can now give less aid and still feel good about it! However, perhaps the basic premise of this concern is unfounded. There may be convergence towards the Western model, but this is very selective and uneven (Chapter 4). Moreover, even the high-flyers among the NICs still have a long way to go to match the levels of economic and social well-being in the West. Although South Korea may have almost as many fax machines per 100 business telephones as the United States, GNP per capita is still only one-third of the OECD average; and despite its spectacular urban centres, the country still has a persistent informal sector, squatter housing and a large external debt, all of which have contributed to recent instability.

Although attention in recent years has focused on these growing contrasts within the Third World itself (and this is a valid concern), it has also masked the more important fact that global contrasts too are continuing to widen. In particular, there has been much concern that a large number of countries, particularly in Africa, have not only failed to exhibit any signs of development but have actually deteriorated, saddled as they now are with the spiralling debts of poverty and harsh structural adjustment programmes. In this context, convergence theory could be seen as a myth. Indeed, it is arrogant to assume that the process of economic and cultural transfer is one-way. The West has not merely exported capitalism to the developing world, capitalism itself was built up from resources transferred to the West from those same countries. Similarly, acculturation is not simply the spread of Gucci and McDonald's around the world. In almost every developed country, clothes, music and cuisine, together with many other aspects of our daily lives,

are permeated with influences from Asia, Africa and Latin America, such as bamboo furniture, curries and salsa music.

The Third World in the 1990s

The extension of the world recession into the 1990s has meant that fragmentation of interests has continued, and weaker communities at both local and global levels have faced increasing difficulties. One response to this has been the emergence of regional economic blocs in the image of the European Union, such as NAFTA and APEC, all of which are designed to protect their member states and which cut across the traditional boundaries of the three worlds. Of course, this conceptualisation has suffered an even greater blow by the apparent demise of the Second World with the break-up of the Soviet Empire and the, admittedly uneven, democratisation and capitalisation of Eastern Europe. If the Second World no longer exists, can there be a Third World? In this etymological sense, there is little justification for retaining the term, particularly since early commonality of non-alignment and poverty has also long been fragmented. Many commentators in the 1990s, particularly those who form part of the antidevelopment school, have suggested that it is time for the term to be abandoned. Sachs (1992: 3), inelegantly but forcefully states that 'the scrapyard of history now awaits the category "Third World" to be dumped.' Corbridge (1986: 112) too has joined 'with others in questioning the current validity of the term the Third World'. J. Friedmann (1992) also rejects the term in favour of a focus on people rather than places, preferring to identify and build policy around the disempowered. And yet, despite such strong condemnation, the term persists in common usage, even by some of those who have criticised its validity.

So why does the term persist in this way when the Second World has all but gone and the developing countries continue to fragment in their interests, among themselves and within themselves? Perhaps, as Norwine and Gonzalez (1988) have remarked, some regions are best defined or distinguished by their diversity. An analogous situation in the biophysical world is the tropical rainforest. 'More diverse in flora and fauna than any other terrestrial biogeographic type, a rainforest is nonetheless one organic whole, consisting of many disparate parts, yet far greater than the sum of them' (Norwine and Gonzalez, 1988: 2).

Despite the variations in the nature of the Third World that we have noted in this review, most people in most developing countries continue to live in grinding poverty (Table 1.3) with little real chance of escape. This is the unity that binds the diversity of the casual labourer in India, the squatter resident in Soweto, or the street hawker in Lima. All are victims of the unequal distribution of resources that this world exhibits. Moreover, this unity is not merely one of pattern or distribution, but of fundamental processes that are linked to the past, present and probable future roles of these states within the world economy, as exploited suppliers of resources, human as well as physical. Still it holds true, there is a unity provided by colonisation, decolonisation and antipathy but lack of resistance to imperialism (socialist as well as capitalist), something noted by Mao Tse Tung, Peter Worsley and John Toye.

In this sense, the concept of the Third World is an 'extremely useful figment of the human imagination. . . . The Third World exists whatever we choose to call it. The more difficult question is how can we understand it' (Norwine and Gonzalez, 1988: 2–3) and change it according to priorities set out by its own inhabitants. Most of those students of development who continue to use the term *Third World* must realise, therefore, that it is not simply a semantic or geographical device (Killick, 1990), but a concept that refers to a persistent process of exploitation through which contrasts at global, regional and national level are growing wider. No matter what abstract conceptualisation we use to structure our development debates – three worlds, two worlds, nation states, cities or whatever – we must not forget that we are discussing human beings. Their welfare and how to improve it must be the focus for our debates, rather than the sterile question of what label is politically correct.

Development and geography

In a recent paper, T.G. McGee (1997) has drawn our attention to some of the implications of the shifts into post-modern development, particularly those which emphasise globalisation as 'a variable geometry of production or consumption, labour, capital management and information – a geometry that denies the specific meaning of place outside its position in a network whose shape changes relentlessly' (Castells in T.G. McGee, 1997: 8). For some time, since Alvin Toffler's *Future Shock* (1970), commentators have been arguing for the end of geography. Toffler himself based his arguments on increased flows of people, goods and information that dissolve difference and distinctions. Apart from geographical differences still being very evident in the world (Chapter 4), Toffler also ignored the fact that linkages and flows between places are largely the province of the geographer. More recently, Richard O'Brien (1991) has also claimed an end to geography on the basis that location matters much less for economic development than it has done in the past.

Although the recent development of technologies does 'challenge conventional notions of distance, boundaries and movement . . . geography matters . . . because global relations construct unevenness in their wake *and* operate through the pattern of uneven development laid down.' (Allen and Hamnett, 1995: 235). Thus, the choice for new investment is often conditioned by the facilities and resources that are already there, perpetuating inequalities. Not surprisingly, therefore, many would argue (Massey & Jess, 1995) that place is more fundamental than ever, since the realities of

Table 1.3 Human development indicators.

	Life expectancy (years)	Adult literacy (% of total)	GNP (US$ per capita)	Daily calorie supply (per capita)
Sub-Saharan Africa	50.9	55.0	555	2096
Arab States	62.1	53.0	1 725	2820
South Asia	60.3	48.8	390	2356
East Asia	68.8	81.0	825	2751
Southeast Asia and the Pacific	63.7	86.0	1 089	2541
Latin America and the Caribbean	68.5	85.9	2 966	2757
All developing countries	61.5	68.8	970	2546
Least developed countries	51.0	46.5	210	2027
Eastern Europe and CIS	69.2	97.6	1 992	–
Industrialised countries	74.3	98.3	16 394	–

Source: UNDP (1996).

development within the Third World are represented by an unevenness and by a constantly shifting fusion and conflict between the global and local, usually filtered through national or regional agency.

It is true that at one level the processes of globalisation are creating super-regions, such as the European Union, NAFTA (North American Free Trade Agreement) and APEC (Asia-Pacific Economic Cooperation), all with varying degrees of cohesion, together with their global and regional megacities, such as London or Tokyo. But at other levels and in other localities in the interstices of the global network, neglect, ignorance and even resistance combine to produce patterns of development which are strongly geared to place and history, and which must be studied as such in order for development to be fully understood (Chapter 4). Such places can even occur within the megacities themselves, as migration throws together people whose roles in the development process are very different.

Although many geographers have re-emphasised the importance of place in this globalisation dimension of development, T.G. McGee (1997: 21) would argue that most geographers continue 'to interrogate the development project from within the modernist project [in] the liberal belief that good research can provide workable solutions'. What constitutes the heart of this approach is that geographical investigation is rooted in an empiricism which focuses on the interaction of society and environment, on networks and flows of people and goods, on uneven and unequal development and, most important in these contexts, on the nature of local places. All of these factors, according to McGee, place development geography firmly in the humanist tradition. What needs to happen now is for the local not only to become the object of the exercise but also the medium, with local input into the development process itself. Only in this way will our preconceived ideologies or images of development be changed (Massey, 1995).

CHAPTER 2

Understanding colonialism

Introduction

The literature on colonialism is scarcely less prolific than, and at times as opaque, as the literature on development. In many of the more accessible texts, however, there is an unfortunate tendency to equate colonialism with the expansion of capitalism in the nineteenth and twentieth centuries, implying that it primarily comprises an economic process. Clearly there is an essential, and at times overwhelming, economic impetus to colonialism, but to construct a framework of analysis based on such a simple equation would be uninformative. Colonialism is a political process, and the establishment of colonies long predates the genesis and subsequent globalisation of European capitalism. Right through this period, colonies were acquired for motives other than the economic imperative for material resources, labour or markets. As in Roman times, otherwise barren or unpromising territory was annexed for strategic reasons: to protect the periphery of pre-existing colonies, to control important military routes or simply to prevent the expansion of rival European powers. Although some of these lands eventually proved to have some economic value, the original motivation was often quite different.

Much of the development literature in fact conceptualises the global expansion of capitalism as imperialism rather than colonialism (Box 2.1), although even here there is a debate about when this process began. For some neo-Marxists, imperialism begins with the division of Africa at the Treaty of Berlin in 1885,

---

BOX 2.1 The meanings of colonialism and imperialism

Colonialism and imperialism are not interchangeable ideas, although in the context of this analysis they overlap considerably. Some observers do not directly acknowledge this, and as a result produce analytical frameworks that are only partially successful. Anthony King (1976: 324), in the index of his seminal text on colonial urbanisation states simply 'see colonialism' under the entry for imperialism. Consequently, his definition of colonialism is only partial: 'the establishment and maintenance for an extended time, of rule over an alien people that is separate and subordinate to the ruling power'.

Blauner is more comprehensive in his approach, defining colonialism as 'the establishment of domination of a geographically extended political unit, most often inhabited by people of a different race and culture, where this domination is political and economic and the colony exists subordinated to and dependent on the mother country' (Wolpe, 1975: 231). How-

ever, once again the specific mode of domination and exploitation is left unidentified and, as with so many analyses of colonialism, the focus is on internal processes at the expense of outward linkages.

Part of the problem is that urban colonialism *per se* is a very long-established process, dating back to Greek cities such as Miletus. It is therefore in itself an inadequate term to describe the political economy of capitalism as it has expanded since the sixteenth century. This process is more correctly known as imperialism and continues today, albeit in a different form. Some have called this an informal imperialism, one more related to the establishment of economic hegemony, compared to the more political emphasis within formal imperialism. The problem with this distinction is that it is not chronologically sequential, as is implied, since establishing a trading hegemony was also characteristic of the mercantile period.

Box 2.1 continued

There are other inconsistencies in the various uses of imperialism. One of the most obvious, but not necessarily a fault, is that the term is employed in two distinct ways: 'a technical sense – to define the latest stage in the evolution of capitalism – and a colloquial sense – to describe the relationships between metropolitan countries and underdeveloped countries' (Bell, 1980: 49). These need not be incompatible, although difficulties of reconciliation between the two approaches have certainly led to contradictions in the chronology of imperialism. Marxist (Leninist) analysts believe that this monopoly stage of capitalism began only around the start of the twentieth century, at least that is what Bell

(1980) argues. Barratt-Brown (1974), on the other hand, has extended consideration of imperialism to roughly the last four hundred years.

Barratt-Brown's chronology is preferable largely because it permits a broader definition of imperialism to be used, one that refers to 'both formal colonies and privileged positions in markets, protected sources of materials and extended opportunities for profitable employment of labour' (Barratt-Brown, 1974: 22). This permits us to examine the way in which the expansion of imperialism affected, and was affected by, the parallel colonisation process in the Americas, Asia and Africa from the early sixteenth century to the decolonisation phases of the 1950s and 1960s.

---

reinforced by the assertion that the term *imperialism* was first coined in the nineteenth century by Napoleon III. Most, however, are of the opinion that imperialism began in the late fifteenth and early sixteenth centuries with the rise to prominence of the European nation state, although the commercial underpinning to the feudal system had also stimulated some colonialism before this period (Blaut, 1993).

At some point in the eighteenth or nineteenth century, depending on whom you read, development

itself becomes an identifiable process in its own right, underpinning and fusing with imperialism and the expansion of capitalism (Dixon and Heffernan, 1991). Throughout this period, colonies continued to be founded for a variety of often complex motives and in many different forms. The particular nature of colonialism varied not only with the motives but also with contemporary political economies and cultures of both the metropolitan power and colonised territory (Box 2.2). Indeed, there has in recent years

---

BOX 2.2   Precolonial sub-Saharan Africa

Before the arrival of European traders from the sixteenth century onwards, there were many substantial and sophisticated communities throughout Asia, Africa and Latin America. Although we must recognise the achievements of these societies as much as we can, given the distorted picture communicated by early European trader-colonists (Blaut, 1993), we must not romanticise 'traditional' societies as a result of remorse. Many such societies were not repositories of simple communism and were rather unpleasant places to live for most of their inhabitants, with a substantial slave trade in which indigenous chiefs willingly participated. Change was sometimes for the better. In sub-Saharan Africa, the 'traditional' indigenous crops, sorghum and millet, were nutritionally inferior to the manioc and maize introduced from the Americas by European traders.

Organisationally, most societies in the immediate pre-European period were (semi) subsistence but this did not preclude either trade or large, powerful

states from emerging. Essentially these societies were structured on two levels. At the local level was a patriarchal agrarian community, where land was allocated and used on a more or less equitable basis and in which the redistribution of surplus took place on a social rather than a market basis. Reciprocity and obligation brought status. At the elite level, status was inherited or was enhanced by wealth accumulated by raids, confiscation or conquest outside the community. Although there was some trade in valuable items such as gold or salt, these were dominated by the small group of elites. Pre-European, precapitalist societies were therefore ravaged by constant wars, fought on behalf of the elites in order to control the production or trade of valuable items or to control people. Before the arrival of European capitalism, life was not comfortable for most Africans. Unfortunately, it did not change for the better with the arrival of the market economy and the ensuing revaluation of commodities.

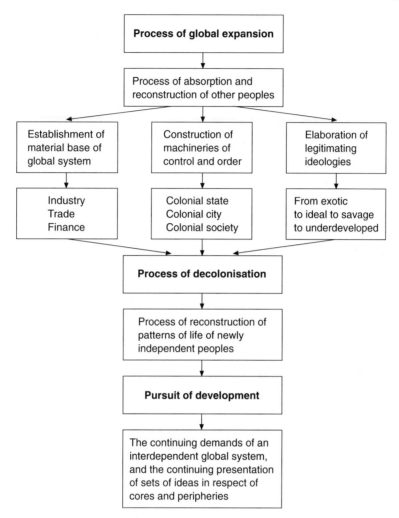

```
┌─────────────────────────────────────┐
│   Process of global expansion        │
└─────────────────────────────────────┘
                  ↓
┌─────────────────────────────────────┐
│   Process of absorption and          │
│   reconstruction of other peoples    │
└─────────────────────────────────────┘
```

Establishment of material base of global system	Construction of machineries of control and order	Elaboration of legitimating ideologies
Industry Trade Finance	Colonial state Colonial city Colonial society	From exotic to ideal to savage to underdeveloped

```
┌─────────────────────────────────────┐
│   Process of decolonisation          │
└─────────────────────────────────────┘
                  ↓
┌─────────────────────────────────────┐
│   Process of reconstruction of       │
│   patterns of life of newly          │
│   independent peoples                │
└─────────────────────────────────────┘
                  ↓
┌─────────────────────────────────────┐
│   Pursuit of development             │
└─────────────────────────────────────┘
                  ↓
┌─────────────────────────────────────┐
│   The continuing demands of an       │
│   interdependent global system,      │
│   and the continuing presentation    │
│   of sets of ideas in respect of     │
│   cores and peripheries              │
└─────────────────────────────────────┘
```

Figure 2.1 Principal processes of colonialism. Adapted from P.W. Preston (1996).

been a tendency to define colonialism as much by the means it employed as by the impetus behind it. Thus 'colonialism is often defined as a system of government which seeks to defend an unequal system of commodity exchange' (Corbridge, 1993a: 177), whereas Said (1979, 1993) maintains that colonialism existed in order to impose the superiority of the European way of life on that of the Oriental, a colonisation of minds and bodies as much as that of space and economies and 'much harder to transcend or throw off' (Corbridge, 1993a: 178).

This myriad of possibilities does not mean to say that some common ground cannot be discerned or that broad phases of colonial development cannot be identified. Indeed, in order to make some sense of the colonial discourse, we need a framework within which

we can address it, identify its principal processes and set out the legacies that persist within current society, in both developed and developing countries. P.W. Preston (1996: 140) has identified several dimensions to what he terms 'the process of absorption and reconstruction of other peoples' (Figure 2.1). He does not elaborate on the links between his three parallel processes and suggests a rather simplistic sequence of ways in which Europeans represented the non-European world: first as exotic cultural equals, then as representatives of innocence and noble savagery during the Age of Enlightenment, subsequently as the uncivilised savages of the nineteenth century who had to be controlled, then improved and eventually guided to independence. Although this sequence is not untrue, it suggests a set of ideologies that were uniform over

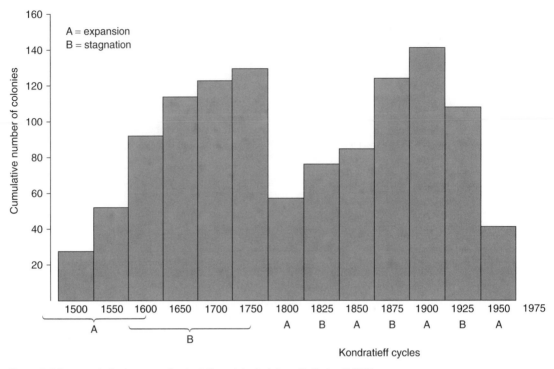

Figure 2.2 Long and short waves of colonialism. Adapted from P. Taylor (1985).

space and through time. The reality was of course very different, particularly over the long period from the early sixteenth to the early nineteenth century. The Enlightenment, as we have noted in Chapter 1, is a period in which the concept of development and the role of the 'enlightened' within this was crystallised. The romantic notion of the noble savage was applied rather sparingly during this period and was very much influenced by the nature of the non-European society encountered. Even Cook on the same journey around the Pacific could both admire the Polynesians and despise Australian Aborigines according to a particular set of British values.

The chronological sequence suggested by Preston, despite its oversimplicity, presents an approach which has been used by many other analysts of colonialism, particularly by world system advocates, such as Wallerstein (1979). P. Taylor (1985), in particular, has set out very clearly a sequence of waves or phases in which imperialism and colonialism combine to produce a series of long and short waves or cycles of development. The long waves coincide with major economic systems: feudalism, mercantilism and industrial capitalism. The shorter waves, often termed Kondratieff waves after their founder, are said to fit into the long

waves in roughly fifty-year cycles. All the waves are characterised by phases of growth and stagnation; during the stagnation phases economic restructuring occurs in order to re-establish economic strength. Such restructuring can involve one or more of a variety of actions, from the development of new technologies, through social change to new sources of raw materials or cheap labour. It is alleged that the acquisition of colonies formed part of this restructuring process by giving access to materials, food and labour. The long waves of mercantilism and industrial colonialism can be seen, therefore, to coincide with the rise and fall of major cycles of colonialism (Figure 2.2).

Despite its rigidity, the chronological sequences of world system theory do present a useful framework from which to examine phases of colonial–imperial development and within which the legitimating ideologies, material base and machineries of control and order may be examined. However, as many writers have noted (Kabbani, 1986; Said, 1993), the narratives from which we draw our material to interpret or read colonialism are themselves subject to deeply embedded prejudices that have found expression in both development thinking and development practice, albeit varying through time. In similar fashion, we need to

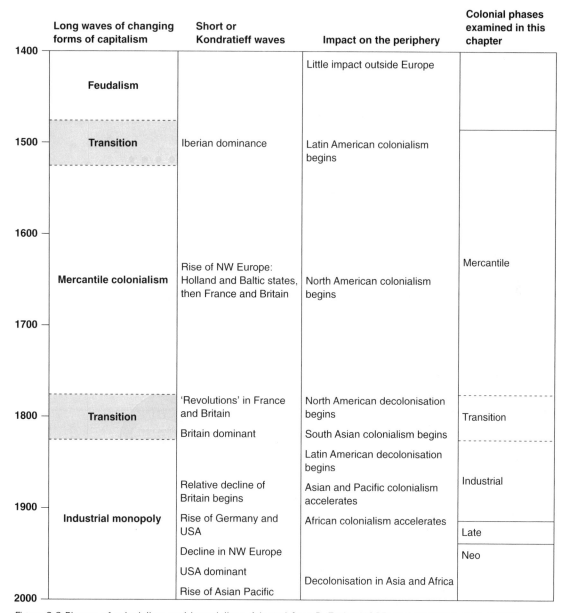

Figure 2.3 Phases of colonialism and imperialism. Adapted from P. Taylor (1985) and Bernstein *et al.* (1992).

be cognisant of the fact that the power exercised within colonialism is not homogeneous; it is often diffuse, fragmented, local and, above all, highly personalised.

In this account, *colonialism* is used to refer to the period from the beginning of the sixteenth century onwards in which economic and political motivations fused together to give spatial expression to the accelerating globalisation of capitalism. The phases of colonialism identified in Figure 2.3 were common to most parts of the non-European world but the chronology, rationale and reactions involved varied enormously. In Clapham's (1985: 13) words, there were 'the Americas, both rich and easy to control; Asia, rich but difficult to control; and Africa, for the most part poor and so scarcely worth controlling'. Latin America and sub-Saharan Africa therefore experienced more intensive plundering activities than either Asia or North America but during very different historical periods (Figure 2.4).

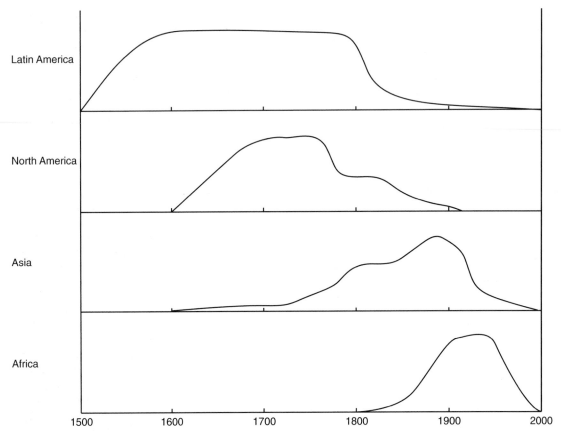

Figure 2.4 Regional colonialism: a chronology of the rise and fall in the numbers of colonies. Adapted from Lowder, 1986.

Indeed, most of both North and South America passed through these phases into independence, and in the case of South America into neocolonialism, before the intensive phase of the colonial project began in Asia or Africa. In parts of sub-Saharan Africa, formal colonies were a relatively short-lived political process, lasting only around half a century, although exploitation was present before and after this period. The phases indentified here are therefore indicative rather than definitive of the major changes that occurred in the expansion of colonial capital.

Phases of colonialism

Mercantile colonialism

The predominant features of this first phase of colonialism were commerce and trade, although in the earliest stages of contact in Latin America it was plunder and conquest which motivated the conquistadors. Furthermore, within North America and the Caribbean, trade and commerce were underpinned by production within the plantation system using slave labour (Blaut, 1993). In Africa, and more especially Asia, the initial contact was structured much more about commodity exchange. In this early period the impact of mercantile colonialism was determined by a variety of factors, including the type of European involvement, the nature of the commodities sought by the Europeans and the strength, culture and organisation of the non-European state.

The last of these varied enormously and European traders found themselves in contact with societies whose ways of life were in material, administrative and spiritual terms far superior to their own. China, for example, thought little of the European goods brought in trade, and as late as 1793 Emperor Chen Lung could condescendingly inform George III's emissary Earl Macartney that our celestial empire possesses all things in prolific abundance and lacks no product within its borders. There is therefore no need to import the manufactures of outside barbarians. . . . But as tea, silk and porcelain . . . are absolute necessities to European nations, and yourselves, we have permitted . . . your wants [to] be supplied and your country [to] participate in our beneficence.

For many years up to this date, European traders were forced to exchange their goods for silks and porcelain only at intermediary ports in Southeast Asia, such as Macao. Here the Europeans were regarded as just one more commercial community and were allocated their own quarter in the flourishing port alongside the Arab, Javanese and Chinese traders (T.G. McGee, 1967).

D.W. Preston (1996) summarises this first phase of colonialism as one in which non-Europeans were regarded as cultural equals, but this occurred only where there were trading goods to be competed for. Not all societies were at their peak when Europeans first encountered them, and the French were distinctly unimpressed by the Angkorian empire of the Khmers, whose 'hydraulic' or irrigation-based society had peaked in the twelfth century and whose extensive temple complexes offered little of interest to the European market. Although inherently precious commodities, such as gold or silver, attracted early Europeans, other exotic commodities such as silk, spices or sugar soon lured many adventurers to the appropriate parts of the world. Initially, trade with these 'distant others' was a high-risk enterprise into which vast sums were invested and from which huge profits were realised. To varying degrees these trading adventurers were accompanied by other kinds of Europeans, such as missionaries, emissaries or even scientists, curious about the non-European world.

The mercantile phase of colonialism in Asia and Africa lasted for some considerable time without extensive European settlement and with no uniform sign of the dominant–subordinate relationship which was to come later. In the Americas, the situation was quite different, with intensification of trade in the seventeenth and eighteenth centuries accompanied by much more extensive settlement from Iberia, France and Britain. Moreover, in North America and the Caribbean, the colonisers were heavily involved in the production process, something which did not occur in Asia and Africa until much later. However, as trade with these two continents grew in both volume and value, so it became more organised in its structure, usually within the context of the trading company. In Asia it was the Dutch that began this trend in the

seventeenth century, and soon the other European nations had their own East India companies too.

The acceleration in the scale and organisation of mercantile colonialism not only expanded profits but also involved increased European commitment to a physical presence in the trading region where commodities were to be assembled, stored and protected (Plate 2.1, overleaf). In order to acquire both commodities and protection, Europeans involved themselves increasingly in local politics, making alliances and inciting conflicts, all of which had enormous repercussions. In much of Africa the slave trade dramatically increased the power of Arab traders and local chiefs well away from the coastal area, intensifying conflicts because of the rewards that could be achieved through the sale of prisoners into slavery.

Although at this juncture the Europeans sustained only a relatively small physical presence in much of the non-European world outside the Americas, this varied enormously. Moreover, even with limited European settlement, change had occurred on an extensive scale by the end of the eighteenth century. The extended trading networks had increasingly drawn many parts and peoples of the non-European world into the capitalist system. European goods, particularly weaponry, European values and ideas, religious and secular, had penetrated most regions. Even where direct impact was still relatively limited, change occurred; for example, in Siam where the present Chakri dynasty was established in the mid eighteenth century through a series of reformist, modernising monarchs who sought to resist the Europeans by becoming more like them. For the great mass of peasants in Asia or Africa, however, life seemed to continue as it had for thousands of years, but their activities, whether subsistence or market-oriented, had over the long mercantile colonial period been subtly linked to a fledgling world economy, the core of which lay in Europe.

The mercantile colonial period merged into the era of industrial colonialism in a highly differentiated transition period. In North America, the United States had decolonised itself and was preparing to become an enthusiastic and powerful metropolitan power in its own right. In Latin America and the Caribbean, colonial production and trade were beginning to be challenged from within, if not by indigenous peoples. In Asia, the East India companies were going bankrupt as their shift into commodity production, in order to ensure quantity and quality of supplies, had escalated their costs of administration and protection. In Europe itself, political revolutions and continental-scale war consumed state resources and attracted the

Plate 2.1 Macao: remnants of Portuguese presence during the mercantile colonial period (photo: David Smith)

individual adventurers who had underpinned much of the mercantile colonialism. But above all, Europe offered new and lucrative profits for the reinvestment of accumulated merchant capital in its accelerating industrial transformation. Even so, the nineteenth century continued to resound to colonial concessions related to trade and commerce in parallel with the broader changes wrought by state-structured colonialism. Thus, the trading islands of Penang, Singapore and Hong Kong were acquired by Light, Raffles and Elliott on behalf of the crown rather than their companies, as were the treaty ports in China.

But if mercantile colonialism had begun to fade, its impact was already fuelling the Industrial Revolution and the renewed burst of colonialism which began in the nineteenth century. The fortunes that had been made from plunder, from commodity trade, and particularly from the triangular trade, were underpinning the accelerating industrial age. As Blaut (1993) argues, the point is not just that profits had been made but that they were in the hands of a new breed of entrepreneur rather than the old elite. The mercantile colonial period not only created new money, it had been accompanied by a social and political revolution, the combination of which gave Britain and other European powers a strong platform from which to launch into a more spatially extended and economic-

ally intensive form of colonialism. In short, by the late eighteenth and early nineteenth century 'capitalism arose as a world-scale process: as a world system. Capitalism became concentrated in Europe because colonialism gave Europeans the power both to develop their own society and to prevent development from occurring elsewhere' (Blaut, 1993: 206).

Industrial colonialism

Two changes characterised the colonialism of the nineteenth and twentieth centuries. The first was related to the dynamics of capitalism itself. Although commerce and trade still made money for the merchants of Liverpool, Bristol and London, the manufacturers themselves were eager to find methods of expanding production, or at least stabilising costs and extending their profits. Two obvious ways were to seek expanded and/or cheaper sources of raw materials and to find new markets overseas. A further development was to expand the production of cheap food overseas, thus lowering the costs of labour production in Europe by keeping wages down. All of these and more, although markets took a while to develop, were made available in the restructured colonies of the nineteenth century, colonies which were established and organised by the state rather than the company, although business and the state worked together through their representatives to transform production, consumption and cultures. The key to this process was territorial acquisition: before 1870 annexation and occupation tended to follow resource exploitation; after 1870 they tended to precede it.

Although the needs of capitalism may have been the driving force behind the industrial colonialism of the nineteenth century, the rationale for the colonial project itself was provided by a consolidation of the ideology of justifiable intervention and occupation of what had become either 'uncivilised savages' or traditional groups whose history was ignored and whose societies and activities were seen as either static or disintegrating (Box 2.3).

Science, reason and, above all, organisation for most nineteenth century thinkers elevated Europeans to their superior position and placed them above the brutality, poverty and imminent death of the peoples in their occupied lands. For Porter (1995) there were 'master metaphors' provided by physics (stability, equilibrium) and biology (constituent parts functioning for the whole) which shaped the ideologies of both colonialism and development. These gave rise to a modernist theme, a universal process of change which is clear and predetermined (Porter, 1995). Fine or sympathetic motives could therefore be written into this process underpinned by a parent–child metaphor (Manzo, 1995), often expressed vividly by the image of the 'mother country' and her fledgling colonies.

BOX 2.3   The scramble for Africa

Although colonialism expanded rapidly throughout the nineteenth century, the speed of Africa's partition was new. Not all the impetus came from outside the continent, although much of it did; some processes were chronologically or spatially specific.

External processes

1 Between the first and second industrial revolutions (i.e. the shift from coal and iron to oil, electricity and steel), Europe went through a deep recession. As rates of profit fell, European firms began to seek new material sources, new markets and new investment opportunities on an extensive scale, partly to forestall other European rivals.
2 The last quarter of the nineteenth century saw the newly united countries of Italy and Germany using colonialism to sidestep internal tensions. As Africa was the largest uncolonised area, it became the site of a national scramble for territory and prestige, with France seeking compensation for her defeat by Germany within Europe. In this process, governments were supported by a popular imperialism created and sustained by a jingoistic media boom in newspapers, journals and books.
3 These processes were facilitated by a technological revolution, particularly in transport, where steamships, railways and telegraphic links accelerated both decision making and physical advances into Africa. New armaments, such as machine-guns, facilitated this process by smaller and smaller European forces.
4 Once acquired, many colonies were also seen as healthy places for the surplus European population that technological advances in medicine and hygiene were beginning to produce. Geographers played an important part in this deterministic process.

Internal processes

1 The acceleration and intensification of European capitalism brought about a breakdown of existing relationships between traders and African societies, goading many of the African societies into reaction and providing excuses for further European invasion.
2 Sub-imperialism occurred when European settlement became extensive and decision making was wrested from the metropolitan centre by ambitious local individuals or groups, forcing retrospective recognition of highly personalised adventurism. Cecil Rhodes provided the most blatant example of such actions.

Box 2.3 continued

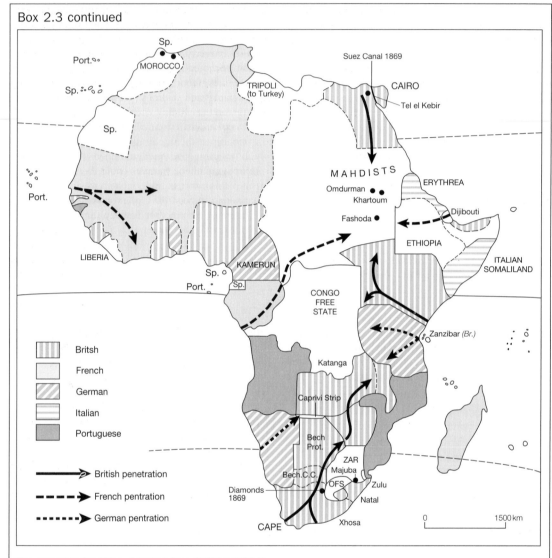

The scramble for Africa. Adapted from Griffiths (1994).

3 Some African groups facilitated and accelerated colonial process by 'inviting' Europeans to 'collaborate' against other groups. Often, however, such invitations were manufactured.

Taken individually, no single reason explains the sudden scramble for Africa, but the conjuncture of many factors in the 1870s gave rise to a spiral of European ambition and nationalism that once started, proved difficult to stop.

We should be careful not to overemphasise the power of ideology and discourse within this process; colonialism had at its heart the economic drive for profit. Every Sir Alfred Milner had his Cecil Rhodes whispering or bellowing into his ear about the returns to investment that will follow annexation and control of yet another piece of territory occupied by traditional people not using it to its full potential. To be sure, there was also prestige for the 'mother country', annexation would be one in the eye for other European rivals and would promote pastures new for grazing missionaries of the true church; but the underlying impetus was usually greed.

Whatever their composition, ideologies need to be translated into action and for colonialism this was through the elaboration and enablement of first, its material base, i.e. production, trade and finance; and second, the establishment of the administrative machinery of control and order. There is no necessary sequence in this process, annexation and the provision of administrative structures could follow economic interests, as in the Transvaal, or could precede them as in the French occupation of Indo-China. There was, however, clearly an expanded role for the state *vis-à-vis* the trading company in the administrative system. In contrast to the mercantile period, the main medium of exploitation was not the trading concessions, although they continued to be squeezed out of 'independent' states, but the acquisition of land on which to organise the mechanics of production. The colonial state then established the infrastructure of legal, transport, administrative and police systems through which the pursuit of wealth and order could be controlled. It is no coincidence that Sir Harry Johnson (Box 2.4) in his address to the Royal Geographical Society in 1895 attributed the transformation of Mlanje to the fact that the natives 'above all, are trained to respect and to value settled and civilised government' (Crush, 1995: 2).

But if the colonial state was an administrative state, it was usually a productive state too, since it was regarded as right and proper for the metropolitan state, metropolitan companies and metropolitan individuals to secure a profitable return on their investment.

The spatial expression of economic exploitation was experienced for the most part in the rural areas in which the export commodities were produced. This varied substantially according to the nature of the commodity, local customs and the metropolitan power involved. In some areas, agricultural restructuring occurred through the creation of large-scale plantations; in others local producers were encouraged to amalgamate their holdings. Both processes resulted in large-scale landlessness, creating labour pools for the new commercial holdings. Others were shifted into agricultural or mining industries by new poll taxes that forced farmers into wage labour to meet these demands, often ruining prosperous and well-organised indigenous systems. Where local labour proved to be 'inadequate' for commercial agriculture, workers were often imported from elsewhere in the country (as in Vietnam where the French shifted workers from north to south) or from overseas (as in Malaya where the British brought in workers from India). However, labour also moved 'voluntarily', recruited through family or kinship systems (e.g. from south China to the Malayan tin mines).

The new agricultural systems often meant that, over large areas, the range of crops produced was narrowed to those commodities required by metropolitan industries. Colonies thus became associated with the production of one or two items, being forced to import whatever else was needed. Needless to say, metropolitan firms were in control of both directions of trade. Although some of these commodities were new introductions to the colonies, such as rubber or coffee in various Southeast Asian countries, more traditional crops continued to play an important role, e.g. coconuts. Local food crops too became an important

---

BOX 2.4   The nineteenth-century logic of colonialism

The discourse of colonialism which first justified then ratified colonial intervention is well expressed by Jonathan Crush in his book *Power of Development*, where he caricatures the transformation of Mlanje through the eyes of Sir Harry Johnson, its fictional governor (Crush, 1995: 1–2). His first description is of Mlanje in 1895 before colonialism extends its benign hand to that unfortunate land:

> In the Mlanje District there was practically chaos . . . throughout all this country there was absolutely no security for life and property for natives, and not over-much for the Europeans. . . . Everything had got to be commenced.

This picture of scorned opportunities was blamed on disinterested local tribes and evil-minded slave traders and was contrasted by Sir Harry Johnson with the scene after just three years of British rule as a placid paradise where

> a planter gallops past on horseback . . . long rows of native carriers pass in Indian file, carrying loads of European goods, or [if prisoners] out mending roads under the superintentence of some very businesslike policeman of their own colour . . . the most interesting feature in the neighbourhood of these settlements at the present time is the coffee plantation which, to a certain extent is the cause and support of our prosperity.

export crop. Siamese rice was exported to many other Asian countries, largely through British firms, where it helped lower the cost of labour reproduction, particularly in the cities (Dixon, 1998).

As a result of the drastic economic, social and demographic changes of industrial colonialism, the last quarter of the nineteenth century also witnessed the acceleration of market potential for Western manufactured products. Indeed, in Pacific Asia it was the purchasing power of the Chinese and Japanese markets that was as important as access to their products in encouraging the Western powers in their almost frantic attempts to gain trading concessions. In the colonies themselves, the initial markets for Western goods were confined to wealthy expatriate and indigenous elites. But the quality and price of these goods and the demonstration effect of purchases by the wealthy, soon resulted in imported commodities dominating the expenditure pattern of all social groups, even those in rural subsistence, thus further destroying the indigenous artisan economy and increasing dependency on the West.

In many areas, however, where indigenous manufacturing posed a real threat, local industries were quickly suppressed, as in the Indian textile towns (Blaut, 1993). Particularly poignant in these circumstances was the re-export of cheap food to the growing markets among the urban and rural poor. Their diet of flour, sugar and tea often had colonial origins but was processed (and value-added) in Europe, thus facilitating a double exploitation of the colonial poor (through their labour in growing the crops, and through their subsequent purchase of it at exorbitant prices). The corollary of this situation is that manufacturing was relatively limited during the industrial colonial phase. Any manufacturing that existed was largely concerned with the preliminary processing of primary products, such as rice milling or tin smelting. Most of the more sophisticated processing (and creation of profits) occurred within the larger parts of the metropolitan country. This is the era during which the big dockside manufacturing plants for tobacco and sugar proliferated in Liverpool, Glasgow and London.

However, it would not be correct to assume that colonial cities were simply points of control and administration. Although few were centres of production, commercial activity, from the manufacture of small consumer goods to the retailing of imported products, was very extensive. Much of this activity was in the hands of non-Europeans. This is not the same as saying they were the prerogative of local entrepreneurs because almost all of the colonial powers in East Africa and in Pacific Asia, other than the Japanese, made a point of encouraging or permitting immigrant groups,

usually Chinese or Indian, to infiltrate and monopolise local commerce. In this way, a convenient demographic, cultural and economic buffer was placed between the colonised and the colonialists. Discontent on the part of indigenous populations with the cost of living was therefore often directed against those immediately available rather than those ultimately responsible.

It cannot be emphasised too much, therefore, that the period from 1850 to 1920 saw a massive restructuring of urban systems (Drakakis-Smith, 1991). Colonial production may have been based in the countryside but colonial political and economic control was firmly centred on the city. Usually just one or two centres were selected for development, giving rise to the urban primacy which remains characteristic of many developing countries today. Within these cities, despite the numerical dominance of the indigenous populations, most of the land space was given over to European activities. Spacious residential and working areas were paralleled by extensive military cantonments, all separated from the usually cramped, crowded indigenous city by railway lines, parks or gardens. Little face-to-face contact took place between the colonisers and the colonised, except within a dominant–subordinate relationship. It was a situation that seemed as though it would go on for ever, but the First World War intervened and widespread changes ensued. Within a generation the political world-order of 1914 was totally undermined and the sun began to set rapidly over the colonial empires.

Late colonialism

A fundamental change occurred in the ethos of colonialism after 1920. Put simply, the 'heroic' age of creating empires gave way to a more prosaic phase of imperial governance. The key to this change is the concept of trusteeship which had permeated the formation of the League of Nations and which elevated to a high priority the well-being and development of colonial peoples. In practice this did not necessarily mean indigenous colonial peoples. Indeed, prevailing anthropological theory conveniently explained that such progress was impossible for 'backward' and 'traditional' societies which did not hold in proper esteem social values such as democracy or the business ethic. Not until after 1945 did metropolitan governments seriously consider fairer representation for indigenous interests, but this was too little, too late.

Between the wars, therefore, colonial government was dominated by bureaucrats, both in metropolitan capitals and overseas, striving on behalf of the colonies with little appreciation either of indigenous aspirations or of the changing world economy in which they were situated. The world wars and the intervening depression severely disrupted colonial economies, with investment from Europe being limited and commodity prices falling steadily. Much of the capital sustaining growth in this interwar period was American, or, in Asia, from the overseas Chinese community. Thus in the Dutch East Indies 80 per cent of all domestic trade was controlled by Nanyang Chinese whereas the largest rubber plantation was owned by a US tyre company. The declining profitability of commodity exports helped cause a shift in the nature of government investment during this period, with increasing amounts of capital being invested in infrastructure – roads, utilities or railways.

During this economic recession official metropolitan and colonial ties grew closer. Thus, by 1930 some 44 per cent of British trade was with the empire. This was reinforced by demographic changes, with increasing numbers migrating away from the European recession to the perceived opportunities of the colonies, often encouraged enthusiastically by their governments. Although most British migrants went to the settler colonies, such as Canada or Australia, most Dutch and French migrants moved to already heavily populated colonies and were forced into a variety of urban occupations rather than farming as in earlier decades. Many unqualified migrants took up relatively low paid work in retail or office locations. The effect of this growing European presence was complex. For the small educated indigenous group, it made personal advancement even more difficult, sending many off to Europe for further education and sharpening their political and organisational skills. The huge peasant population, suffering dreadfully from the recession, were not organised enough to threaten more than the local representatives of the system. But the growing urban indigenous population posed more of a problem – not an overtly political problem in the late colonial period, rather one associated more with social issues.

For the majority of European colonists, the interwar years seemed like 'a golden age', despite the economic vicissitudes. Even those with more modest incomes had status and privilege relative to the indigenous population which compensated somewhat. This was the languorous era from which most contemporary images of colonialism are drawn in the media, particularly in the cinema, e.g. *Out of Africa*. The luxurious way of life was facilitated by the shift in the balance of administrative power from the metropolitan centre to the colonies, certainly in the British Empire, so that salaries, privileges and jobs remained secure and rewarding, notwithstanding the economic situation. This shift in power also enabled the administrators to more or less ignore trusteeship as far as indigenous populations were concerned, favouring the colonial settlers in numerous ways. Urban planning became a distorted version of European concepts, so the garden city movement in Africa and Asia was largely used to further segregate European and indigenous population by swathes of greenery, particularly recreational areas such as golf-courses and racecourses. Despite the health risks, burgeoning indigenous populations were often crammed into crumbling 'old quarters'. In 1911 in Delhi, on the eve of the emergence of the new imperial capital, the old Mughal city of Shahjahanabad contained almost a quarter of a million people in its 2.5 square miles (King, 1976).

The Second World War destroyed the myth of European invincibility, particularly in Asia as a result of Japanese victories, although ironically the eventual success of the European allies owed not a little to the colonial or imperial forces that served under the metropolitan flag and to the resources that the colonies continued to supply. In the postwar period, however, decolonisation was strongly encouraged by the United States, partly because it wished to extend its own sphere of economic influence through a kind of informal imperialism or neocolonialism. Not all the metropolitan powers responded to this call or to the call of the colonial peoples for independence. Britain found it easier than most to withdraw because of its less formal and more decentralised administrative systems, although there were one or two places such as southern Rhodesia where settler resistance saw the colonial sun set rather more slowly than elsewhere. For France and the Netherlands, however, decolonisation was an altogether more complex and bloody affair, with settlers resentful of being abandoned and unwilling to return to a war-ravaged Europe putting up fierce resistance. Brutal and cruel decolonisation wars resulted in Algeria, the East Indies and Indo-China. Eventually, however, independence came to all except a few small territories, but their development paths were strongly affected not only by the neocolonial forces that quickly moved into the political and economic vacuum but also by the considerable legacies that colonisation had bequeathed to these nascent states.

The legacies of colonialism

Some historians would argue that colonialism was too varied and, in some instances, too brief to judge whether it has left a beneficial or detrimental legacy.

Certainly the local textures of colonialism were immensely complex but as a component of a broader, changing global process, there were immense overall repercussions. Politically, the national units that emerged between the 1940s and 1970s essentially comprised the territorial divisions of the colonial era and often had limited correlation with environmental geographies, precolonial structures or with contemporary cultural and ethnic patterns. Thus the Malay world of Southeast Asia was divided between three countries, Malaysia, Indonesia and the Philippines, in accordance with the territorial spread of the British, Dutch and Spanish-American empires. Even non-Malay areas were incorporated into these territories if they had been part of the colonial territory. Thus Indonesia has held on to Melanesian Irian Jaya despite the cultural and ethnic contradictions and is currently attempting to flood the area with Javanese migrants as part of its accumulation process. As Clapham (1985: 20) has remarked, 'There is still no more striking, even shocking, reminder of the impact of colonialism in Africa than to cross an entirely artificial frontier and witness the instant change of language . . . that results.' Moving from Zambia (English) to Angola (Portuguese) illustrates this comment very vividly.

The demographic legacy of colonialism was also considerable. During the late colonial period, the rate of indigenous population growth began to accelerate, partly due to transfers of medical technology, improvements in health care and hygiene and wider food security. Not that they were provided on either a widespread or enthusiastic basis by the colonial authorities, but improvements did occur and fertility did rise to the extent that the newly independent nations inherited an accelerating population growth. Perhaps even more important, however, was the mixing of populations that occurred as a result of labour movements, whether forced, contracted or voluntary. Add to this the multiple ethnic groups that were already in the artificial colonial states which became independent territories, and the consequence has been an ethnic melting-pot that has simmered and bubbled almost everywhere during the post-colonial period, at best considerably hampering the development process, at worst resulting in appalling acts of expulsion or genocide, as in East Africa.

In material terms too, the legacies of colonialism have been substantial, particularly in the form of transport and communications links. It is often claimed that the road and rail networks of many contemporary states owe their origins to the colonial era. Unfortunately, most of these lines of communication reflect the economic needs of the colonial powers and do not necessarily coincide with the contemporary needs of independent states. In Zimbabwe the principal road and rail links still connect the main areas of European settlement and economic activity on the high veld, and exit through what was British South Africa (Figure 2.5). As a result, transportation in the communal areas of the country is much inferior, hindering development, and has been a priority of the hard-pressed government since independence.

Transport and communication links were, in a way, subordinate to the urbanisation process which, as we have seen, was either established for the first time or completely restructured during the colonial period to favour one or two major ports that functioned as crucial connecting points between the colonial and global economies. Many instances of exaggerated urban primacy in the contemporary world have their origins in the colonial economy. Within these cities too, there is often a considerable physical legacy of colonial triumphalism in architectural form, planning layout and infrastructure provision (Plates 2.1 and 2.2 pages 32 and 40) (King, 1990).

Colonial cities were points of administration, rather than production, and some would argue that the most useful legacy of colonialism is the administrative, legal and judicial systems that were established by the metropolitan powers. There is certainly some truth in this as far as orderly and efficient administration was concerned. But the elitism inherent within these systems still continues in their successors, particularly the use of European languages in the highest echelons of the state. However, fairness was rarely a characteristic of colonial administrative and legal systems, and here again some most unfortunate legacies have persisted. Thus Singapore, like many other countries, retained and uses legislation on detention without trial; elsewhere representation of populations in parliament is often restricted by outmoded franchise conditions. In similar fashion, building regulations and standards retained from the colonial period have long prevented more effective housing policies being pursued within contemporary Third World cities.

What underpinned almost all these consequences of colonialism were its economic activities and here the direct legacy has also been substantial. The narrowing of production into one or two commodities persisted in the great majority of countries, even to the same commodities being grown. This posed enormous problems for those countries, the great majority of which watched commodity prices continue to fall and failed to diversify their economies. And the spatial

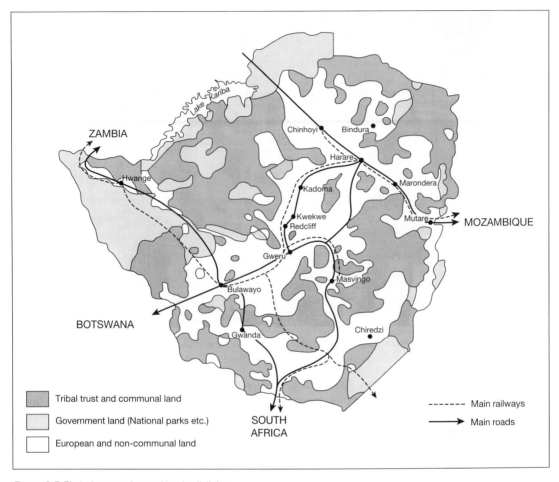

Figure 2.5 Zimbabwe: main road and rail links.

concentrations of these activities continues to cause problems of regional inequality and imbalance, often reinforced by post-colonial urbanisation trends. For some twenty-five years after the end of the Second World War, the economies of the newly independent states of Asia and Africa remained virtually unchanged under the development strategies of neoclassical advisers such as Lewis or Rostow (Chapter 3). These economies were still reliant on the export of a narrow range of primary commodities, possibly with some diversification into import substitution industries. For the most part, these economies were still controlled from the outside through the medium of tied aid or the activities of transnational corporations (TNCs). Although metropolitan powers continued to be linked in this way with their former colonies, new international players were equally dominant, particularly the United States (Table 2.1, page 41).

From the 1970s this situation begins to change, with a more sustained and substantial outflow of investment funds from Europe and North America as industrial profits begin to decline in those areas. The various reasons for this are often related to the rising costs of labour and environmental protection. New financial systems, particularly artificial international currencies, such as Eurodollars, and the windfall profits from the OPEC oil price rise, all facilitated the investment of money overseas in new manufacturing plants in the Third World. The usual term to describe this post colonial phenomenon is the new international division of labour (NIDL), in which the low-cost labour-intensive parts of the manufacturing process are siphoned off to the developing world where costs are lower. A detailed critique of this fragmentation of the production process and the role played by TNCs can be found later in this volume. However, at this

Plate 2.2 Grand French colonial architecture in Dakar, Senegal (photo: Tony Binns)

point and as a link to Chapter 3, it might be useful to examine the concept of NIDL as it has become almost synonymous with the post-colonial period.

NIDL is perhaps the third international division of labour. The first comprised the production and extraction of primary commodities in the colonies and their manufacture in the metropolitan countries; the second involved the shift of some industry into the newly independent countries under import substitution policies; the third began to expand in the 1970s and involves the fragmentation of the manufacturing process and the shift of a large proportion of this to developing countries, largely through the medium of TNCs. It is important to realise that all three international divisions of labour are currently in operation; in particular, large amounts of TNC investment continue to sustain primary resource production in countries such as Papua New Guinea, almost all of which is exported to advanced capitalist economies.

The latest international division of labour came about through a conjuncture of changes: reduced profitability in Europe, largely through increased production costs, cheap production costs in developing countries, encouragement given to urban-industrial growth in the Third World by international development agencies, facilitating developments in communications technology and the parallel increasing mobility and flexibil-

ity of financial services. The result was new levels of extraction of surplus value created by superexploitation of Third World labour, which accrues few skills and with limited backward linkages into the local economy.

The problem with NIDL is that it overemphasises the inevitability of this exploitation and fails to credit peripheral social formations with any autonomy to manipulate foreign investment to maximise local advantage. The concept is also undermined in its global applicability by the fact that NIDL was initially very selective, with just six countries, usually identified as the four Asian tigers together with Mexico and Brazil, receiving the brunt of TNC investment, largely because TNCs rely on more than just cheap labour for efficient and profitable production. It can be argued that the range of countries which are industrialising through the medium of foreign investment is widening. In particular, analysts point to the rapid rise of countries such as Malaysia, Thailand and Indonesia. But these countries receive much, if not most, of their investment from within the Asia-Pacific region, notably from the four little tigers who used NIDL to retain surplus value and generate regional investment. It is clear in this context that national governments have not been overwhelmed or displaced by Western TNCs but act in concert with indigenous equivalents to

Table 2.1 Foreign direct investment by TNCs, 1960–1985.

	1960			1975			1980			1985		
	Value	% total	% GDP	Value	% total	% GDP	Value	% total	% GDP	Value	% total	% GDP
Developed market economies	67.0	99.0	6.7	275.4	97.7	6.7	535.7	97.2	6.7	693.3	97.2	8.0
United States	31.9	47.1	6.2	124.2	44.0	8.1	220.3	40.0	8.2	250.7	35.1	6.4
United Kingdom	12.4	18.3	17.4	37.0	13.1	15.8	81.4	14.8	15.2	104.7	14.7	23.3
Japan	0.5	0.7	1.1	15.9	5.7	3.2	36.5	6.6	3.4	83.6	11.7	6.3
German Federal Republic	0.8	1.2	1.1	18.4	6.5	4.4	43.1	7.8	5.3	60.0	8.4	9.6
Switzerland	2.3	3.4	26.9	22.4	8.0	41.3	38.5	7.0	37.9	45.3	6.4	48.9
Netherlands	7.0	10.3	60.6	19.9	7.1	22.9	41.9	7.6	24.7	43.8	6.1	35.1
Canada	2.5	3.7	6.3	10.4	3.7	6.3	21.6	3.9	8.2	36.5	5.1	10.5
France	4.1	6.1	7.0	10.6	3.8	3.1	20.8	3.8	3.2	21.6	3.0	4.2
Italy	1.1	1.6	2.9	3.3	1.2	1.7	7.0	1.3	1.8	12.4	1.7	3.4
Sweden	0.4	0.6	2.9	4.7	1.7	6.4	7.2	1.3	5.8	9.0	1.3	9.0
Other	4.0	5.9	3.1	8.5	3.0	1.7	17.4	3.2	1.9	25.6	3.6	3.3
Developing countries	0.7	1.0	–	6.6	2.3	–	15.3	2.8	2.8	–	19.2	–
Centrally planned economies of Europe	–	–	–	–	–	–	–	–	–	1.0	0.1	–
Total	67.7	100.0	–	282.0	100.0	–	551.0	100.0	–	713.5	100.0	–

Source: UN Centre on Transnational Corporations (1988) *Transnational Corporations in World Development: Trends and Prospects*, United Nations, New York.

promote joint interests overseas. What has emerged, therefore, is a regional division of labour (RDL) operating both within and in conflict with NIDL. Nascent RDLs exist in other areas of the developing world too, such as the Middle East or South Asia or even southern Africa but the extended world recession of the 1980s and 1990s will slow their emergence considerably.

Often confused with NIDL, largely because it first surfaced at about the same time was the new international economic order (NIEO). This concept mainly derived from the United Nations and was underpinned by a mounting concern with the failure of modernisation strategies (Chapter 3) to achieve much for most of the people in the Third World. Mounting poverty, growing national debts and the pessimistic environmental predictions of the Club of Rome all resulted

in the declaration by the UN in 1994 of its intention to establish an NIEO, although we must also be aware of the self-interest of Western nations threatened by recession, oil price hikes and inflation. The five principal areas of concern identified were trade reform, monetary reforms, debt relief, technology transfer and regional cooperation. In fact, very little was achieved for the lasting benefit of the developing nations, not least because of their growing diversity of interests (Chapter 1), and eventually the NIEO became suffused into the structural adjustment programmes of the New Right. In effect, the NIEO became a justification for TNC investment in the Third World in the name of development, and as such was a factor in the internationalisation of the division of labour which paralleled it.

CHAPTER 3

Strategies of development

Introduction

A major characteristic of the interdisciplinary field of development studies since its establishment in the 1940s has been a series of sea-level changes in thinking about the process of development itself. This search for new conceptualisations of development has been mirrored by changes in development practice in the field. Thus, there has been much debate and controversy about development, with many changing views as to its definition, and the strategies by means of which, however defined, it may be pursued.

In short, the period since the 1950s has seen the promotion and application of many varied geographies of development and there is much vitality in the field (Apter, 1987; Corbridge, 1986; Slater, 1992a, 1992b, 1993; Schuurman, 1993; O'Tuathail, 1994; Hettne, 1995; M.P. Cowen and Shenton, 1996; Brohman, 1996; Leys, 1996; P.W. Preston, 1987, 1996; Streeten, 1995). The purpose of this chapter is to provide an introduction to the different approaches to development which have been proposed, and followed, since around 1940, both in theory and practice. These different approaches reflect, of course, the changing paradigms (or supramodels) of development that were outlined in Chapter 1. Many of these paths to development will be further elaborated in Parts II and III.

A major theme in this account will be to illustrate that ideas about development have long been highly controversial and contested. This argument is further highlighted in the conclusion to the chapter, which sets out briefly some of the relationships between development theory and the societal conditions of modernity and post-modernity. The linked contention that development theory and development studies have currently reached an impasse or deadlock is also considered as part of this account.

Theories, strategies and ideologies of development

The account provided in this chapter is very wide and eclectic. Accordingly, a broad definition of paths to development is adopted at the outset. To use Hettne's (1995) nomenclature, the chapter reviews selected aspects of development theories, development strategies and development ideologies. Before doing so, these three basic terms must be defined.

Development theories may be regarded as sets of apparently logical propositions, which aim to explain how development has occurred in the past, and/or should occur in the future. Development theories can either be normative, when they generalise about what should be the case in an ideal world, or positive in the sense of dealing with what has been the case (Figure 3.2). The arena of development theory is primarily, although by no means exclusively, to be encountered in the academic literature. On the other hand, development strategies can be defined as the practical paths to development which may be pursued by international agencies, states in the First, Second and Third Worlds, non-government organisations and community-based organisations, in an effort to stimulate change within particular nations and regions and continents.

Different development agendas will reflect different goals and objectives. These goals will reflect social, economic, political, cultural, ethical, moral and even religious influences. Thus, what may be referred to as different development ideologies may be recognised. Chapter 1 stressed how, both in theory and in practice, early perspectives on development were almost exclusively concerned with promoting economic growth. Subsequently, however, the predominant ideology within the academic literature changed to emphasise political, social, ethnic, cultural, ecological

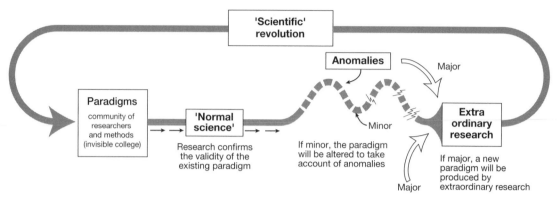

Figure 3.1 Scientific revolutions: picturing Kuhn's model of their structure.

and other dimensions of the wider process of development and change.

Perhaps the only sensible approach is to follow Hettne (1995) and to employ the idea of development thinking, in the body of this chapter. The expression *development thinking* may be used as a catch-all phrase indicating the sum total of ideas about development, including pertinent aspects of development theory, strategy and ideology. Thus, the present chapter takes a very broad remit in presenting an overview of development strategies.

Such an all-encompassing definition is necessary due to the nature of thinking about development itself. Development thinking has shown many sharp twists and turns during the twentieth century. The various theories that have been produced have not commanded attention in a strictly sequential manner. In other words, as a new set of ideas about development has come into favour, earlier theories and strategies have not been totally replaced. Rather, theories and strategies have tended to stack up upon one another, coexisting in what can sometimes be described as a very convoluted manner. Thus, in discussing development theory, Hettne (1995: 64) has drawn attention to the 'tendency of social science paradigms to accumulate rather than fade away.'

This characteristic of development thinking as a distinct field of enquiry can be considered in a more sophisticated manner, using Thomas Kuhn's ideas on the structure of scientific revolutions (Figure 3.1). Kuhn (1962) argued that scientific disciplines are dominated at particular points in time by communities of researchers and their associated methods, and they define the subjects and the issues deemed to be of importance within them. He referred to them as 'invisible colleges', and he noted that they serve to define and perpetuate research which confirms the validity of the existing paradigm (or supramodel). Kuhn referred to this as 'normal science'. Kuhn noted that a fundamental change occurs only when the number of observations and questions confronting the status quo of normal science becomes too large to be dealt with by means of small changes. However, if the proposed changes are major, and a new paradigm is adopted, a scientific revolution can be said to have occurred, linked to a period of so-called extraordinary research.

In this model, therefore, scientific disciplines basically advance by means of revolutions in which the prevailing normal science is replaced by extraordinary science, and ultimately a new form of normal science develops. In dealing with social scientific discourses, it is inevitable that the field of development theory is characterised by evolutionary rather than revolutionary change. Evidence of the persistence of ideas in some quarters, years after they have been discarded elsewhere, will be encountered throughout this chapter, and indeed through this book. Given that development thinking is not just about the theoretical interpretation of facts, but rather about values, aspirations, social goals, and ultimately that which is moral and ethical, it is understandable that change in development theory leads to the parallel evolution of ideas, rather than revolution. Hence conflict, debate, contention and positionality are all inherent in the discussion of development strategies and associated plural and diverse geographies of development.

There are many ways to categorise development thinking through time. Broadly speaking, it is suggested here that four major approaches to the examination of development theory can be recognised, and these are shown in Figure 3.2. This follows the framework recently suggested by Potter and Lloyd-Evans (1998). The four approaches are (i) the classical–traditional

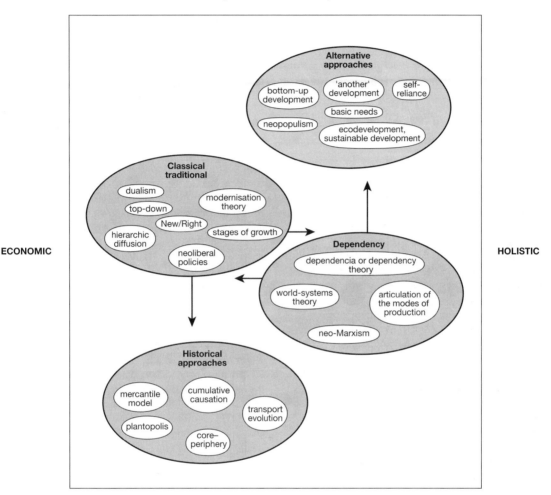

NORMATIVE
THEORY
(what should be the case)

Alternative
approaches

bottom-up
development

'another'
development

self-
reliance

basic needs

neopopulism

ecodevelopment,
sustainable development

Classical
traditional

dualism

modernisation
theory

top-down

New/Right

hierarchic
diffusion

stages of growth

neoliberal
policies

Dependency

dependencia or dependency
theory

world-systems
theory

articulation of
the modes of
production

neo-Marxism

ECONOMIC   HOLISTIC

Historical
approaches

mercantile
model

cumulative
causation

transport
evolution

plantopolis

core–
periphery

POSITIVE
THEORY
(what has been the case)

Figure 3.2 Development theory: a framework for this chapter.

approach, (ii) the historical–empirical approach, (iii) the radical–political economy–dependency approach and (iv) bottom-up and alternative approaches. Following the argument presented in the last section, each of these approaches may be regarded as expressing a particular ideological standpoint, and can also be identified by virtue of having occupied the centre stage of the development debate at particular points in time.

However, each approach still retains currency in some quarters. Hence, in development theory and academic writing, left-of-centre socialist views may well be more popular than neoclassical formulations, but in the area of practical development strategies, the 1980s and 1990s have seen the implementation of neoliberal interpretation of classical theory, stressing the liberalisation of trade, along with public sector cutbacks, as a part of structural adjustment programmes (SAPs), aimed at reducing the involvement of the state in the economy. Notwithstanding this caveat, the account which follows uses these four divisions to overview the leading theories which have been used to explain and promote the development process.

Classical–traditional approaches: early views from the developed world

The classical approach to the study of development derives from neoclassical economics and totally dominated thinking for a period close on forty years. Summarised in simple terms, these approaches argue that developing countries are characterised by a dualistic structure. Hettne (1995) notes the strong role of dichotomous thinking in early anthropology, where comparisons were made between backward and advanced societies, the barbarian and the civilised, and the traditional and the modern. The fundamental dualism exists between what is seen as a traditional, indigenous, underdeveloped sector on the one hand, and a modern, developed and Westernised one on the other. It follows that the global development problem is seen as a scaled-up version of this basic dichotomy. Seminal works include those of Meier and Baldwin (1957), Perloff and Wingo (1961), Schultz (1953), Perroux (1950), Myrdal (1957) and Hirschman (1958).

The general economic development model of the American economist A.O. Hirschman forms a convenient starting point for discussion of the approach. Hirschman (1958), in his volume, *The Strategy of Economic Development*, advanced a notably optimistic view in presenting the neoclassical position (Hansen, 1981). Specifically, Hirschman argued that polarisation should be viewed as an inevitable characteristic of the early stages of economic development. This represents the direct advocacy of a basically unbalanced economic growth strategy, whereby investment is concentrated in a few key sectors of the economy. It is envisaged that the growth of these sectors will create demand for the other sectors of the economy, so that a 'chain of disequilibria will lead to growth.' The corollary of sectorally unbalanced growth is geographically uneven development and, Hirschman specifically cited Perroux's (1955) idea of the natural growth pole.

The forces of concentration were collectively referred to by Hirschman as polarisation. The crucial argument, however, was that eventually, development in the core will lead to the 'trickling down' of growth-inducing tendencies to backward regions. These trickle-down effects were seen by Hirschman as an inevitable and spontaneous process. Thus, the clear policy implication of Hirschman's thesis is that governments should not intervene to reduce inequalities, for at some juncture in the future, the search for profits will promote the spontaneous spin-off of growth-inducing industries to backward regions. Hirschman's

approach is therefore set in the traditional liberal model of letting the market decide. The process whereby spatial polarisation gives way to spatial dispersion out from the core to the backward regions has subsequently come to be known as the point of 'polarisation reversal' (Richardson, 1977, 1980).

The full significance of these ideas concerning polarised development extends beyond their use as a basis for understanding the historical processes of urban-industrial change, for in the 1950s and 1960s they came to represent an explicit framework for regional development policy (J. Friedmann and Weaver, 1979). Thus, the doctrine of unequal growth gained both positive and normative currency in the first postwar decade and the path to growth was actively pursued via urban-based industrial growth. The policies of non-intervention, enhancing natural growth centres, and creating new induced subcores became the order of the day. As J. Friedmann and Weaver (1979: 93) observe, the 'argument boiled down to this: inequality was efficient for growth, equality was inefficient,' so that 'given these assumptions about economic growth, the expansion of manufacturing was regarded as the major propulsive force.'

Hirschman's ideas can be seen as part of a wider modernisation theory, which was in vogue during the 1950s and 1960s. The paradigm was grounded on the view that the gaps in development which exist between the developed and developing countries can gradually be overcome on an imitative basis. The emphasis was placed on a simple dichotomy between development and underdevelopment. Thereby developing countries would inexorably come to resemble developed countries, and 'in practice, modernization was thus very much the same as Westernization' (Hettne, 1995: 52). The modernisation thesis was largely developed in the field of political science, but was taken up by a group of geographers in the late 1960s (Soja, 1968, 1974; P.R. Gould, 1970; Riddell, 1970), although sociologists also spent some considerable time working along these lines.

In such work, sets of indices which were held to reflect modernisation were mapped and/or subjected to multivariate statistical analysis to reveal the 'modernisation surface'. For example, using such an approach, P.R. Gould (1970) examined what he regarded as the modernisation surface of Tanzania (Box 3.1). One of the classic papers written in the mould was by Leinbach (1972), who investigated the modernisation surface in Malaya between 1895 and 1969, using indicators such as the number of hospitals and schools per head of the population, together with the incidence of

BOX 3.1   Modernisation and development in Tanzania

Tanzania became independent in 1961 after a British and German colonial history (Hoyle, 1979). The area was occupied by Germans in the 1880s, and after the First World War it became British-administered Tanganyika. Like many former colonies, the population was very concentrated along the narrow coastal region (see figure below). The other major urban nodes such as Morogoro, Iringa and Mbeya formed a corridor running in a south-westerly direction from Dar-es-Salaam on the Indian Ocean coast.

During the era when modernisation thinking was is vogue, 'islands' of development linked by major

transport lines were recognised by geographers such as Gould (1969, 1970), Hoyle (1979) and Safier (1969). Traditionally, the settlement pattern had comprised dispersed villages, although strong urban concentration around Dar-es-Salaam occurred during the colonial period, with Hoyle (1979) referring to it as an 'hypertropic cityport' (O'Connor, 1983).

Lundqvist (1981) identified four main phases of development planning in Tanzania between 1961 and 1980. The period from 1961 to 1966 was indeed seen as the legacy of the colonial era, during which such planning as was carried out was sectoral rather than

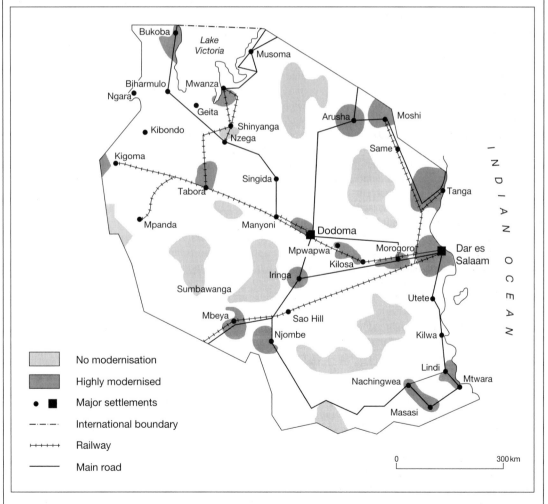

Tanzania in the 1970s: settlements and the modernisation surface. Adapted from Gould and White (1974).

Box 3.1 continued

regional in scope, as a result of which, infrastructure remained concentrated in the principal towns and urban–rural disparities were maintained. Thus, one could talk about a highly polarised 'modernised–non-modernised' development surface, which largely reflected colonial penetration.

However, subsequent to this, development in Tanzania has been far more complex than this simplistic overview implies. Thus, the principal policy efforts to reduce urban–rural differences can be identified as giving rise to the second and most important development phase, lasting from 1967 to 1972, and witnessing the emergence of a strong commitment to rural-based development, linked to strong principles of traditional African socialism.

These policies were based on the Arusha Declaration of 1967, which attacked privilege and sought to place strong emphasis on the principles of equality, cooperation, self-reliance and nationalism. Such ideas were put into practice in the second five-year plan, 1969–74. The major policy imperative was ujamaa villagisation, which was regarded as the expression of 'modern traditionalism', i.e. a twentieth-century version of traditional African village life. The word *ujamaa* is Swahili for familyhood. The intention was to concentrate scattered rural populations and, by this process of villagisation, to provide the services required for viable settlements. Reducing rural-to-urban migration was a major goal, along with lessening the dependence on major cities such as Dar-es-Salaam. Ujamaa villages were envisaged as cooperative ventures, by means of which initiative and self-reliance would be fostered, along socialist lines. In addition, efforts were also made to spread urban development away from Dar-es-Salaam toward nine selected regional growth centres. In overall terms, President Nyerere regarded these policies as a distinct move away from a slavish imitation of Western-style planning and development, based on uncritical 'modernisation'.

Despite having received much praise from certain quarters, the policies adopted in Tanzania have been viewed with considerable scepticism by others, especially those from a committed Marxist perspective. During the third phase, from 1973 to 1968, enforced movement to development villages occurred. Furthermore, by the fourth stage, starting in 1978, industry and urban development were once again being upgraded at the expense of ujamaa villages and rural progress. Thus, the fourth five-year plan, 1982–83 to 1985–86, gave priority to industrial development, and by this juncture the ujamaa concept appeared to have all but fallen from the consciousness of both planners and politicians alike.

---

postal and telegraph facilities and road and rail densities. This modernisation approach served to emphasise that core urban areas and the transport corridors running between them are the focus of dynamic change (Leinbach, 1972). In 1895 early growth was almost exclusively related to the west coast, specifically centering on Kuala Lumpur, with a clear inland island around Ipoh. By 1955 the so-called modernisation surface had penetrated to the east of the nation along two 'corridors' (Figure 3.3).

The process involved in the Malaysian case is shown as an ideal-typical sequence in the sequence of four boxed diagrams in the lower half of Figure 3.4 (page 50). It essentially represents the diffusion downwards of 'development' from the largest to the smallest settlement. Thus, from a critical perspective, J. Friedmann and Weaver (1979: 120) observed that the approach only succeeded in 'mapping the penetration of neo-colonial capitalism'.

The hallmark of this work was that it posited that modernisation is basically a temporal–spatial process. In such a vision, underdevelopment is seen as something which can be overcome, principally by the spatial diffusion of modernity. A number of studies argued that growth occurs within the settlement system from the largest urban places to the smallest in a basically hierarchical sequence. This is shown in the upper part of Figure 3.4 (page 50). Foremost among the proponents of such a view was Hudson (1969), who applied the classic ideas of Hagerstrand (1953) to the settlement or central place system. Hudson argued that innovations can travel through the settlement system by a process of contagious spread, where there is a neighbourhood or regional effect of clustered growth. This was close to Schumpeter's (1911) general economic theory, in which he argued that the essence of development is a volume of innovations. Opportunities tend to occur in waves which surge after an initial

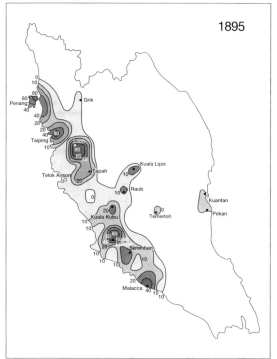
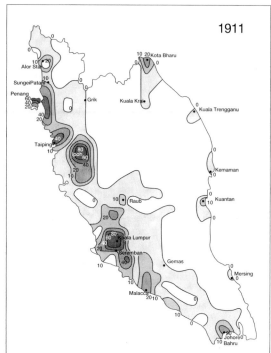
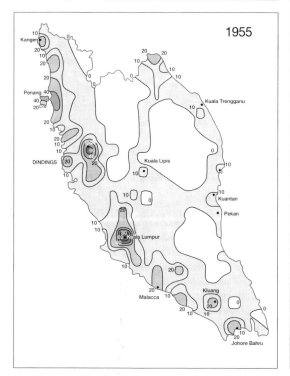
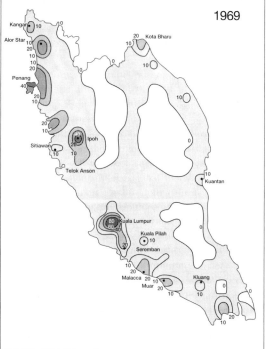

Figure 3.3 The modernisation surface for Malaya, 1895–1969. Adapted from Leinbach (1972).

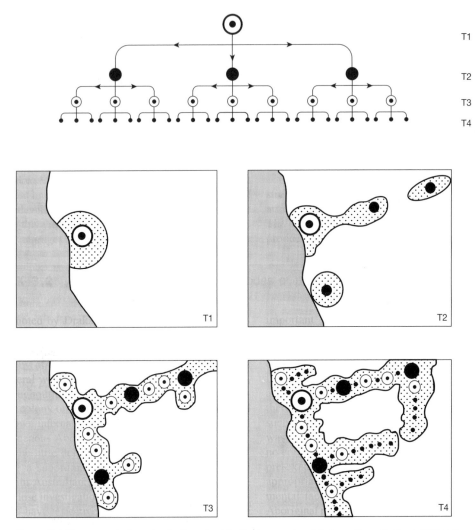

Figure 3.4 The spread of modernisation: hypothetical examples (a) down through the settlement system from the largest places to the smallest places and (b) over the national territory.

innovation. Thus, Schumpeter argued that development tends to be 'jerky' and to occur in 'swarms'.

Alternatively, Hudson noted that diffusion can occur downwards through the settlement system in a progressive manner, the point of introduction being the largest city. Pedersen (1970) argued the case for a strictly hierarchical process of innovation diffusion, an assertion which seemed to be borne out by some historical–empirical studies carried out in advanced capitalist societies such as the United States (Borchert, 1967) and England and Wales (B.T. Robson, 1973). Pedersen drew a very important distinction between domestic and entrepreneurial innovations; entrepreneurial innovations were the instrument of urban growth, not domestic innovations. In another frequently cited paper of the time, Berry (1972) also argued strongly in favour of a hierarchical diffusion process of growth-inducing innovations; this was seen as the result of the sequential market-searching procedures of firms, along with imitation effects. But notably, Berry's analysis was based on the diffusion of domestic as opposed to entrepreneurial innovations, namely of television receivers. In other words, it dealt with what was happening to consumption rather than production. Furthermore, the critique of modernisation has to accept that even larger firms are currently coming to dominate

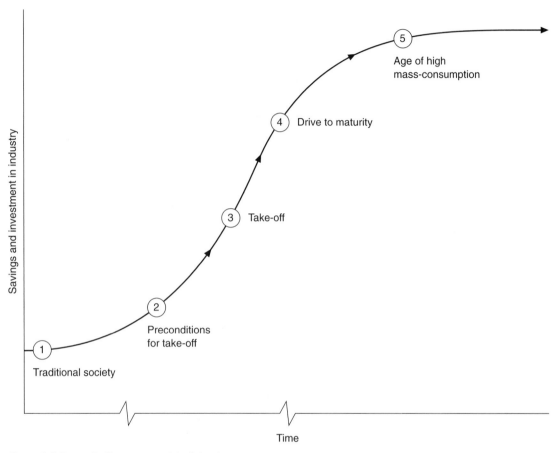

Savings and investment in industry

5 Age of high
mass-consumption

4 Drive to maturity

3 Take-off

2 Preconditions
for take-off

1

Traditional society

Time

Figure 3.5 Rostow's five-stage model of development.

the world capitalist system, a major development that is detailed in Chapter 4.

All of these approaches, involving unequal and uneven growth, modernisation, the diffusion of innovation and hierarchic patterns of change may be grouped together and regarded as constituting the top-down paradigm of development (Stöhr and Taylor, 1981), which advocates the establishment of strong urban-industrial nodes as the basis of self-sustained growth. Such an approach is premised on the occurrence of strong trickle-down effects, by means of which it is believed that modernisation will inexorably be spread from urban to rural areas (Figures 3.3 and 3.4, pages 49 and 50). As with modernism, all such approaches 'had a great appeal to a wider public due to the paternalistic attitude toward non-European cultures' (Hettne, 1995: 64). These approaches, together with modernisation, reflected the desire of the United States of America to order the postwar world, and were used

to substantiate the logic of 'authoritative intervention', as noted in Chapter 1 (P.W. Preston, 1996).

Such models, including Rostow's (1960) classic *Stages of Economic Growth*, see urban-industrial nodes as engines of growth and development. Rostow's work can be seen as the pre-eminent theory of modernisation in the early 1960s (P.W. Preston, 1996). Rostow envisaged that there were five stages through which all countries have to pass in the development process: the traditional society, preconditions to the take-off phase, take-off, the drive to maturity, and the age of mass consumption, as depicted in Figure 3.5. Rostow's stage model encapsulates faith in the capitalist system, as expressed by the subtitle of the work: a non-communist manifesto for economic growth. For Rostow, the critical point of take-off can occur where the net investment and savings as a ratio to national income grow from 5–10 per cent, thereby facilitating industrialisation.

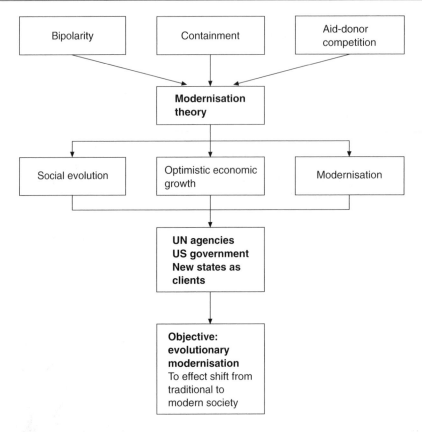

Figure 3.6 A schematic view of modernisation theory. Adapted from
P.W. Preston (1996).

Although Rostow's work can be regarded as a derivation of Keynesian work, its real significance derives from the simple fact that it seemed to offer every country an equal chance to develop (P.W. Preston, 1996). In particular, the 'take-off' period was calibrated at twenty or so years, long enough to be conceivable but short enough not to seem oppressive and unattainable. The salience of the Rostowian framework was that it purported to explain the advantages of the Western development model. Further, in the words of P.W. Preston (1996: 178), the 'theory of modernisation follows on from growth theory but is heavily influenced by the desire of the USA to combat the influence of the USSR in the Third World' (Figure 3.6).

One of the most influential theories of the 1950s was authored by W. A. (Arthur) Lewis (1950, 1954), an economist of West Indian origin, but who was working at the University of Manchester at the time. Lewis set out the foundations of modernisation theory

when he maintained that the juxtaposition of a backward traditional sector with an advanced modern sector meant that an 'unlimited supply of labour' existed for development. Thus, the argument ran, this duality means that industry can expand rapidly if industrialisation is financed by foreign capital. This led to the so called policy of industrialisation by invitation (Plates 3.1 and 3.2). It was ironic that a St Lucian economist should use the metaphor of a snowball, arguing that once the process started to move, it would develop its own self-sustaining momentum, like a snowball rolling downhill, an essentially similar argument to Walt Rostow's.

Indeed, all such formulations place absolute faith in the existence of a linear and rational path to development, based on Western positivism and science, and the possibility that all nations can follow this in an unconstrained manner. Modernism was very much an urban phenomenon from 1850 onwards (Harvey, 1989). Universal or high modernism became hegemonic after

Plate 3.1 Part of an industrial estate in Bridgetown, Barbados
(photo: Rob Potter)

Plate 3.2 United States-owned baseball factory in Port au Prince, Haiti
(photo: Sean Sprague, Panos Pictures)

1945. Thus, the top-down approach was strongly associated with the 1950s, through to the early 1970s.

Taken together, many writers refer to these theories as representing 'Eurocentric development thinking', i.e. development theories and models rooted in Western European history (Hettne, 1995; Slater, 1992a, 1992b). Via such approaches, during the 1950s and 1960s, development was seen as a strengthening of the material base of society, principally by means of industrialisation (Plate 3.3, overleaf). Inevitably, the history of the first industrial state was taken as the model which should be followed, not only by the rest of Europe, but ultimately by the rest of the world, for 'it is quite natural that the original recipe for

Plate 3.3 A heavily polluted town with the Pannex refinery in the background, Mexico
(photo: Ron Giling, Panos Pictures)

development given by the developed countries should emanate from their own experiences and prejudices' (Hettne, 1995: 37). It is notable that all the early theories of development were authored by men, and that virtually all of them were of Anglo-European origin.

In conclusion, however, it would be wrong to give the impression that the focus on top-down, urban-industrial growth, associated with the quest for modernisation, was associated with a single and unified path. Four more or less distinct development strategies making up the early Western tradition are identified and described by Hettne (1995):

The liberal model, the one implicitly discussed through much of the above account, stresses the importance of the free market and largely accepts the norm based on English development experience during the Industrial Revolution. Such views are presently gaining currency once again in the form of structural adjustment programmes enforced by the International Monetary Fund (IMF), USAID and the World Bank as the condition of loans given to sort out the economic situation. These neoliberal polices will receive attention in many sections later in this text.

Keynesianism departs from the liberal tradition by virtue of arguing that the free market system

does not self-regulate effectively and efficiently, thereby necessitating the intervention of the state in order to promote growth in capitalist systems. Since the 1930s, Keynesianism has been the dominant development ideology in the industrialised capitalist world, especially in countries with a social democratic tendency (Hettne, 1995).

State capitalist strategy refers to an early phase of industrial development in continental Europe, principally tsarist Russsia and Germany. The approach advocated the development of enforced industrialisation based primarily on agrarian economies, in order to promote nationalism and for reasons pertaining to national security.

The Soviet model represents a radical state-oriented strategy inspired by Stalin's five-year mandatory economic development plans. The approach regarded modernisation as the goal, to be achieved by means of the transfer of resources from agriculture to industry. The agricultural sector was collectivised, and heavy industry was given the highest priority. The state completely replaced the market mechanism.

The common denominator linking these four approaches is an unswerving faith in the efficacy of urban-based industrial growth, although some approaches in the early

Plate 3.4 Urban-based modernisation: the CBD of Johannesburg, South Africa
(photo: Tony Binns)

modernisation phase also emphasised resource-based development strategies (Plate 3.4). Notwithstanding the variations noted above, Hettne (1995) comments on the fact that through time the role of the state has generally been central to Western development strategies. This was certainly true of the Keynesian, state capitalist and Soviet models. However, the recent orthodoxy has seen the dismantling of the welfare state and the reduced scope of government in the form of neoliberalism. Arguing that the modern welfare state together with trades unions and state bureaucracies have destroyed the market system, Milton Friedman (1962) and others argued against Keynesianism and promoted the New Right.

In this way, the New Right neoliberal theorists have been seen as giving rise to a counter-revolution, celebrating the unrestrained power of the unregulated free market, arguing both for its economic efficiency and its role in liberating the social choices of the individual. There can be little doubting the fact that the general pro-market position of the New Right has served to inform the policies of the World Bank, the IMF and the United States government over the 1980s and beyond. Thus, as detailed in many of the chapters of this book, the World Bank and the IMF have pressed for economic liberalisation, the elimination of market imperfections and market-inhibiting social

institutions, plus the redefinition of planning regulations in Third World countries. Some sections of the New Right have gone so far as to argue that the 'Third World only exists as a figment of the guilty imaginations of First World scholars and politicians (P.W. Preston, 1996: 260).

In Britain, neoliberalism was witnessed in the form of Thatcher's popular capitalism, and in America, in the guise of Reaganism. Both of them saw the extension of the market into new fields such as hospitals, schools, universities and other public sector establishments. On the global scale, the 1980s also saw 'liberalisation' being advocated in the Third World, especially by the monetarist school, which advanced an extreme *laissez-faire* approach, as in the so-called global Reaganomics. Using the example of the newly industrialising countries (NICs), particularly those of Asia, countries were advised to liberalise their economies, encourage entrepreneurship and to seek comparative cost advantage. Through the 1980s and into the early 1990s, monetarism became the firm policy of the World Bank and the International Monetary Fund. However, P.W. Preston (1996: 260) among others has argued that 'the schedule of reforms inaugurated by the New Right have not generally proved to be successful. It is also true to say that the rhetorically important attempt to annex the development experiences of Pacific Asia to the position of the New Right has been widely ridiculed by development specialists.'

Historical approaches: empirical perspectives on change and development

Another way in which scholars and practitioners can seek to generalise about development is by empirical or real-world observations through time. By definition, this approach will give rise to descriptive–positive models of development (see Figure 3.2, page 45), but some think these frameworks have a real role to play in the discussion of development, specifically for grounding theory in the historical realities of developing nations. Although such approaches deal primarily with the pre-independence period, it can be argued that they still have relevance to contemporary patterns and processes of development and change.

In contrast to Hirshmann, the Swedish economist Gunnar Myrdal (1957), although writing at much the same time, took a noticeably more pessimistic view, maintaining that capitalist development is inevitably marked by deepening regional and personal income and welfare inequalities. Myrdal followed the arguments of the vicious circle of poverty in presenting his theory of 'cumulative causation'. Thereby, it was argued that once differential growth has occurred, internal and external economies of scale will perpetuate the pattern.

Such a situation is the outcome of the 'backwash' effect, whereby population migrations, trade and capital movements all come to focus on the key growth points of the economy. Increasing demand, associated with multiplier effects and the existence of social facilities also serve to enhance the core region. Although 'spread' effects will undoubtedly occur, principally via the increased market for the agricultural products and raw materials of the periphery, Myrdal concluded that, given unrestrained free market forces, these spread effects would in no way match the backwash effects. Myrdal's thesis leads to the advocacy of strong state policy in order to counteract what is seen as the normal tendency of the capitalist system to foster increasing regional inequalities.

The view that, without intervention, development is likely to become increasingly polarised in transitional societies was taken up and developed by a number of scholars towards the end of the 1960s and the beginning of the 1970s. As such, they ran counter to the conventional wisdom of the time. These works were based mainly on empirical studies which encompassed an historical dimension. Perhaps the best-known example is provided by American planner John Friedmann's (1966) core–periphery model. From a purely theoretical perspective, Friedmann's central contention was that 'where economic growth is sustained over long time periods, its incidence works toward a progressive integration of the space economy' (J. Friedmann, 1966: 35). This process is made clear in the much-reproduced four-stage ideal-typical sequence of development shown in Figure 3.7.

The first stage, independent local centres with no hierarchy, represents the precolonial stage and is associated with a series of isolated self-sufficient local economies. There is no social surplus product to be concentrated in space and an even and essentially stable pattern of settlement and development is the result.

In the second stage, a single strong centre, it is posited that as the result of some form of 'external disruption' – a euphemism for colonialism – the former stability is replaced by dynamic change. Growth is envisaged to occur rapidly in one main region and urban primacy is the spatial outcome. Social surplus product is strongly concentrated. The centre (C) feeds on the rest of the nation, so the extensive periphery (P) is drained. Advantage tends to accrue to a small elite of urban consumers, who are located at the centre. However, Friedmann regarded this stage as inherently unstable.

The outcome of this instability is the development of a single national centre with strong peripheral sub-centres. Over time, the simple centre–periphery pattern is progressively transformed to a multinuclear one. Subcores develop (SC1, SC2), leaving a series of intermetropolitan peripheries (P1 to P4). This is the theoretical representation of the point of polarisation reversal when development starts to be concentrated in parts of the former periphery, albeit on a highly concentrated basis.

The fourth and final stage, which sees the development of a functionally interdependent system of cities, was described by Friedmann as 'organised complexity' and is one where progressive national integration continues, eventually witnessing the total absorption of the intermetropolitan peripheries. A smooth progression of cities by size and a smooth progression of national development are envisaged as the outcomes.

The first two stages of the core–periphery model describe directly the history of the majority of developing countries. Indeed, it often appears not to be appreciated that, in the first stage, the line along which

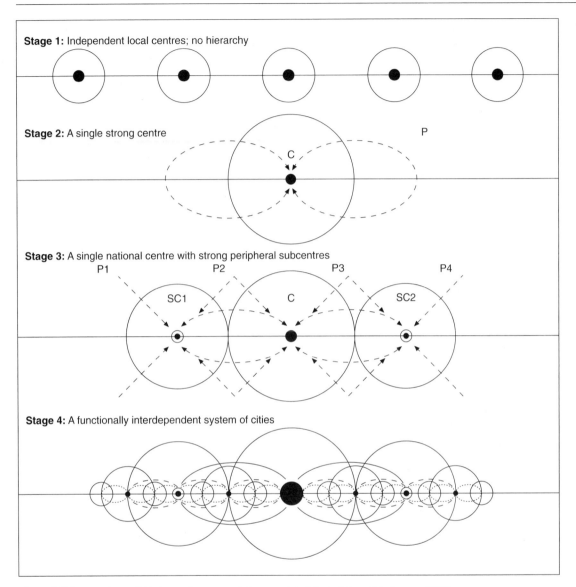

Stage 1: Independent local centres; no hierarchy

Stage 2: A single strong centre

Stage 3: A single national centre with strong peripheral subcentres

Stage 4: A functionally interdependent system of cities

Figure 3.7 An overview of Friedmann's core–periphery model. Adapted from Friedmann, 1966.

the small independent communities are drawn represents the coastline. The occurrence of uneven growth and urban concentration in the early stages of growth is seen as being the direct outcome of exogenic forces. Thus, Friedmann commented that the core–periphery relationship is essentially a colonial one, his work having been based on the history of regional development in Venezuela.

The principal idea behind the centre–periphery framework is that early on, factors of production will be displaced from the periphery to the centre, where marginal productivities are higher. Thus, at an early stage of development, nothing succeeds like success. However, the crucial change is the transition between the second and third stages, where the system tends toward equilibrium and equalisation. Friedmann's model is one which suggests that, in theory, economic development will ultimately lead to the convergence of regional incomes and welfare differentials.

But at the very same time as he was presenting the simplified model as a template, Friedmann observed that in reality there was evidence of persistent

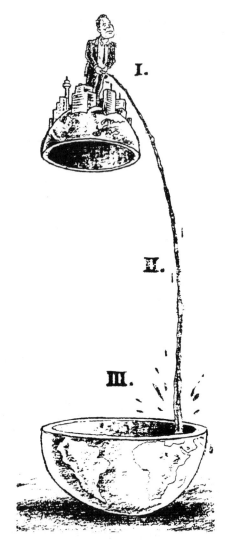

Trickle-down is the notion that making the North, the cities and elites in the South, and the rich generally, richer will ultimately produce trickle down benefits for the world's less advantaged. We hesitated in publishing this cartoon (left) because we thought it could offend people we don't wish to offend. If you are offended (and we hope you are not), think of the greater obscenity of those who profess, act and benefit from the notion – in a world of increasing inequity and poverty.

Figure 3.8 Trickle-down: cartoon interpretations of the trickle from First World to Third World (from *Te Amokura, Na Lawedua*, The New Zealand and Pacific Islands Development Digest), and from rich to poor within a single country (reprinted with kind permission from *Private Eye*).

disequilibrium. Thus, in a statement which appeared alongside the model, Friedmann (1966: 14) observed that there was 'a major difficulty with the equilibrium model: historical evidence does not support it.' Despite this damning caveat, many authors have represented the model as a statement of invariant truth, ignoring Friedmann's warning that 'disequilibrium is built into transitional societies from the start' (J. Friedmann, 1966: 14). Effectively, Friedmann was maintaining that, without state intervention, the transition from the second stage to the third stage will not occur in developing societies; in this respect, he was in agreement with

Myrdal's prescriptions that development will become ever more concentrated in space, with polarisation always tending to exceed the so-called trickle-down effect. Figure 3.8 depicts a somewhat irreverent view of this argument.

Writing just a few years after the appearance of Friedmann's much-cited model, American geographer Jay E. Vance (1970) noted that it was with the development of mercantile societies from the fifteenth century onward that settlement systems started to evolve along more complex lines. The main development came with colonialism, for continued economic growth required greater land resources. Frequently, this requirement was initially met by local colonial expansion via trading expeditions. By the seventeenth and eighteenth centuries, however, this need was increasingly fulfilled by distant colonialism, the transoceanic version of local colonialism. The implications of these historical developments have been well summarised by Vance (1970: 148):

> The vigorous mercantile entrepreneur of the seventeenth and eighteenth centuries had to turn outward from Europe because the long history of parochial trade and the confining honeycomb of Christaller cells that had grown up with feudalism left little scope there for his activity. With overseas development, for the first time the merchant faced an unorganised land wherein the designs he established furnished the geography of wholesale-trade location. By contrast, in a central-place situation (such as that affecting much of Europe and the Orient), to introduce wholesale trade meant to conform to a settlement pattern that was premercantile.

During the period of mercantilism, ports came to dominate the evolving urban systems of both the colony and the colonial power. In the colony, once established, ports acted as gateways to the interior lands. Subsequently, evolutionary changes occurred that first saw increasing spatial concentration at certain nodes, then lateral interconnection of the coastal gateways and the establishment of new inland regions for expansion. The settlement pattern of the homeland also underwent considerable change, for social surplus product flowed into the capital city and the principal ports, thus serving to strengthen considerably their position in the urban system.

These historical facets of trade articulation led Vance (1970) to suggest what amounted to an entirely new model of colonial settlement evolution; one that was firmly based on history. This is called the mercantile model, and its main features are summarised in Figure 3.9 (overleaf). The model is illustrated in five stages. In each of these, the colony is shown on the left of the figure, and the colonial power on the right.

The first stage represents the initial search phase of mercantilism, involving the search for economic information on the part of the prospective colonising power. The second stage sees the testing of productivity and the harvest of natural storage, with the periodic harvesting of staples such as fish, furs and timber. However, no permanent settlement is established in the colony. At the third stage, the planting of settlers who produce staples and consume the manufactures of the home country occurs. The settlement

system of the colony is established via a point of attachment. The developing symbiotic relationship between the colony and the colonial power is witnessed by a sharp reduction in the effective distance separating them. The major port in the homeland becomes pre-eminent. The fourth stage is characterised by the introduction of internal trade and manufacture in the colony. At this juncture, penetration occurs inland from the major gateways in the colony, based on staple production. There is rapid growth of manufacturing in the homeland to supply the overseas and home markets. Ports continue to increase in significance. The fifth and final stage sees the establishment of a mercantile settlement pattern and central-place infilling occurs within the colony; there emerges a central-place settlement system with a mercantile overlay in the homeland.

The mercantile model stresses the historical–evolutionary viewpoint in examining the development of national patterns of development and change. The framework offers what Vance sees as an alternative and more realistic picture of settlement structure, based on the fact that in the seventeenth and eighteenth centuries, mercantile entrepreneurs turned outward from Europe. Hence the source of change is external to developing countries. In contrast, the development of settlement patterns and systems of central places in the developed world was based on endogenic principles of local demand, thereby rendering what is essentially a closed settlement system (Christaller, 1933; Lösch, 1940).

The hallmark of the mercantile model is the remarkable linearity of settlement patterns, first along coasts and especially in colonies, and secondly, along the routes which developed between the coastal points of attachment and the staple-producing interiors. These two alignments are also given direct expression in Taaffe, Morrill and Gould's (1963) model of transport development in less developed countries, as shown in Figure 3.10a (page 61). The model was based on the transport histories of West African nations such as Nigeria and Ghana, plus Brazil, Malaya, and East Africa. Hoyle's (1993) application of the framework to East Africa is shown in Figure 3.10b (page 61).

In plantation economies such as those of the Caribbean, a local historical variant of the mercantile settlement system is provided by the plantopolis model. A simplified representation of this is shown in Figure 3.11 (page 62). The first two stages are based on Rojas (1989), although the graphical depiction of the sequence and its extension to the modern era have been effected by Potter (1995a). In the first stage,

THE OVERSEAS COLONY

THE COLONIAL POWER

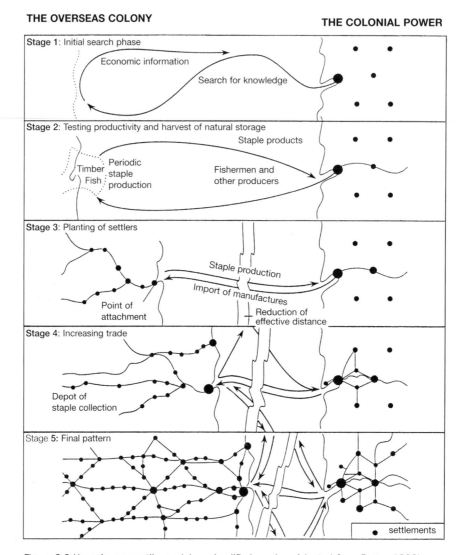

Figure 3.9 Vance's mercantile model: a simplified version. Adapted from Potter 1992b.

plantopolis, the plantations formed self-contained bases for the settlement pattern, such that only one main town was required for trade, service and political control functions. Following emancipation, the second stage, small, marginal farming communities – clustered around the plantations, practising subsistence agriculture and supplying labour to the plantations – added a third layer to the settlement system. The distribution of these communities would vary according to physical and agricultural conditions.

Figure 3.11 (page 62) suggests that, in the Caribbean, the modern era has witnessed the extension of this highly polarised pattern of development; this is the third stage. The emphasis is placed on extension, for this may not in all cases amount to intensification *per se*. This has come about largely as the result of industrialisation and tourism being taken as the twin paths to development. In 1976 Augelli and West (1976: 120) commented on what they regarded as the disproportionate concentration of wealth, power and social status in the chief urban centres of the West Indies. As shown in Figure 3.11, such spatial inequality is sustained by strong symbiotic flows between town an country.

The virtues of the mercantile and plantopolis models are many. Principally, they serve to stress that the evolution of most developing countries amounts to a

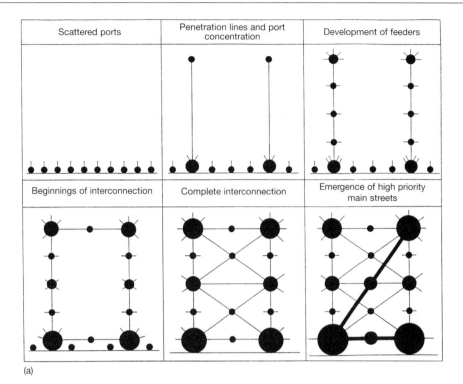

(a)

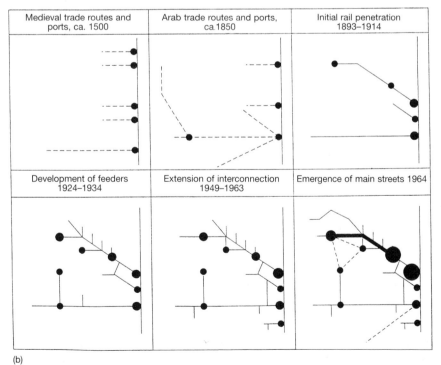

(b)

Figure 3.10 The Taafe, Morrill and Gould model of transport development and its application to East Africa. Adapted from Hoyle (1993).

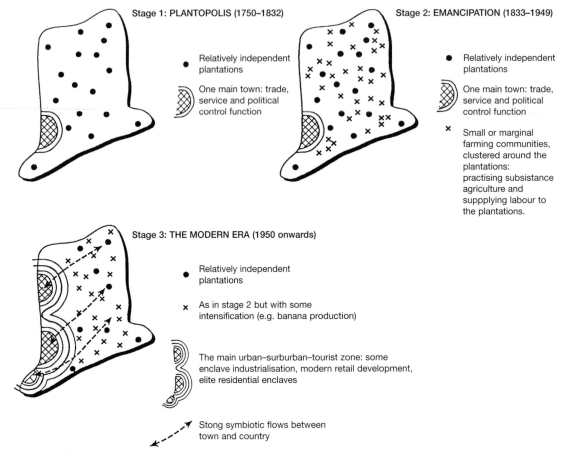

Stage 1: PLANTOPOLIS (1750–1832)

• Relatively independent plantations

One main town: trade, service and political control function

Stage 2: EMANCIPATION (1833–1949)

• Relatively independent plantations

One main town: trade, service and political control function

× Small or marginal farming communities, clustered around the plantations: practising subsistance agriculture and suppplying labour to the plantations.

Stage 3: THE MODERN ERA (1950 onwards)

• Relatively independent plantations

× As in stage 2 but with some intensification (e.g. banana production)

The main urban–surburban–tourist zone: some enclave industrialisation, modern retail development, elite residential enclaves

Stong symbiotic flows between town and country

Figure 3.11 The plantopolis model and its extension to the modern era. Adapted from Potter (1995).

highly dependent form of development. Certainly we are reminded that the high degree of urban primacy and the littoral orientation of settlement fabrics in Africa, Asia, South America and the Caribbean are all the direct product of colonialism, not accidental happenings or aberrant cases, hence the comment that modernisation surfaces merely chart neocolonial penetration.

According to these models, ports and other urban settlements became the focus of economic activity and of the social surplus which accrued. A similar but somewhat less overriding spatial concentration also applies to the colonial power. Hence a pattern of spatially unequal or polarised growth emerged strongly several hundred years ago with the strengthening of this symbiotic relationship between colony and colonial power. The overall suggestion is that, due to the requirements of the international economy, far greater levels of inequality and spatial concentration are produced than may be socially and morally desirable.

Radical dependency approaches: the Third World answers back?

A major advance came with what some refer to as the indigenisation of development thinking; that is, the production of ideas purporting to emanate from, or at least relate to, the sorts of conditions that are encountered in the Third World, rather than ideas emanating from European experience. As Slater (1992a, 1992b) has recently averred, the core can learn from the periphery, and it does need to learn. The empirically derived mercantile and plantopolis models reviewed above can be regarded as graphical depictions of the outcome of the interdependent development of the world since the 1400s. The dependencia, or dependency school, specifically took up this theme as a rebuttal of the modernisation paradigm. Although some consider that dependency theory developed as a voice from the Third World, others have maintanied that its most cogent formulation represents Eurocentric development thinking by virtue of the origins of its leading author, a German-born economist.

The dependency school, the origins of which can be traced back to the 1960s, became a global force in the 1970s. It had its origins in the writings of Latin American and Caribbean radical scholars known as structuralists, because they focused on the unseen structures which may be held to mould and shape society (Girvan, 1973). The approach was the outcome of the convergence of pure Marxist ideas on Latin American and Caribbean writings about underdevelopment. Essentially, by such a process, the Eurocentric ideas of Marxism became more relevant to the Latin American condition. Marx and Engels had seen capitalism as initially destructive of non-capitalist social forms, and thereafter progressive (P.W. Preston, 1996). Prebisch (1950) and Furtado (1964, 1965, 1969), who wrote about Argentina and Brazil respectively, are the best-known Latin American structuralists. Raul Prebisch stressed the importance of economic relations in linking industrialised and Latin American less developed countries, and observed that these were primarily prescribed in the form of centre–periphery relations. Furtado (1969) argued that present-day Latin American socioeconomc structures were the result of the manner of incorporation into the world capitalist system, an argument which is close to the full dependency line. Other later writers included Dos Santos (1970, 1977) and Cardoso (1976).

As noted by Hettne (1995), opinions differ as to whether the well-developed Caribbean school of dependency should be seen as an autonomous development. It is salient to note that the whole history of the Caribbean is one of dependency, and this issue was central to the so-called New World Group, which first met in Georgetown, Guyana, in order to discuss Caribbean development issues. Key names include George Beckford (1972), Norman Girvan (1973) and Clive Thomas (1989).

The development of a distinctly neo-Marxist line in the 1960s had been strongly associated with Paul Baran (Baran, 1973; Baran and Sweezy, 1968), who focused on the vital topic of how local economic surpluses were realised and subsequently employed. Thus, Baran saw the key to development as disengagement from the deforming impact of the world economy, as capitalism creates and then diverts much of the economic surplus into wasteful and sometimes immoral consumption.

However, the dependency approach proper is particularly associated with the work of Andre Gunder Frank, a Chicago-trained economist, who underwent a rapid and thorough radical conversion (Brookfield, 1975). Frank's key ideas were outlined in 'The development of underdevelopment', an article published in 1966, and the book *Capitalism and Underdevelopment in Latin America*, which appeared in 1967. Although a scholar of European origin working in the United States, Frank had carried out research in Mexico, Chile and Brazil.

Frank (1967) maintained that development and underdevelopment are opposite sides of the same coin, and that both are the necessary outcome and manifestation of the contradictions of the capitalist system of development. The thesis presented by Frank was devastatingly simple. It was argued that the condition of developing countries is not the outcome of inertia, misfortune, chance, climatic conditions or whatever, but rather a reflection of the manner of their incorporation into the global capitalist system. Viewed in this manner, so-called underdevelopment, and associated dualism, were not a negative or void, but the direct outcome and reciprocal of development elsewhere (Figure 3.12, overleaf). The only real alternative for such nations was to weaken the grip of the global system by means of trade barriers, controls on transnational corporations and the formation of regional trading areas (see Plate 3.5, overleaf).

The phrase 'the development of underdevelopment' (Frank, 1966) has come to be employed as a shorthand description of the approach, which has a strong graphical tie-in with the mercantile and plantopolis models. Quite simply, if the development of large tracts of the earth's surface has depended upon metropolitan cores, called metropoles, then the development of cities has also depended principally upon the articulation of capital and the accumulation of surplus value (Figure 3.13, page 65). The process has operated internationally and internally within countries. Viewed in this light, so-called backwardness results from integration at the bottom of the hierarchy of dependence. Frank argued that the more such 'satellites' are associated with the metropoles, the more they are held back, and not the other way around. In this connection, Frank specifically cited the instances of northeast Brazil and the West Indies as regions where processes of internal transformation had been rendered impossible due to such close contact.

Dependency theory represents an holistic view because it describes a chain of dependent relations which has grown since the establishment of capitalism as the

Figure 3.12 Dependent relations according to the *New Internationalist*. Cartoon by Paul Fitzgerald.

Plate 3.5 Organisation for Rural Development poster advocating more domestic production and fewer imports in St Vincent (photo: Rob Potter)

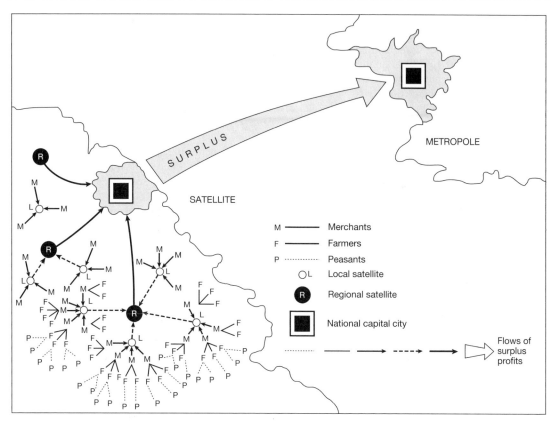

Figure 3.13 Dependency theory: a graphical depiction. Adapted from Potter (1992b).

dominant world system, so its expansion is regarded as coterminous with colonialism and underdevelopment. The chain of exploitative relations witnesses the extraction and transmission of surplus value via a process of unequal exchange, extending from the peasant, through the market town, regional centre, national capital, to the international metropole, as shown in Figure 3.13. The terms of trade have always worked in favour of the next higher level in the chain, so that social surplus value is progressively concentrated (Harvey, 1973; Castells, 1977). By such means, dependency theorists argue that the dominant capitalist powers, such as England and then the United States, encouraged the transformation of political and economic structures in order to serve their interests. According to this view, colonial territories were organised to produce primary products at minimal cost, simultaneously becoming an increasing market for industrial products. Inexorably, social surplus value was siphoned off from poor to rich regions, and from the developing world to the developed.

The chief criticism of dependency theory is that it is economistic, seeing all as the outcome of a form of economic determinism, conforming with what Armstrong and McGee (1985: 38–39) have described as the 'impersonal, even mechanical analysis of structuralism'. Furthermore, the theory only appears to deal with class structure and other factors internal to a given nation, insofar as they are the outcome of the economic processes described. Another point of contention is how dependency theory suggests that socialist countries can only advance their lots by delinking from the global economy, whereas the capitalist world system is busily becoming more global. For all these reasons, dependency theory has largely been out of fashion in the First World of late (P.W. Preston, 1996).

Wallerstein (1974, 1980) attempted to get around some of these issues, including the internal–external agency debate, by stressing the existence of a world system (P.J. Taylor, 1986). The essential point is that Wallerstein distinguishes not only between the core nations, which became the leading industrial producers, and the peripheral states, which were maintained as agricultural providers, but also identifies the semiperipheries. It is argued that the semiperipheries play a key role, for these intermediate states are strongly ambitious in competing for core status by increasing their importance as industrial producers relative to their standing as agricultural suppliers. For the semiperipheral capitalist nations, read the NICs. Within the world system since the sixteenth century there have been cyclical periods of expansion, contraction, crisis and change. Hence it is envisaged that the fate of a particular nation is not entirely externally driven, but depends on the internal manner in which external forces have been accommodated.

Frank's ideas are seen by many as being near to the orthodox Marxist view that the advanced capitalist world at once both exploited and kept the Third World underdeveloped. Although many would undoubtedly refute this view as extreme, if one can clear away the moral outrage, it may be argued that elements of the analysis, even if in world-systems form, are likely to provide food for thought for those interpreting patterns of development. As Hettne (1995) notes, dependency theory stressed that the biggest obstacles to development are not a lack of capital or entrepreneurial skill, but are to be found in the international division of labour. As already observed, certainly the graphical representation of pure dependency theory exhibits many parallels with the spatial outcomes of the core–periphery, mercantile and plantopolis models of settlement development and structure.

One final element of radical development theory stressing exploitation is the so-called theory of the articulation of the modes of production (Box 3.2). The basic argument is that the capitalist mode of production exists alongside (and is articulated with)

BOX 3.2 Aborigines, development and modes of production

As noted by Drakakis-Smith (1983), Australia has experienced many different facets of colonialism. For much of the last two hundred years, it was a direct colony of Britain and was exploited as such for its mineral wealth and agricultural produce. However, throughout its history as a White nation state, whether colony or independent, Australia has harboured its own internal process of exploitation – that of its indigenous Aboriginal population by Whites. There can be little doubt that Aborigines do in fact comprise a subordinate, deprived and exploited group within Australian society.

Close investigation reveals how government measures have effectively institutionalised Aboriginal dependency in an unequal relationship which brings benefits principally to the White middle classes. Before the initial settlement by the British in 1788, Aborigines lived in what Meillassoux (1972, 1978) has termed a natural economy where land is the subject rather than the object of labour. The technology was simple but effective, with human energy alone being employed to tap the environmental resources through systems of hunting and collecting. Although life was primarily organised around the collection and consumption of subsistence food, a considerable amount of time was given over to ritual and ceremonial activities and to the production of consumer durables in the form of hunting weapons, tools and items with religious significance.

The mode of distribution was based on sharing the hunted and gathered food, and was therefore strongly related to the mode of production. One of the most important features of the precapitalist Aboriginal economy was an intense involvement with the land, both in physical and spiritual terms, with a strong emphasis on the spiritual. The small numbers, simple lifestyle and apparent lack of political organisation among the Aborigines convinced the British that there was no need for negotiation or treaties with such a 'primitive' people. Accordingly, all land was declared to be Crown land from the outset, so no compensation was paid to the Aborigines and no pre-existing rights were recognised. This legalistic appropriation of the land itself has been a fundamental factor in the subsequent exploitation of the Aboriginal people in Australia. Simply on the grounds that the indigenous population appeared to be disorganised and primitive, the British removed at one administrative stroke both the economic and spiritual basis of Aboriginal society.

For several decades after these developments, Aborigines appeared to be unimportant within the Australian colonial economy. In the first instances, labour was provided by assigned convicts, and later by larger landowners buying out many of the initial smallholders. By the 1830s, however, the rapid development of sheep farming to supply wool for export had led to the encouragement of large-scale labour migration from Britain.

But this rapid expansion of pastoral farming in the second half of the nineteenth century brought Aborigines once more into direct contact with the vanguard of White settlement. Hitherto, they had been virtually ignored – despised, destitute and decimated

---

Box 3.2 continued

by starvation, anomy and disease. By the mid nine-teenth century, pastoral settlement was beginning to push into the central and northern regions of the country, where conditions were harsher and Aborigines more numerous. Thus, Aboriginal labour was for the first time becoming a necessity on those properties where shere size and harsh climate led to a sharp reduction in the enthusiasm of recruited White labour to move north.

Thus, Aboriginal labour, either as station hands or domestics, was extensively used from the 1880s onwards. Bands and families were encouraged to stay on the land after its appropriation, where they received payment in kind for work undertaken on the station. Labour relations were at best paternalistic, but more often than not, the property owners or managers had little interest in the reproduction of Aboriginal labour, considering the supply to be limitless and the individuals unworthy of detailed attention. Living conditions and diet were often totally inadequate

and populations were decimated. As the value of the Aboriginal labour began to be more appreciated, so the Australian government began to establish a series of expanded reserves and settlements.

Australia has changed markedly since 1945. Although it still makes a notable contribution to the production process, the Aboriginal community is now more important as a consumer group for the goods and services of an extensive tertiary system operated almost entirely by Whites. In effect, this comprises a third stage in the institutionalisation of Aborigines within a dependency framework, following the appropriation of their land and labour power.

In the contemporary situation in Aboriginal Australia, therefore, the dominant capitalist mode of production has conserved the Aboriginal precapitalist mode of production largely for its role as a consumer of goods and services. The class position and economic prosperity of the White population is largely dependent on this relationship.

---

non-capitalist and precapitalist modes of production. The capitalist system replaces the non-capitalist system where profits are to be accued. However, the precapitalist form is left intact wherever profits are unlikely to be made and it is therefore advantageous to do so. A frequently cited example is housing, where the poor are left to provide their own folk or vernacular homes whereas the formal sector provides for the middle and upper classes (T. McGee, 1979; Drakakis-Smith, 1981; Burgess, 1990, 1992; Potter, 1992a, Potter, 1994; Potter and Conway, 1997). A more specific example of the operation of the modes of production approach was apartheid in South Africa, where the so-called homelands preserved the traditional mode of production in order to conserve labour, which was then allowed to commute to 'White South Africa'. Hence, according to this radical perspective, dualism is a product of the contradictions of the capitalist system, not some form of aberration. Similarly, according to this view, underdevelopment can be described as a stalemate in the process of articulation.

Bottom-up and alternative approaches: perspectives on 'another' development

The somewhat inelegant and uninformative expression *another development* has been used to denote

the watershed in thinking which has characterised the period since the mid 1970s (Hettne, 1995; Brohman, 1996). The concept was born at the Seventh Special Session of the United Nations General Assembly and the allied publication by the Dag Hammarskjöld Foundation '*What Now?*'. The session stressed the need for self-reliance to be seen as central to the development process, and for the emphasis to be placed on endogenous (internal) rather than exogenous (external) forces of change. It also came to be increasingly suggested that development should meet the basic needs of the people. At the same time, development needed to be ecologically sensitive and to stress more forcefully the principles of public participation (Potter, 1985).

Thus, from the mid 1970s, a growing critique of top-down policies, especially growth-pole policy, argued that such approaches had merely replaced concentration at one point in space with concentrated deconcentration at a limited number of new localities. The assertions that there is only one linear path to development, and that development is the same thing as economic growth came at long last to be seriously challenged. Liberal and radical commentators averred that top-down approaches to development were acting as the servants of transnational capital (J. Friedman and Weaver, 1979). In a similar vein, yet other commentators argued that what had been achieved in the past was economic growth without development, but with increasing poverty (Hettne, 1995).

Plate 3.6 Environmental costs of development; pollution near Witbank, Eastern Transvaal, South Africa (photo: Eric Millex, Panos Pictures)

In their book *Territory and Function*, John Friedmann and Clyde Weaver (1979) presented the important argument that development theory and practice to that point had been dominated by purely functional concerns relating to economic efficiency and modernity, with all too little consideration being accorded to the needs of particular territories, and to the territorial bases of development and change (Plate 3.6).

Since 1975 a major new paradigm has come to the fore, which involves stronger emphasis being placed on rural-based strategies of development. As a whole, this approach is described as development from below. Other terms used to describe the paradigm include agropolitan development, grassroots development and urban-based rural development. In the context of wider societal change, such developments can be related to the rise of what is called neo-populism. Neopopulism involves attempts to recreate and re-establish the local community as a form of protection against the rise of the industrial system (Hettne, 1995: 117). One manifestation of neopopulism has been the rise of the green ideology as a global concern, allied to green politics.

The provision of basic needs became a major focus during the early 1970s. The idea of basic needs originated with a group of Latin American theorists, and was officially launched at the International Labour Organisation's World Employment Conference, which was held in 1976. P.W. Preston (1996) argues that the pessimistic view of the Club of Rome's Limits to Growth (Chapters 5 and 6), was the motivating force behind basic needs strategies. Thus, the approach stressed the importance of creating employment over and above the creation of economic growth. This was because the economic growth that had occurred in developing countries seemed to have gone hand in hand with increases in relative poverty. Development, it appeared, was failing to improve conditions for the poorest and weakest sectors of society. The argument ran that what was needed was redistribution of wealth to be effected alongside growth. During this period, the basic needs approach was accepted and adopted by a range of international agencies, not only the International Labour Organisation (ILO) but also UNEP and the World Bank. However, it has to be recognised that many basic needs approaches used the aegis of the poor in cheap basic needs programmes in place of greater state commitment to poverty alleviation. This is why the World Bank became such a rapid convert to the approach. In these circumstances it has to be acknowledged that the practical implementation of basic needs had little to do with socialist principles *per se*.

The principal idea is that basic needs, such as food, clothing and housing, must be met as a clear first priority within particular territories. In the purest form, it is argued that this can only be achieved by nations becoming more reliant on local resources, the communalisation of productive wealth, and closing up to outside forces of change. This aspect of development theory is known as selective regional and territorial closure.

In simpler terms, it is argued that Third World countries should try to reduce their involvement in processes of unequal exchange. The only way round the problem is to increase self-sufficiency and self-reliance. It is envisaged, however, that later the economy can be diversified and non-agricultural activities introduced. But it is argued that, in these circumstances, urban locations are no longer likely to be mandatory, and cities can in this sense be based on agriculture. Thus, J. Friedmann and Weaver (1979: 200) comment that 'large cities will lose their present overwhelming advantage'.

Clearly, such approaches are inspired by, if not entirely based on, socialist principles. Classic examples of the enactment of bottom-up paths to development have been China, Cuba, Grenada, Jamaica and Tanzania under the *ujaama* policies inspired by African socialism. As Hettne (1995) notes, self-sufficiency has frequently been perceived as a threat to the influence of superpowers, as in the case of tiny Grenada (Brierley, 1985a, 1985b, 1989; Potter, 1993a, 1993b; Potter and Welch, 1996). This example is more fully explored in Box 3.3.

Walter Stöhr (1981) provides an informative overview of development from below. In particular, his account stresses that there is no single recipe for such strategies, as there is for development from above. Development from below needs to be closely related to specific sociocultural, historical and institutional conditions. Simply stated, development should be based on territorial units and should endeavour to mobilise their indigenous natural and human resources. More particularly, the approach is based on the use of indigenous resources, self-reliance and appropriate technology, plus a range of other possible factors, as shown in Table 3.1.

Bottom-up strategies are varied, with alternative paths to development being stressed. They share the characteristic of arguing that development and change should not be concentrated at each higher level of the settlement and social systems, but should focus on the needs of the lower echelons of these orders. It is this characteristic which gives rise to the term *bottom-up*, for such strategies are in fact often enacted by strong state control and direction from the political 'centre'.

Perhaps the major development since the 1970s has been the emergence of environmental consciousness in the arena of development (see Plate 3.5). Central to this evolving concern was the Brundtland Commission on Environment and Development which reported in 1987 (WCED, 1987). Even more important was the so-called Earth Summit held in Rio in the summer of 1992. This United Nations Conference on Environment and Development (UNCED) brought together some 180 nations. It was at this stage that principles of environmental sustainability became a political issue in the development debate.

Ecodevelopment, now known as sustainable development, has become the development paradigm of the 1990s, stressing the need to preserve the natural biological sytems that underpin the global economy (Redclift, 1987; Elliott, 1994); see Chapter 6. Sustainability constitutes the ecological dimension of territorialism discussed previously (Hettne, 1995). Territory it is argued should be considered before function, and developing countries should not look to developed nations for the template on which to base their development. Rather, they should look toward their own

Table 3.1 Stöhr's criteria for the enactment of 'development from below'.

Broad access to land

A territorially organised structure for equitable communal decision-making

Granting greater self-determination to rural areas

Selecting regionally appropriate technology

Giving priority to projects which serve basic needs

Introduction to national price policies

External resources only used where peripheral ones are inadequate

The development of productive activities exceeding regional demands

Restructuring urban and transport systems to include all internal regions

Improvement of rural-to-urban and village communications

Egalitarian societal structures and collective consciousness

Source: Based on Stöhr (1981).

BOX 3.3   Paths to development: the case of Grenada

The experience of Grenada in the eastern Caribbean is useful in demonstrating that alternative paths to development do not have to be revolutionary in the Marxist political sense (Potter and Lloyd-Evans, 1998). In March 1979, Maurice Bishop, a UK-trained lawyer, otherthrew what was regarded as the dictatorial and corrupt regime of Eric Gairy. Maurice Bishop led the New Jewel Movement (NJM), the principal theme of which was anti-Gairyism allied with anti-imperialism. The movement also expressed its strong commitment to genuine independence and self-reliance for the people of Grenada (Brierley, 1985a and b; Kirton, 1988; Hudson, 1989, 1991; Potter, 1993a, 1993b; Ferguson, 1990).

On the eve of the revolution, Grenada suffered from a chronic trade deficit, strong reliance on aid and remittances from nationals based overseas, dependence on food imports and very substantial areas of idle agricultural land. After the overthrow of Gairy, the NJM formed the People's Revolutionary Government (PRG), the movement taking a basic human needs approach as the core of its development philosophy. The PRG stated its intention of preventing the prices of food, clothing and other basic items from rocketing, along with its wish to see Grenada depart from its traditional role as an exporter of cheap produce. The Government also set up the National Cooperative Development Agency in 1980, the express aim of which was to engage unemployed groups in villages in the process of 'marrying idle hands with idle lands'.

Between 1981 and 1982, two agro-industry plants were completed, one producing coffee and spices, the other juices and jams. A strong emphasis was placed on encouraging the population to value locally grown produce together with local forms of cuisine, although the scale of this task was clearly appreciated by those concerned (Potter and Welch, 1996). The PRG also pledged itself to the provision of free

medicines, dental care and education. Finally, it was an avowed intention of the People's Revolutionary Government to promote what Bishop referred to as the New Tourism, a term which is now widely employed in the literature. New Tourism meant the introduction of what the party regarded as sociologically relevant forms of holiday-making, especially those which emphasised the culture and history of the nation, and which would be based on local foods, cuisine, handicrafts and furniture-making (Patullo, 1996). Such forms of tourist development, it was argued, should replace extant forms based on overseas interests and the exploitation of the local environment and sociocultural history.

The salient point is that, throughout the period, 80 per cent of the economy of Grenada remained in the hands of the private sector, and a trisectoral strategy of development that encompassed private, public and cooperative parts of the economy was the declared aim of the PRG. In this sense, the so-called Grenadian Revolution was nothing of the sort. The economy of Grenada grew quite substantially from 1979 to 1983, at rates of between 2.1 and 5.5 per cent per annum. During the period, the value of Grenada's imported foodstuffs fell from 33 to 27.5 per cent. Even the World Bank commented favourably on the state of the Grenadian economy during the period from 1979 (Brierley, 1985a).

For many it was a matter of great regret that Maurice Bishop was assassinated in October 1983, and the island invaded by United States military forces, because this saw the end to the four-year experiment in alternative development set up in this small Commonwealth nation (Brierley, 1985a). This deprived other small dependent Third World states of the fully worked-through lessons of grassroots development that Grenada seemed to be in the process of providing.

---

ecology and culture (Box 3.3). In this context too, it is recognised that development does not have a universal meaning. Strongly allied to this, the need for emancipatory views on women and development, and ethnicity and development has started to receive the attention that it merits. It is in this sense that sustainable development means more than preserving

natural biological systems. There is the assumption of implicit fairness or justice within sustainable development, so the poor and disadvantaged do not have to degrade or pollute their environments in order to be able to survive on a day-to-day basis (Drakakis-Smith, 1996; Lloyd-Evans and Potter, 1996; Eden and Parry, 1996).

The much quoted definition of sustainable development was provided by the Brundtland Commission, as development 'that meets the needs of the present without compromising the ability of future generations to meet their own needs' (WCED, 1987: 43). This is a far cry from the unilinear, Eurocentric functional perspectives advanced during the 1960s and 1970s, and demonstrates the wide-ranging changes that have occurred in development theory and geographies and policies of development over the past forty years. However, as noted at the outset of this chapter, the promotion of sustainable development is occurring in a context where neoliberal economic policies are also being promulgated, and many would stress the incompatability of these two forces of change at their extremes.

Development theory, modernity and post-modernity: concluding remarks

Abandoning the evolutionary and deterministic modernisation paradigm opens up several post-modern options for future development, and some of them have been reviewed in the previous section on bottom-up and alternative conceptualisations of development.

These trends in development thinking can be linked with the idea that, globally speaking, we are entering the post-modern age, associated with the rise of a knowledge-based post-industrial economy. This theme is briefly addressed here, but receives further, more explicit attention in relation to globalisation trends and development in the next chapter. Some aspects of post-modernity in relation to development ideologies have already been covered in Chapter 1.

At a straightforward level, post-modernity involves moving away from an era dominated by notions of modernisation and modernity (Plates 3.7a and 3.7b and Figure 3.14, overleaf). It is therefore intimately associated with development theory and practice. It involves the rejection of modernism and a return to premodern and vernacular forms, as well as the creation of distinctly new post-modern forms (Urry, 1990; Soja, 1989; Harvey, 1989). It can be seen as a reaction against the functionalism and austerity of the modern period in favour of a heterogeneity of styles, drawing on the past and on contemporary mass culture (Plate 3.8, page 73). As argued earlier in this chapter, there is much in the idea that the whole ethos of the modern period privileged the metropolitan over the provinces, the developed over the developing worlds, North America over the Pacific Rim, the professional over the general populace and men over women. The

Plate 3.7a Modern Hong Kong towers above traditional sampans (photo: Rob Potter)

Plate 3.7b Modern high-rise apartments in Havana, Cuba (photo: Rob Potter)

Figure 3.14 Modernism destroying the urban fabric. Cartoon by J.F. Batellier.

post-modern world in contrast potentially involves a diversity of approaches, which may serve to empower 'other' alternative voices and cultures. A strong emphasis on bottom-up, non-hierarchical growth strategies, which endeavour to get away from the international sameness, depthlessness and ahistoricism of the modern epoch, can be seen as part and parcel of the post-modern world (Table 3.2). The accent can

Table 3.2 Some polar differences between modernism and post-modernism that have relevance to development.

Modernism	Post-modernism
Form	Antiform
Conjunctive, closed	Disjunctive, open
Purpose	Play
Design	Chance
Hierarchy	Anarchy
Finished work	Performance, happening
Synthesis	Antithesis
Centring	Dispersal
Root, depth	Rhizome, surface
Interpretation, reading	Against interpretation, misreading
Narrative	Antinarrative
Master code	Idiolect
Determinancy	Indeterminancy
Transcendence	Immanence

Source: Adapted from Harvey (1989).

potentially be placed on growth in smaller places rather than bigger, in the periphery not the core.

However, although post-modernism may in certain respects be seen as an optimistic, liberating force associated with small-scale non-hierarchical development and change, there is another distinct facet to the trend of post-modernism. As well as the rejection of the modern and a hankering for the premodern, there is the establishment of 'after the modern'. This is frequently interpreted as 'consumerist post-modernism', involving the celebration of commercialism, commercial vulgarity, the glorification of consumption, and the related expression of the self (Cooke, 1990). As will be explored in Chapter 4, these are trends which are of interest in relation to what is happening in parts of the developing world, an example being the promotion of international tourism as a major plank of development (Plate 3.8) (Chapter 4). It is this aspect of post-modernism which gives rise to the suggestion that it maintains significant affinitites with both right- and left-wing lines of thought and associated policy prescriptions (Jones, Natter and Schatzki, 1993).

Such a condition is related to a conflation of trends in which aspects of art and life, high and low culture are fused together, or 'pastiched'. Images, signs, hoarding and advertisements are potentially more important than 'reality'. Mass communications lead to mass image creation (Robins, 1989, 1995; Massey, 1991). History and heritage may be rewritten and re-interpreted in order to meet the needs of international business. This may all lead to further external control, exploitation and neocolonialism. Many of these features can be interpreted in terms of the over-consumptive lifestyles which were offered to members of the upper white-collar strata of society in the

Plate 3.8 Resonances of post-modernity in a tourist setting: tourists and members of a traditional 'Tuk' band in Barbados (photo: Rob Potter)

Plate 3.9 Images of development ideology in Cuba: 'Revolution Yes!'
(photo: Rob Potter)

Reagan–Thatcher era. Several of these themes will be further explored in the next chapter.

In this regard, rather than being seen as a freeing and enabling force, post-modernism may alternatively be interpreted as essentially the logical outcome of late capitalism (Harvey, 1989; Jameson, 1984; Dann and Potter, 1994; Potter and Dann, 1994, 1996; Sidaway, 1990; Cuthbert, 1995; Kaarsholm, 1995). The role of TNCs in the promotion of tourism in the Caribbean may be seen as a clear example of this, where advertising and promotion campaigns may be interpreted as being aimed directly at increasing both the environmental and social carrying capacities of the nation. Such developments have interesting implications in contexts where nations themselves are still endeavouring to modernise. Indeed, the situation may give rise to all sorts of ambiguities (Potter and Dann, 1994; Austin-Broos, 1995; Masselos, 1995; G.A. Thomas, 1991).

The development of post-modern trends which influence developing societies, in particular, via the activities of TNCs and tourists, is of further interest in the new world-order following the collapse of the communist world in 1989. We are certainly entering a noticeably less certain, less monolithic and unidirectional world. Hence the already wide diversity of development, with many varied geographies of development, seems likely to get more complicated, rather than less, over the coming years. For example, some authorities are now talking of the tripolarity of development, with the Americas, Europe and Pacific Asia each presenting a particular version of industrial cap-

italism (P.W. Preston, 1996). The recent problems being faced by Asian economies are another sign of increasing volatility and dynamism.

This view is reflected in what has been called an impasse in development studies (D. Booth, 1985; Preston, 1985; Slater, 1992a; Corbridge, 1986; Schuurman, 1993). The world has become a more complex place since the collapse of communism in 1989, making the division into the First, Second and Third Worlds much more complex (Plate 3.9). Similarly, Hettne (1995) argues that the rise of neoconservatism in the global political realm, and monodisciplinary trends in the academic world have both presented development thinking with fundamental challenges. Some have pointed to what they regard as an impasse in theorising development itself, although this seems unduly pessimistic given the range of ideas considered in this chapter. Hettne (1995) also refers to the failure of development in practice as contributing to self-criticism, pessimism and 'development fatigue', especially in relation to the ultimate relevance of Western-developed research and ideas.

Notwithstanding these justifiable concerns, it is axiomatic that 'development', defined as change for the better or for the worse (Brookfield, 1975), will proceed in each and every corner of the globe. This being the case, there will continue to be the need for the generation and discussion of realistic, although challenging, sets of ideas concerning the process of development, and the conditions that are to be encountered in developing countries themselves.

CHAPTER 4 Globalisation and development

An era of global change

Over the last fifteen years or so, one of the major trends has been that the world in which we live has become ever more global in character and orientation. This trend has been witnessed in increasing actual and potential interactions between different parts of the globe. There are at least three distinct aspects to this process of global change. Firstly, the world is effectively shrinking in terms of the distances that can be covered in a given time period, due to faster and more efficient transport. Secondly, better communications such as satellite television mean that we hear about what is happening elsewhere in the world more swiftly than we ever did before. The global web now has far more connections than it had in the past (Allen, 1995). Thirdly, global corporations and global marketing activities are resulting in the availability of standardised products (Robins, 1995) and television programmes throughout the world. In addition, not only do we live in a world of Big Macs, Coca-Cola and Levi jeans, we are also witnessing the emergence of global financial markets. This is associated with a dramatic acceleration in the speed of financial flows and transactions, with the possibility that money can now exist in a purely electronic form, so that it takes only seconds to send sums from one part of the world to another (Leyshon, 1995: 35).

Following this argument, Allen (1995) has recognised three broad strands to globalisation: the economic, the cultural and the political. In respect of economic globalisation, distance has become less important to economic activities, and large corporations subcontract to branch plants in far distant regions, effectively operating within a 'borderless' world. The stereotype of cultural globalisation suggests that as Western forms of consumption and lifestyles spread across the globe, there is an increasing convergence of cultural styles on a global norm, with that norm being codified and defined by the global capitalist system (Plate 4.1). In the arena of political globalisation, internationalisation is regarded as leading to the erosion of the former role and powers of the nation state.

Plate 4.1 Globalisation in the Third World – modern advertising poster provides a contrast to the surrounding streets in Lagos, Nigeria (photo: Mark Edwards, Still Pictures)

Following the overview of basic development theory presented in the previous chapters, the present account focuses on the question of what development means in a contemporary context that is dominated by processes of globalisation and global change. One of the important questions to be addressed is whether globalisation is in fact a new process in the first place. Does globalisation mean the entire world is becoming more uniform? Does it also mean there is a chance that the world will become progressively more equal over time? If not, is it the case that such a process of accelerated homogenisation will come about in the near future? In short, does globalisation mean that change and development will trickle down, and that this will occur more speedily than in the past? These are just a few of the basic issues that will be addressed in this final chapter of Part I.

Globalisation and the shrinking world

Over the past forty years, there has been much talk about the world becoming a global village, and the associated 'compression' or 'annihilation' of space by time, in the context of what is known as a shrinking world. The phrase 'annihilation of space by time' is commonly attributed to Karl Marx (Leyshon, 1995: 23). Leyshon (1995) credits Marshall McLuhan (1964) with the first use of the expression 'Global Village', noting that the world was becoming compressed and electronically contracted, so that 'the global is no more than a village'. McLuhan went on to observe that due to the evolving electronic media, humans were beginning to participate in village-like encounters, but at a global scale, thereby cogently anticipating the development of electronic mail and the Internet (Chapter 8).

The main aspects of this change were outlined at the start of the chapter. Firstly, the world has effectively become a smaller place than it was fifty years ago, in terms of the time it takes to travel around it. At the global level, this is illustrated by the much reproduced representations of the world shown in Figure 4.1. The best average speed of horse-drawn coaches and sailing ships was about 10 miles per hour in the period between 1500 and 1840. In 1830 the first railway was opened between Liverpool and Manchester, and the first telegraph system was patented. By 1900 a global telegraph system had been established, based on submarine cables, giving rise to the world's first ever global communications system. By the end of the period 1850–1940, steam trains averaged 65 mph and steamships around 36 mph. But, as shown by Figure 4.1, the real change came after 1950, with propeller-driven aircraft travelling at 300–400 mph. After the 1960s, commercial jet aircraft took speeds into the 500–700 mph range. As a result of these progressive changes, the earth has effectively been shrunk to a fraction of its effective size of some five hundred years ago (Figure 4.1 and Plate 4.2 page 78).

In the 1960s there was an exponential increase in the number of scheduled international flights globally; whereas at the national level, large-scale highway construction proceeded in North America and Europe in association with rapidly increasing levels of car ownership. Between 1950 and 1960, domestic television was disseminated, followed by the exploration of space and the launch of communication satellites (Leyshon, 1995). At the beginning of the 1970s, Janelle (1973) referred to the 'thirty-minute world', this being the time it would take for an intercontinental missile to travel from its launch site to its target on the other side of the world.

The process by means of which improvements in transport technologies have effectively moved places within the settlement system *vis-à-vis* one another was described by Janelle (1969) as time–space convergence. To give a national example, in 1779 it took four days or 5,760 minutes to travel the 330 miles which separate Edinburgh from London. By the 1960s, the time taken to travel between them had effectively been reduced to less than 180 minutes by plane, so the two places had been 'converging' on one another at the rate of approximately 30 miles per year.

Of course, the world is also shrinking in another sense, in that we are all potentially increasingly aware of what is happening in other far distant places, without the need to move from our home localities. This is now achieved via the mass media. As Leyshon (1995: 14) notes, it is 'in the area of news and current affairs that television's ability to shrink space is best illustrated,' as was all too clearly demonstrated by the coverage of the Gulf War by cable news early in 1991. But cable news depends on relatively sophisticated and expensive technologies, so the relatively rich have also tended to become the information-rich (Leyshon, 1995), a point to which we shall return at several junctures in this chapter. There are other development-related implications to this set of changing circumstances in that they invite a redefinition of our ethical and moral responsibilities in relation to people who live far away from us. This 'responsibility to distant others' (Corbridge, 1993b; Smith, 1994; Potter, 1993b) is not unrelated to the observation that

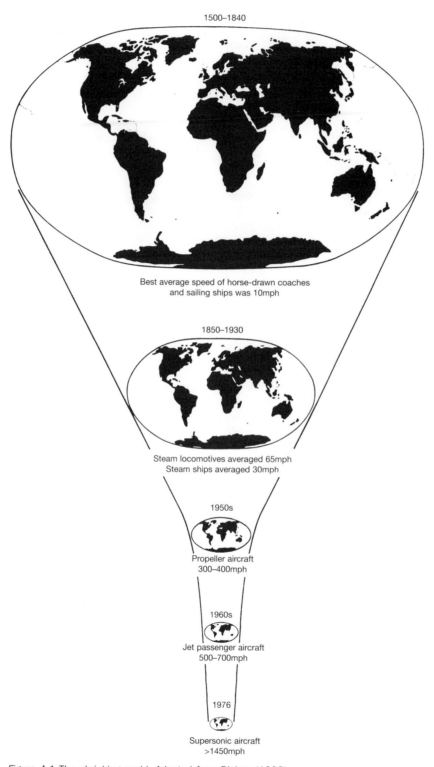

1500–1840

Best average speed of horse-drawn coaches
and sailing ships was 10mph

1850–1930

Steam locomotives averaged 65mph
Steam ships averaged 30mph

1950s

Propeller aircraft
300–400mph

1960s

Jet passenger aircraft
500–700mph

1976

Supersonic aircraft
>1450mph

Figure 4.1 The shrinking world. Adapted from Dicken (1998).

Plate 4.2 Shrinking world: long- and short-haul aircraft at Seychelles international airport (photo: Tony Binns)

the global mass media frequently only refers to developing countries when reporting natural disasters, social disturbances, poverty, mass starvation and other crises and mishaps. Some writers, especially those concerned with Africa, have observed how this is leading to the notion that Africa is literally 'bad news', gradually desensitising the relatively wealthy from the real daily plight of Africans (Harrison and Palmer, 1986; Milner-Smith and Potter, 1995). Such a view leads to the implication that the Third World is literally a disaster zone, and serves to emphasise its status as something quite separate, representing 'the Other'.

It is all too easy to conclude that as the world shrinks, all parts of the global village share in the benefits of global development. However, this leads to a vital argument, for the places that share in development are those which are already well connected on the network. Places which are eccentric to it, or which are off the network altogether, are by definition massively disadvantaged. This is a fundamental point, and pursuing it at the subglobal level makes a very telling point about the developmental impact of globalisation. As well as relative distances being reduced by the process of global development, distances to other places can *increase* in relative terms, even within the overall pattern of a shrinking world.

Figure 4.2 gives a specific example of this. The figure shows the Pacific Basin. Figure 4.2(a) shows the conventional cartographic projection, whereas Figure 4.2(b) has been redrawn according to travel times between places by scheduled airline in the mid 1970s.

The figure is adapted from Haggett (1990) and Leyshon (1995). At first glance, North America has 'moved' closer to Asia, and Australia has 'drifted' north towards Asia. But if we look in a little more detail, we find that places like Tokyo, San Francisco and Sydney have indeed 'moved' closer to one another.

Even more detailed scrutiny of Figure 4.2(b) reveals that some places have in fact become more 'distant' from one another. This is evident from the fact that South America has 'trailed behind' North America in its 'convergence on' Asia. But if we look at specific places in more detail, they appear to have 'moved' quite substantially relative to one another. In particular, it is noticeable that poorer and less frequent air transport links mean that Papua New Guinea appears to have moved to the south of Australia, away from Asia, whereas the Trust Territories of the Pacific appear to have moved north, apparently now existing outside the Pacific Basin altogether. This clear example of overall time–space convergence shows that the process is far from homogeneous. In fact, it is sufficiently inhomogeneous as to produce instances of what may be called relative time–space divergence.

The idea that globalised improvements in transport and communications are leading to the intensification of the functional importance of certain places or nodes is confirmed if we look at world airline networks in the 1990s. Keeling (1995) produced a map showing the magnitude of the international air connections between major cities. The map, reproduced here as Figure 4.3 shows the number of outward and return non-stop

(a)

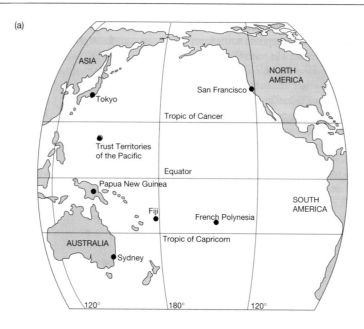

(b)

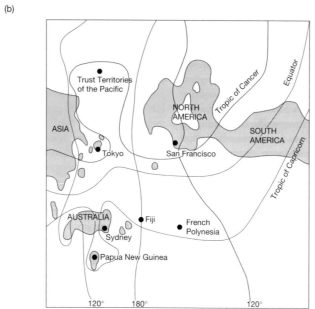

Figure 4.2 Time–space convergence and divergence: (a) the
conventional projection of the Pacific; (b) time–space map of the Pacific
based on travel times by scheduled airline in 1975. Adapted from
Haggett (1990) and Leyshon (1995).

flights per week from various nodes. The data on which
the diagram is based is reproduced in Table 4.1. The
outcome illustrates all too clearly the predominance
of three global cities, London, New York and Tokyo,
and the role these cities play as dominant global hubs.
Together these three cities receive a staggering 36.5 per
cent of total global non-stop flights to the world airline
network's twenty dominant cities. Beyond these three

Table 4.1 Total number of non-stop flights per week to the major world cities, 1992.

City	Notation on map	Global	Regional	Domestic
London	L	775	3 239	1 063
New York	N	644	634	8 837
Paris	P	565	2 264	1 436
Tokyo	T	538	401	1 814
Frankfurt	FRA	482	1 376	771
Miami	MIA	311	1 389	2 146
Cairo	CAI	277	34	114
Los Angeles	LAX	245	419	7 150
Bangkok	B	231	483	307
Singapore	SIN	221	831	0
Hong Kong	HKK	154	713	0
Sydney	SYD	144	89	1 541
Rio de Janeiro	RIO	93	44	933
Moscow	MOW	87	400	1 430
Bombay	BOM	64	111	313
São Paulo	SAO	64	97	1 418
Buenos Aires	BUE	52	336	414
Johannesburg	JNB	40	108	450

Source: Adapted from Keeling (1995).

cities, Paris, Cairo, Singapore, Los Angeles and Miami appear as secondary global hubs, and Johannesburg, Moscow, Bombay, Bangkok, Hong Kong, Sydney, São Paulo, Rio de Janeiro and Buenos Aires as secondary hubs. Looking at the flows in Figure 4.3 shows just how marked is the concentration. Globalisation is leading to strong local concentration within continents; it can be seen as increasing 'differences'.

Far more commonly experienced is the phenomenon of time–space compression noted by Harvey (1989). The capitalist system demands efficiency and leads to the economic logic of reducing barriers to movement and communications over space, as time costs money. This leads to the progressive acceleration in the pace of life that seems to be universally experienced in the 'modern' world, and which seems to affect countries whether developed or developing, although perhaps in contrasting and locally specific manners.

Globalisation and development

From the ideas in the previous section, there have emerged two generalised views concerning the relationships between globalisation and patterns of development.

The first view is the familiar claim that, to all intents and purposes, places around the world are fast

becoming if not exactly the same then certainly very similar. This view dates from the 1960s belief in the process of modernisation (Chapters 1 and 3). Such a perspective tacitly accepts that the world will become progressively more 'Westernised', or more accurately, 'Americanised' (Massey and Jess, 1995). The approach stresses the likelihood of social and cultural homogenisation, with key American traits of consumption being exemplified by the 'coca-colonisation', or 'coca-colaisation', and the Hollywoodisation, or Miamisation of the Third World, replete with McDonald's golden arches.

The second more realistic view of globalisation is implicit in the account which resulted from consideration of the transformed map of the Pacific Basin. It presents almost the reverse view, stressing that rather than uniformity, globalisation is resulting in greater flexibility, permeability, openness, hybridity, plurality and difference, both between places and between cultures (Massey and Jess, 1995; Massey, 1991; Robins, 1995; Potter, 1993a; 1997). Following on from this perspective, far from leading to a uniform world, globalisation is viewed as being closely connected with the process of uneven development, and the perpetuation and exacerbation of spatial inequalities.

This view of globalisation argues that, by such a process, localities are being renewed afresh. This is particularly so in respect of economic change, where production, ownership and economic processes are

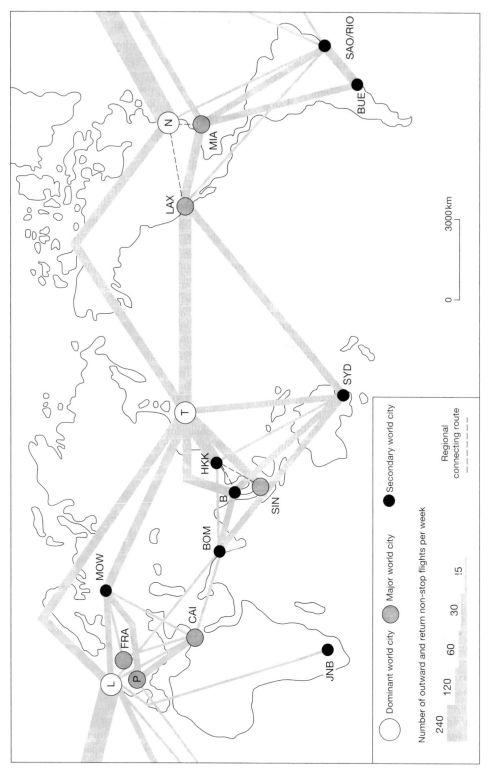

Figure 4.3 Dominant flows on the world airline network. Adapted from Keeling (1995).

highly place- and space-specific. Even in regard to cultural change, it may be argued that although the hallmarks of Western tastes, consumption and lifestyles, such as Coca-Cola, Disney, McDonald's and Hollywood are available to all, such worldwide cultural icons are reinterpreted locally, and take on different meanings in different places (Cochrane, 1995). This view sees fragmentation and localisation as key correlates of globalisation and post-modernity. Again, all of this can be seen as an extension of the lessons that are to be drawn from considering the transformed map of the Pacific Basin shown in Figure 4.2 (page 79).

A further major point substantiates this view. Evidence shows that globalisation is anything but a new process; it has been operating for hundreds of years. The process of globalisation can be seen as having started with the age of discovery (Allen, 1995; Hall, 1995). This argument has been clearly summarised by Stuart Hall (1995: 189):

> Symbolically, the voyage of Columbus to the New World, which inaugurated the great process of European expansion, occurred in the same year as the expulsion of Islam from the Spanish shores and the forced conversion of Spanish Jews in 1492. This . . . [is] as convenient a date as any with which to mark the beginnings of modernity, the birth of merchant capitalism as a global force, and the decisive events in the early stages of globalization.

This perspective highlights how globalisation has always been intimately connected with power differentials and changes in culture. In this particular connection, early globalisation was associated with the conquest of indigenous populations, great rivalries between the major European powers in carving up colonial territories, and the eventual establishment of the slave trade, as detailed in Chapter 2. Globalisation has always been associated with increasing differences between peoples and places, rather than with evenness and uniformity. And globalisation has been a gradual, as well as a partial and uneven process, which has spread heterogeneously across the globe. This overarching theme is now addressed in the contemporary context, firstly in relation to economic aspects of globalisation, and then in relation to cultural change.

Global shifts: industrialisation and economic globalisation

The pursuit of unequal development as a matter of policy came to affect the newly independent, formerly colonial territories in the 1960s, as noted in Chapter 3. It was perhaps inevitable that in seeking to progress during the post-colonial era, developing countries should associate development with industrialisation. This was hardly surprising given that the conventional wisdoms of development economics stressed so cogently this very connection (Potter and Lloyd-Evans, 1998, Chapter 3). For many Third World countries, decolonisation afforded political independence and promoted the desire for the economic autonomy to go with it. In the words of J. Friedmann and Weaver (1979: 91), such nations

> took it for granted that western industrialised countries were already developed, and that the cure for 'underdevelopment' was, accordingly, to become as much as possible like them. This seemed to suggest that the royal road to 'catching up' was through an accelerated process of urbanisation.

In the early phase, the trend toward industrialisation in developing countries was closely associated with the policy of import substitution industrialisation (ISI). This represented an obvious means of increasing self-sufficiency, as such nations had traditionally imported most of their manufactured goods requirements in return for their exports of primary products such as sugar, bananas, coffee, tea and cotton. During the era of import substitution industrialisation, key industrial sectors for development were those which were relatively simple and where a substantial home market already existed; for example, food, drink, tobacco, clothing and textile production (Plate 4.3). By now most developing countries have followed this path toward import substitution industrialisation, although as Dickenson et al. (1996) observe, few have progressed much beyond it. An exception, perhaps, is Taiwan, where between 1953 and 1960 the ISI policy was put into practice, focusing on textiles, toys, footwear, agricultural goods and the like. During this era, manufacturing output increased by 11.7 per cent per annum. Only after the 1960s did Taiwan develop export-oriented manufacturing, and after 1980 there came technologically advanced, high-value added manufacturing, so as to stay ahead in the industrialisation stakes. The development of heavy industries, such as steel, chemicals and petrochemicals, along the lines of the Soviet model, has not been possible for most Third World nations.

Such a policy – which might seem attractive when following Rostow's (1960) linear model of development (Chapter 3) – requires a level of population and effective demand not normally present in most

Plate 4.3 Import substitution industrialisation – a brewery in Ouagadougou, Burkina Faso
(photo: Jorgen Schytte, Still Pictures)

developing countries. Furthermore, the competition from developed nations, along with capital and infrastructural shortages and problems of lumpy investment, technological transfer and capital rather than labour intensity, also militate against such heavy industrial development. An exception, however, is provided by India, which has achieved a high level of industrial self-sufficiency since 1945 (Johnson, 1983) and is now around the thirteenth industrial producer in the world.

However, from the 1960s onwards, a number of developing countries embarked upon policies of light industrialisation by means of making available fiscal incentives to foreign companies (Potter and Lloyd-Evans, 1998, Chapter 3). This policy of so-called industrialisation by invitation was strongly recommended by Caribbean-born economist Sir Arthur Lewis (1950, 1955). Reviewed at length in Chapter 3, it involved the establishment of branch plants by overseas firms, with the products being exported back to industrialised countries.

The approach became closely associated with the setting up of free trade zones (FTZs) and export processing zones (EPZs). The FTZ is an area, usually located in or near to a major port, in which trade is unrestricted and free of all duties (Plate 4.4, over-

leaf). The EPZ is normally associated with the provision of buildings and services, and amounts to a specialised industrial estate. Firms locating on EPZs frequently pay no duties or taxes whatsoever, and may well be exempt from labour and other aspects of government legislation. The approach is often known as enclave industrialisation.

According to Hewitt, Johnson and Wield (1992), the first EPZ established in a developing country was at Kandla in India in 1965, and this was quickly followed by further developments in Taiwan, the Philippines, the Dominican Republic and on the United States–Mexico border. In the latter case, during the 1960s, Mexico enacted legislation permitting foreign, especially American, companies to establish 'sister plants', called *maquiladoras*, within 19 kilometres of the United States borders, for the duty-free assembly of products destined for re-export (Figure 4.4, overleaf). By the early 1990s, more than two thousand such assembly and manufacturing plants had been established, producing electronic products, textiles, furniture, leather goods, toys and automotive parts. In aggregate, the plants generated direct employment for over half a million Mexican workers (Getis, Getis and Fellman, 1994; Dicken, 1998).

Plate 4.4 Ice cream salesman outside the Koggala free trade zone, Sri Lanka (photo: Ron Giling, Still Pictures)

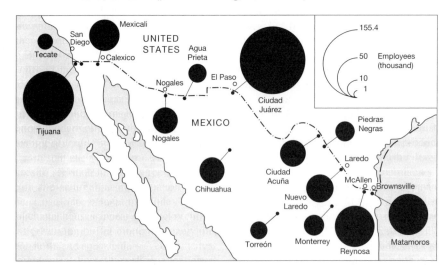

Figure 4.4 The principal maquiladora centres on the United States–Mexico border. Adapted from Dicken (1998).

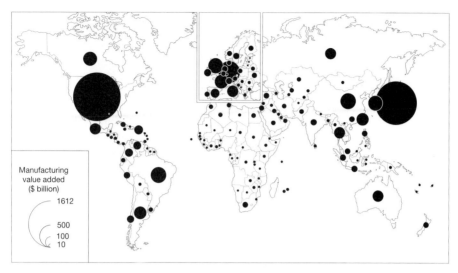

Figure 4.5 The global distribution of manufacturing, 1994, Adapted from Dicken (1998).

By 1971, nine countries had established EPZs; this increased to twenty-five by 1975 and fifty-two by 1985. In 1985 it was estimated that there were a total of 173 such zones around the world, which together employed 1.8 million workers. Frequently, programmes of industrial development have been strongly urban-based, as in the case of Barbados, where ten industrial estates were established, all within the existing urban envelope (Potter, 1981). Recently, data processing and the informatics industry have become very important on one of the central Bridgetown industrial parks (Clayton and Potter, 1996).

By such means, developing countries have increased their overall level of industrialisation. From 1938 to 1950, developing nations experienced a 3.5 per cent growth rate of manufacturing per annum, and from 1950 to 1970, this annual rate increased to 6.6 per cent (Dickenson *et al.* 1996). But this growth has been minuscule compared with the growth of the urban population, and in many instances it has been based on a non-existent prevailing level of industrial activity. Furthermore, industrial growth has been characterised by two additional features. The first has been its highly unequal global distribution, and in the postwar period this has been associated with major changes in the global distribution of industrial production.

This process is illustrated in Table 4.2. The traditional industrial nations, such as the United States, the United Kingdom and France along with other developed countries, and latterly Germany, have all shown marked declines in their percentage share of world industrial production since 1948. This has gone hand in hand with rising industrial production in Japan, which

Table 4.2 Changes in the global distribution of industrial production, 1948–1984.

Country/region	Percentage of world industrial production		
	1948	1966	1984
United States	44.4	35.2	28.4
United Kingdom	6.7	4.8	3.0
West Germany	4.6	8.1	5.8
France	5.4	5.3	4.4
Japan	1.6	5.3	8.2
Other developed countries	14.7	12.5	11.1
Centrally planned economies	8.4	16.7	25.4
Newly industrial countries	4.9	5.7	8.5
Other less developed countries	9.1	6.5	5.4
Total	100.0	100.0	100.0

Source: Chandra, 1992.

by 1985 had increased its share of the world total to 8.2 per cent. Since 1948, industrial production has also risen sharply in the centrally planned economies. A key feature has been the increasing importance of the newly industrialised countries (NICs), which by 1984 accounted for 8.5 per cent of global production.

However, the remaining less developed nations have shown a declining proportion of total manufacturing production, from 9.1 per cent in 1948 to 5.4 per cent in 1984 (Table 4.2). The emergence of the NICs such as China, Brazil, India, South Korea, Mexico and Taiwan is shown by their inclusion among the top twenty-five industrial nations in 1986 (Table 4.3);

Table 4.3 The world 'league table' of manufacturing production, 1994.

Rank	Country	Manufacturing value added (US$ million)	Percentage of world total
1	United States	1 611 763	26.9
2	Japan	1 257 761	21.0
3	Germany	692 191	11.6
4	France	268 611	4.5
5	United Kingdom	243 653	4.1
6	South Korea	159 172	2.7
7	Brazil	154 425	2.6
8	China	139 031	2.3
9	Italy	128 486	2.2
10	Canada	100 322	1.7
11	Argentina	88 366	1.5
12	Spain	81 196	1.4
13	Taiwan	73 295	1.2
14	Australia	64 417	1.1
15	Switzerland	60 111	1.0
			———
		Total	85.8

Source: Dicken (1998).

see also Dickenson (1994) and Courtney (1994). The global distribution of manufacturing production is shown in Figure 4.5 (page 85), and although the United States, Western Europe and Japan between them account for three-quarters of total production, the importance of the Asian tigers, together with Brazil and Mexico, is clear from the figure. The process of change in the world distribution outlined above has been referred to by Dicken (1998) as global shift, whereby economic activity is becoming increasingly internationalised or globalised. An example of the complex changes that are occurring is given in Box 4.1.

The second characteristic feature of post-1948 industrial change has been the rise of transnational corporations (TNCs), which now represent the most important single force creating global shifts and changes in production (Dicken, 1998). TNCs can be traced back to the late nineteenth century; to begin with they focused on agricultural, mining and extractive activities, but in the period since 1950 they have become increasingly associated with manufacturing (Jenkins, 1987, 1992; Dicken, 1998).

In 1985 the United Nations identified six hundred TNCs operating in the fields of manufacturing and mining, each of which had annual sales in excess of US$1 billion. These corporations between them generated more than 20 per cent of the total production in the world's market economies. Roughly 40 per cent of total world trade now takes place between the subsidiaries and parent companies of TNCs (Corbridge, 1986; Hettne, 1995). During the 1960s, the foreign output of TNCs was growing twice as fast as world gross national product. By 1985, developing countries accounted for 25 per cent of total foreign direct investment (FDI). The largest share of this was in Latin America and the Caribbean (12.6 per cent of the world total), followed by Asia (7.8 per cent), Africa (3.5 per cent), with other areas accounting for 1 per cent (Dicken, 1998). However, much FDI is still related to the resource production sector, and a considerable amount of it now emanates from Third World NICs (Figure 4.6, page 88).

These contributory trends of globalisation are far too complex to be dealt with adequately by means of traditional models of development. In particular, the hierarchical model of change linked to modernisation, put forward by Hudson (1969), Pederson (1970) and Berry (1972) and reviewed in Chapter 3, may be reinterpreted as far too simplistic. Given the omnipotence of TNCs, new production, innovations, capital and social surplus are not likely to trickle down the national space economy in a step-by-step manner, from the top to the bottom. Ownership and production are likely to be much more concentrated, an important theme which is picked up in the next section of this chapter. Furthermore, it follows that the decision to base production in one developing nation rather than another will have considerable impact on the geography of development and change, especially when it is remembered that many TNCs have annual turnovers that greatly exceed the gross national products of some small and impoverished developing nations.

It was just this sort of patterning that was identified by Pred (1973, 1977) in his historical examination of the growth and development of the United States. Pred noted that the growth of the mercantile city was based on circular or cumulative causation, linked to multiplier effects. In addition, Pred argued that the growth of large cities was based on their interdependence, so that large city stability has been characteristic. However, Pred maintained that key innovation adoption sequences were not always hierarchic, frequently flowing from a medium-sized city up the urban hierarchy, or from one large city to another large city. Pred (1977) looked at the headquarters of TNCs in postwar America, and stressed the close correspondence with the uppermost levels of the world system. Thus, growth within the contemporary global system is increasingly

---

BOX 4.1   Globalisation and the production of athletic footwear

The footwear industry is labour-intensive but it is also highly dynamic. In a paper published in 1993, Barff and Austen show how sales tripled in the United States over the preceding ten-year period. And they show the industry's volatility, measured in spatial and geographical terms. In 1989, US market leader Nike Inc. had about 2 per cent of its shoes made by Chinese-based subcontractors. Just four years later, in 1993, almost 25 per cent of Nike athletic shoes came from Chinese factories (Barff and Austen, 1993).

Although characterised by such dynamism, the majority of athletic footwear production continues to occur in Southeast Asia. The three US companies which account between them for over 60 per cent of sales in the United States have the vast majority of their production based there. However, the details of this pattern are quite volatile, and many producers of athletic footwear have developed a complex set of long- and short-term subcontracting agreements with other firms that change from year to year as a result of factory improvements, market fluctuations, and technological change (Donaghue and Barff, 1990).

On the other hand, several athletic footwear firms still manufacture in the United States. In particular, the cheapest sport shoes continue to be produced in the United States, whereas the more complex, expensive models tend to be manufactured in Asia. Barff and Austen (1993) show that, in order to understand this complex global geography, one must move beyond the basic consideration of international labour-cost differentials.

By means of case studies, the authors demonstrate that domestic production involves very different labour processes from those of production based in other countries. As in many sectors of the economy, domestic producers gain advantage by carrying smaller inventories via faster lead times. However, the best explanation for the globalised pattern of differential production centres on the nature of shoes themselves. The athletic shoes produced in the United States tend to have far fewer stitches in them than those manufactured elsewhere, and this minimises the most expensive component of the production process. Furthermore, the authors explain how tariffs on athletic shoes massively discriminate against imported shoes of a particular construction.

This example of global-scale production therefore demonstrates how processes of globalisation are based on subtle aspects of differentiation between world regions, and suggests that new forms of economic localisation may well be the outcome.

---

linked to the locational decisions of multinational firms and government organisations.

These types of notion have recently been given expression in the concept of the world city or global city. Although nebulous in terms of size definitions, the idea is that certain cities dominate world affairs. At one level, this is a very straightforward and obvious proposition, but its contemporary relevance has been elaborated by J. Friedmann and Wulff (1982) and J. Friedmann (1986). J. Friedmann (1986) put forward six hypotheses about world cities, observing that they are used by global capitalism as 'basing points' in the spatial organisation and articulation of production and markets, and that they act as centres for capital accumulation. Friedmann also suggested that the growth of world cities involves social costs which in fiscal terms the state finds it hard to meet. World cities have large populations, but more important, they have large and/or sophisticated manufacturing bases, they have sophisticated finance and service complexes, and they act as transport and communication hubs, involving TNCs and NGOs (D. Simon, 1992a, 1993); see also Knox and Taylor (1995) and J. Friedmann (1995).

The principal world cities such as New York, Brussels, Paris, London, Amsterdam and Milan are located in the developed world. But Singapore, Hong Kong, Bangkok, Taipei, Manila, Shanghai, Seoul, Osaka, Mexico City, Rio de Janeiro, Buenos Aires and Johannesburg (Plate 4.5, page 89) have all been recognised as part of an emerging network of world cities (J. Friedmann, 1995). This emergence is given spatial expression in Figure 4.7 (page 90). In short, world cities may be seen as points of articulation in a TNC-dominated capitalist global system.

There is the implication, therefore, that uneven development is particularly likely to be associated with developing countries, and that their paths to development in the late twentieth century will be infinitely more difficult than those which faced developed countries. This argument has been reviewed in the case

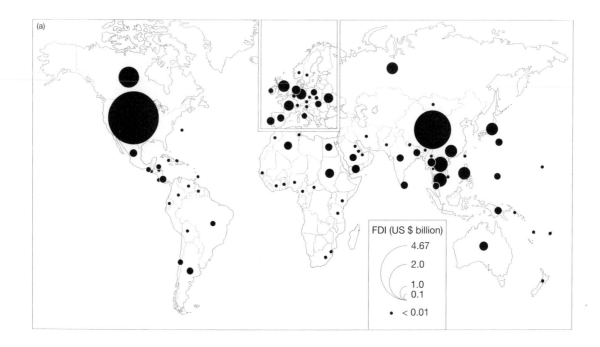

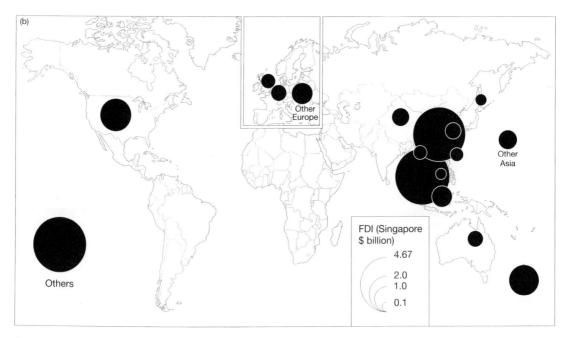

Figure 4.6 Foreign direct investment by (a) South Korea and (b) Singapore. Adapted from Dicken (1998).

Plate 4.5 A world city: the centre of Johannesburg from the air
(photo: Tony Binns)

of poor countries by Lasuen (1973). He started from the premise that, in the modern world, large cities are the principal adopters of innovations, so that natural or spontaneous growth poles become ever more associated with the upper levels of the urban system. Lasuen also observed that the spatial spread of innovations is generally likely to be slower in developing countries, due to the frequent existence of single plant industries, the generally poorer levels of infrastructural provision, and sometimes the lack of political will.

Thus, developing countries facing spatial inequalities have two policy alternatives. The first is to allow the major urban centres to adopt innovations before the previous ones have spread through the national system. The second option is to attempt to hold and delay the adoption of further innovations at the top of the national urban system, until the filtering down of previous growth-inducing changes has occurred. This may sound somewhat theoretical, but these options represent the two major strategies that can be pursued by states. The first option will result in increasing economic dualism, but some would argue, the chance of a higher overall rate of economic growth. On the other hand, the second option will lead to increasing regional equity, but potentially lower rates of national growth. Most developing countries have adopted policies close to the first option of unrestrained innovation adoption, seeking to maximise growth rather than equity. This theme is re-examined in Chapter 9.

Industrialisation in developing countries has been far from characterised by uniformity and homogenisation. In fact, it has been associated with global shifts,

non-hierarchic adoption sequences and the growth of global or world cities. In short, globalisation is leading to increasing differences between regions and places; for example, giving rise to centres, peripheries and semiperipheries at the broadest scale, as noted in Chapter 3 from a theoretical viewpoint. But in reality, global patterns of differentiation and localisation are much more complex than this in the contemporary context. The chapter now turns to consider this argument in further detail.

Global convergence and global divergence

This leads to a major conceptualisation of what is happening to the global system in the contemporary world, and what this means for growth and change in present-day developing countries. The basic argument is that the uneven development that has characterised much of the Third World during the mercantile and early capitalist periods has been intensified post 1945, as a result of the operation of what may be called the dual processes of global convergence and global divergence, terms which originate in the work of Armstrong and McGee (1985); see also Potter (1990). Together these processes may be seen as characterising globalisation.

Divergence relates to the sphere of production and the observation that the places which make up the world system are increasingly becoming differentiated,

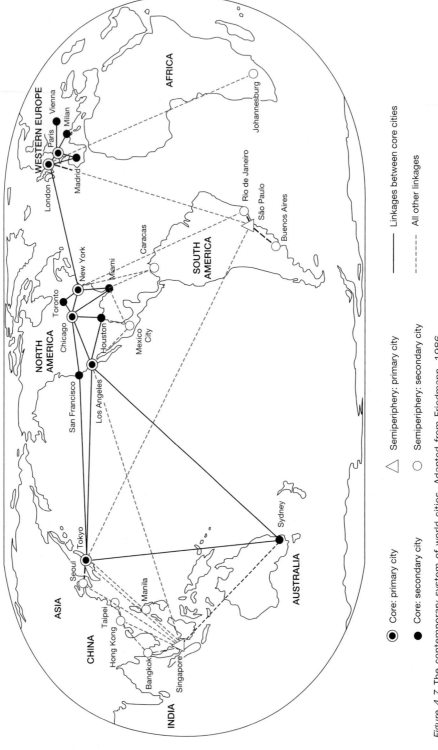

Core: primary city   △ Semiperiphery: primary city   —— Linkages between core cities

● Core: secondary city   ○ Semiperiphery: secondary city   ----- All other linkages

Figure 4.7 The contemporary system of world cities. Adapted from Friedmann, 1986.

i.e. diverse and heterogeneous, as discussed in the first part of this chapter. Starting from the observation that the 1970s witnessed a number of fundamental shifts in the global economic system, not least the slowdown of the major capitalist economies and rapidly escalating oil prices, Armstrong and McGee stressed that such changes have had a notable effect on developing nations.

Foremost among these changes has been the dispersion of manufacturing industries to low labour-cost locations, and the increasing control of trade and investment by TNCs. It is this trend which has witnessed the establishment of Fordist production-line systems in the NICs, whereas smaller-scale, more specialised and responsive, or so-called flexible systems of both production and accumulation have become more typical of advanced industrial nations. In this fashion, productive capacity is being channelled into a limited number of countries and metropolitan centres. Increasing global division of labour, and the increasing salience of TNCs are leading to enhanced heterogeneity or divergence between nations with respect to their patterns of production, capital accumulation and ownership. Thus, the industrialising export economies of Taiwan, Hong Kong and South Korea must be recognised, along with the larger internally directed industrialised countries such as Mexico, raw material exporting nations like Nigeria, and low-income agricultural exporters such as Bangladesh, and so on.

As argued in the previous section, in the contemporary world such changes are highly likely to be non-hierarchic in the sense that they are focusing development on specific localities and settlements. Armstrong and McGee (1985: 41) state that 'Cities are . . . the crucial elements in accumulation at all levels, . . . and the *locus operandi* for transnationals, local oligopoly capital and the modernising state.' It is these features that gave rise to the title of their book, which characterised cities as 'theatres of accumulation' or global palladiums!

Perspectives on cultural globalisation

Other commentators point to what ostensibly appears to be the reverse trend: the increasing similarity which appears to characterise world patterns of change and development (figure 4.8). There is at least one major respect in which a predominant pattern of what may be called global convergence is occurring. This is in the sphere of consumer preferences and habits. Of particular importance is the so-called demonstration effect, involving the rapid assimilation of North American and European tastes and consumption patterns (McElroy and Albuquerque, 1986) (see Plate 4.1).

The influence of the mass media, in particular television, videos, newspapers, magazines and various forms of associated advertising, is likely to be especially critical in this respect. The televising of North American soap operas may well lead to a mismatch between extant lifestyles and aspirations (Miller, 1992; 1994; Potter and Dann, 1996), although there is equally the chance that such events will be reinterpreted and reconstituted from a local perspective. This argument is developed in Box 4.2. A major reality is that such media systems have become truly global in character in the 1990s (Robins, 1995). Potter and Dann (1996) show that the ownership of televisions and radio receivers is almost universal, even among low-income households in Barbados in the eastern Caribbean. The data on which this observation is made are reproduced in Table 4.4 (page 93). It is clear that a surprisingly high proportion of households have a video recorder, some 43.24 per cent in 1990. Video ownership was as high as 27.82 per cent for the occupants of all-wood houses, and 48.26 per cent for the occupants of combined wood and concrete houses, those which are generally in the process of being upgraded. Other aspects of the wider trend of convergence involve changes in dietary preferences, and the rise of the 'industrial palate', whereby an increasing proportion of food is consumed by non-producers (MacLeod and McGee, 1990; Drakakis-Smith, 1990).

Developing cities may be seen as the prime channels for the introduction of such emulatory and imitative lifestyles, which are sustained by imports from overseas along with the internal activities of transnational corporations and their branch plants. These in turn are frequently related to collective consumption, indebtedness and increasing social inequalities. These changes toward homogenisation are ones which are particularly true of very large cities.

Such a view sees globalisation as a profoundly unsettling process both for cultures and the identity of individuals, and it suggests that established traditions are dislocated by the invasion of foreign influences and images from global cultural industries. The implication is that such influences are pernicious and are extremelty difficult to reject or contain (Hall, 1995). Following this line of argument, Hall (1995: 176) has observed that the view is expressed that 'global consumerism, though limited by its uneven geography of power (Massey, 1991), spreads the same thin cultural

---

BOX 4.2   Global mass media, metropolitanisation and cultural change

British anthropologist Daniel Miller examined the popularity of soap operas produced in metropolitan regions of developed countries in a study published in 1992. This phenomenon can be seen as part of the evolution of 'global forms'. Such global forms have received a good deal of attention in relation to shifts in global production, but less evidence has been cited concerning the parallel process in global mass consumption.

Miller was researching on households and culture in Trinidad in the Caribbean, but he observed that 'for an hour a day, fieldwork proved impossible since no-one would speak to me, and I was reduced to watching people watching a soap opera'. The author goes on to note that much of the relevant research has been carried out on the pioneer coloniser of this type of television programme, *Dallas*. However, the programme that was receiving so much attention in Trindad was *The Young and the Restless*. This has been produced since 1973, and has always had a strong emphasis on sexual relations and associated social breakdown.

It is noted that many people went to extreme lengths in order to watch the programme. Those with low income, e.g. a large squatter community, were found to be the most resourceful in gaining access to the programme. Although most householders had neither electricity nor water, many homes had televisions connected to car batteries so they could watch the show. The car batteries were recharged for a small fee per week by those residents who had electricity.

Although the programme has little to do with the environmental context of Trinidad, Miller notes that it was regarded as realistic in portraying key structural problems of Trinidadian culture. In particular, in fashion and style conscious Trinidad, local audiences identified with the clothing worn by the characters. Thus, a retailer observed: 'What is fashion in Trinidad today? *The Young and the Restless* is fashion in Trinidad today.' The programme was also seen to match with the local sense of truth as revealed by exposure and scandal.

The author concludes that 'Trinidad was never, and will never be, the primary producer of the images and goods from which it constructs its own culture,' and 'Trinidad is largely the recipient of global discourses for which the concept of spatial origin is becoming increasingly inappropriate,' however different they may be in terms of the physical environment. But Miller also stresses it would be wrong to assume that such developments mean an end to Trinidadian culture, which has always been derived from here, there and everywhere – Africa, India, France, Jamaica, United States, among others!

Source: Miller (1992).

---

film over everything – Big Macs, Coca-Cola and Nike trainers everywhere' (Plates 4.1 and 4.6).

However, the impact of standardised merchandising is likely to be highly uneven, especially when viewed in terms of social class. Thus, it is the urban elite and the urban upper income groups who are most able to adopt and sustain the 'goods' provided by standardised merchandising; for example, health care facilities, mass media and communications technologies, improvements in transport and the like. It may be conjectured that the lower income groups within society disproportionately receive the 'bads'; for example, formula baby milk and tobacco products. Thus, once again, forces of globalisation may be seen to etch out wider differences on the ground. This heterogenising effect is true within urban areas too, with the residential subdivisions of the rich contrasting with those extensive areas that are inhabited predominantly by squatters and the low-income residents of the city. But the capitalist system must inevitably be recognised as having a vested interest in globalising the expectations of consumption aspirations and tastes.

A direct and important outcome of this suggestion is a strong argument that the form of contemporary development which is to be found in particular areas of the developing world is the local manifestation and juxtaposition of the two seemingly contradictory processes of convergence and divergence at the global scale. In terms of examples, Armstrong and McGee look at the ways in which these trends are played out in Ecuador, Hong Kong and Malaysia. Potter (1993c, 1995a) has examined how well the framework fits the Caribbean, where it has been argued that tourism has a direct effect on trends of convergence and divergence. This is another way of saying, via changes in production and consumption, that globalisation is not leading to uniformity, but to heterogeneity and differences between places. This is also reflected in contexts where

Table 4.4 Availability of household appliances by house type in Barbados, 1990.

Material of outer walls of dwelling	Number of occupied dwellings having household appliances in use (Percentage of total occupied dwellings)											
	Radio	Television	Video recorder	Telephone	Refrigerator	Washing machine	Solar water heater	Other water heating	None of these	All of these	Not stated	Total occupied dwellings
Wood	25 566 (85.21)	21 797 (72.65)	8 347 (27.82)	13 719 (45.72)	21 769 (72.55)	1 424 (4.75)	286 (0.95)	845 (2.82)	1 610 (5.37)	273 (0.91)	854 (2.85)	30 004
Wood and concrete block	13 971 (92.58)	13 968 (92.56)	7 282 (48.26)	12 026 (79.70)	14 210 (94.17)	3 098 (20.47)	1 256 (8.32)	2 204 (14.61)	72 (0.48)	1 120 (7.42)	201 (1.33)	15 090
Wood and concrete	829 (91.40)	843 (92.94)	443 (48.84)	714 (78.72)	843 (92.94)	197 (21.72)	83 (9.15)	148 (16.32)	4 (0.44)	76 (8.38)	14 (1.54)	907
Concrete block	23 615 (92.50)	23 765 (93.09)	14 548 (56.99)	21 569 (84.49)	24 291 (95.15)	11 963 (46.86)	9 400 (36.82)	4 503 (17.64)	96 (0.38)	6 961 (27.27)	394 (1.54)	25 529
Stone	2 130 (88.71)	2 149 (89.50)	1 191 (49.60)	2 085 (86.84)	2 208 (91.96)	1 367 (56.93)	840 (34.99)	702 (29.24)	30 (1.25)	775 (32.28)	71 (2.96)	2 401
Concrete	1 082 (94.09)	1 075 (93.48)	663 (57.65)	994 (86.43)	1 100 (96.65)	555 (48.26)	496 (43.13)	217 (18.87)	6 (0.52)	340 (29.57)	13 (1.13)	1 150
Other	96 (83.48)	84 (73.04)	42 (36.52)	80 (69.57)	91 (79.13)	48 (41.74)	23 (20.00)	25 (21.74)	6 (5.22)	21 (18.26)	5 (4.35)	115
Not stated	7 (46.67)	6 (40.00)	4 (26.67)	7 (46.67)	7 (46.67)	1 (6.67)	1 (6.67)	0 (0.00)	0 (0.00)	1 (6.67)	8 (53.33)	15
Barbados	67 296 (89.48)	63 687 (84.68)	51 194 (43.24)	64 519 (68.07)	18 634 (85.78)	12 384 (24.78)	12 385 (16.47)	8 644 (11.49)	1 824 (2.43)	9 567 (12.72)	1 560 (2.07)	75 211

Source: Potter and Dann (1996); derived from Barbados Population and Housing Census, 1990.

Plate 4.6 McDonald's in Nanjing Road, Shanghai
(photo: Harmut Schwarzbach, Still Pictures)

Figure 4.8 Globalisation and Third World societies: Rip Kirby airs the stereotypical argument. Reprinted by kind permission of Yaffa Advertising and King Features.

the Third World is represented in the First World. Thus, it is necessary to acknowledge that the flow is not one way, and although the North to South flow is dominant, it can be argued that the nature of the South to North flow is becoming of increasing salience.

The emerging system: hierarchic and non-hierarchic change

We are now ready to reconcile a number of important arguments. It can be posited that it is the key traits of Western consumption and demand that are potentially being spread in an hierarchical manner within the global system, from the metropolitan centres of the core world cities to the regional primate cities of the peripheries and semiperipheries, and subsequently down and through the global capitalist system (Plate 4.6). But their actual impact will be highly specific to locality, class and gender. It is interesting to observe that the innovations cited by Berry and others in the 1960s and 1970s as having spread sequentially from the top to the bottom of the urban system of America, were all consumption-oriented; for example, the diffusion of television stations and receivers. But the spread is one of potential, and many real differences are evolving.

In contrast, aspects of production and ownership are becoming more unevenly spread; they are becoming concentrated into specific nodes. This process involves strong cumulative feedback loops. Hence considerable stability is likely to be maintained at selected points within the global system, frequently the largest world cities. Entrepreneurial innovations will be concentrated in space, and are not likely to be spread through the urban system; this argument has parallels with the view that sees dependency theory as the diffusion of underdevelopment rather than development.

The key elements of the argument presented above are summarised in Figure 4.9 (overleaf). On the one hand, the culture and values of the West are potentially being diffused on a global scale. By such means, patterns of consumption are spread through time (T1, T2, T3, etc.), and there is an evolving tendency for convergence on what may be described as the global norms of consumption. The figure recognises that these aspects of global change are primarily expressed hierarchically, and are essentially top-down in nature. In contrast, cities appear to be accumulating and centralising the ownership of capital, and this process is closely associated with differences in productive capabilities. The tendency toward divergence is expressed in a punctiform, sporadic manner, which stresses activities in area (A1, A2, A3). TNCs and associated industrialisation are the most important agents involved in this process. This goes a long way toward explaining the contradictory nature of the post-modern world system.

Cities and urban systems have to be studied as important functioning parts of the world economy. In such a role, cities are agents of concentration and spread at one and the same time. Viewed in this light, the age-old argument as to whether cities are generative or parasitic is naive, simplistic and unfounded (Potter and Lloyd-Evans, 1998). Similarly, it is far too simplistic to ask whether cities spread change in a hierarchical or non-hierarchical manner, for in fact they are doing both simultaneously (Chapter 9). In this regard, it is tempting to argue that the breaking down of rigid hierarchical systems at a global level is very much part of the post-modern world. What we can certainly conclude is that globalisation is much to do with new and perpetuated forms of uneven development.

Globalisation and development: tourism

Tourism is now regarded as the world's leading industry. By 1987, it recorded US$2 trillion sales, and employed an estimated 6.3 per cent of the global workforce, making it the global premier industry (Gale and Goodrich, 1993). Tourism is also quintessentially linked to globalisation and the phenomenon of time–space compression. In the sociocultural realm, tourism is emblematic of globalisation, hyperreality, fantasy and post-modernity. Tourism can also be seen as a productive enterprise that is actively etching out differences betweeen places. In addition, tourism is closely connected with conspicuous consumption and the adoption of outside norms, via the operation of the demonstration effect (Chapter 6).

Many developing nations have adopted specific programmes to promote the growth of tourism as part of their development strategies. One of the first examples, however, is afforded by the First World European nation of Spain, as developed in Box 4.3. Barbados in the Caribbean is a good example from the developing world; the growth of tourism in Barbados dates from the late 1950s. In 1955 there were only 15,000 visitors to the island annually (Potter, 1983). With the advent of jet aircraft, there was dramatic growth in the tourist sector from 1966 to 1972; there was an overall increase in visitors of 165.9 per cent over this seven-year period. There were 210,349 visitors in 1972, this having increased from 79,104 in 1966. It was the heyday of the expansion in tourism, with yearly increases never falling below 14.2 per cent. By 1980, total tourist numbers had reached the dazzling heights of 369,915, compared with a national population of 248,983 (Potter, 1983). By 1992, tourist arrivals to Barbados amounted to 385,472 (Dann and Potter, 1997).

Barbados has, through time, pulled in most of its visitors from Canada, the United States, the United Kingdom and other parts of Europe. This aspect of internationalisation has its downside, as it entails dependency on economic conditions and inevitable fluctuations elsewhere. This openness of the economy was witnessed in the vicissitudes of the mid 1970s, when recessionary tendencies and increasing oil prices had a marked effect on tourist arrivals. The number of visitors actually fell during 1975, and between 1973 and 1976 the increases at no time exceeded 5.6 per cent per annum.

A major aspect of globalisation in relation to tourism in Barbados is the foreign ownership of hotels. In Barbados, as elsewhere in the developing world, foreign ownership and participation in the tourist sector is particularly conspicuous in the larger, up-market tourist sector. Approximately 74 per cent of all class I hotel bed spaces are owned by non-nationals, whereas foreign ownership accounts for over half of all hotel bed spaces (52.6 per cent). Such influence does not end there, however, with foreign ownership being directly responsible for 47 per cent of class I apartments, and a grand total of 44.2 per cent of all bedspaces.

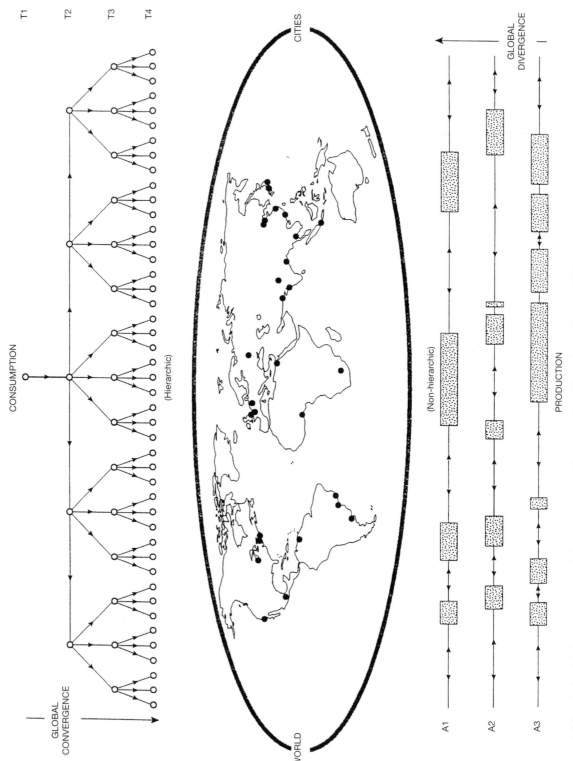

Figure 4.9 Trends in global convergence and divergence: a graphical depiction. Adapted from Potter (1997).

BOX 4.3 Tourism and development: the case of Spain

Since the 1960s, tourism has been the mainstay of Spain's economic miracle. Spain now has a GNP per capita which stands at US$13,650, significantly above both Portugal and Greece (at around US$7,500 per capita). Indeed, in 1995, Spain was the twelfth most powerful industrialised nation in the world (Dicken, 1998). Today Spain is almost synonymous with the words *holiday* and *tourism*, and perhaps even the expression *package holiday*. In 1984 Spain recorded 43 million visitors, and by 1995 this had increased to a staggering 55 million. This should be put alongside the indigenous population figure of 39.1 million. Spain now accounts for some 9 per cent of total world tourists. The number of tourist arrivals had risen from 6 million in 1960.

Tourism in Andalucia, which covers 17 per cent of Spain in the south of the country, accounted for nearly 13 million visitors in 1996 and yielded Pta 1 billion, amounting to some 10.5 per cent of Andalucia's GDP. The area has an exceptionally mild climate, with 3,000 sunshine hours reported per year. Its 812 km of coastline is comprised of more than 300 beaches, of which 63 meet the European Blue Flag criteria for excellence.

The development of tourism in Spain started in the late 1950s. Years of ultra right-wing rule under Franco had seen Spain fall further behind the rest of Europe in economic terms. As a result, General Franco consciously decided to open the Spanish economy up to the outside world. Thus, the peseta was devalued in 1959, internationally making Spain a very cheap place to visit, in terms of accommodation, food and drink. These developments occurred at the same time as advances in cheap jet travel for the masses. The development of tourism in Spain can therefore be seen as an important step in the progress of its industrialisation and internationalisation. It also marked the first direct effort to use tourism as a central plank of development policy.

The speed with which tourism was developed meant there was little or no state or local planning. Effectively, the free market was left to sort things out. This was demonstrated most cogently by the fact there was little or no control over developments. Almost literally overnight, former small fishing villages such as Torremolinos, Fuengirola and Lloret de Mar became large tourist-led cities.

Planning regulations were blatantly flouted during the Franco era, with virtually all land projects being left entirely to the private sector. 'Anarchic growth' is how some have described the type of development that occurred. It has been argued that many early developments in the tourist industry were at the expense of both the environment and the local population (Plate 4.7).

Plate 4.7 High-rise tourism development in Benidorm, Spain
(photo: Panos Pictures)

There is another area of increasing globalisation and that involves the suggestion that Third World locations such as Barbados are playing host to First World post-modern tourists who are seeking to handle the pace of life and 'future shock' (Toffler, 1970) in their own countries. Such tourists are provided with selective glimpses of the simple life of yesteryear in a Third World or developing context. It is a short step to argue that this may represent one of the most lucrative and exploitable trends in the contemporary tourism industry (Dann, 1994; Potter and Dann, 1996, Dann and Potter, 1997).

Another frequently debated series of issues concerns the relations between tourism, globalisation, culture and learning about 'other' areas of the world. One side of the argument has it that tourism is a major positive force in promoting knowledge about far distant places and ways of life. Although this is undoubtedly true at a relatively simple and trivial level, there is another side to the argument which stresses the production of false information about places and the propagation of enduring stereotypes.

In the case of Barbados it can be argued that by putting plantation great houses and sugar factories on the tourist 'consumption' map, the past ignominious history of slavery is being openly discussed in a culturally sensitive, educationally sensitive and progressive manner. However, Potter and Dann (1996) and Dann and Potter (1997) demonstrate that it is undoubtedly part of the post-modern turn to conveniently disregard anything which may be deemed unpleasant and which might reduce enjoyment of the tourist 'product'.

BOX 4.4 Whither the real Barbados?

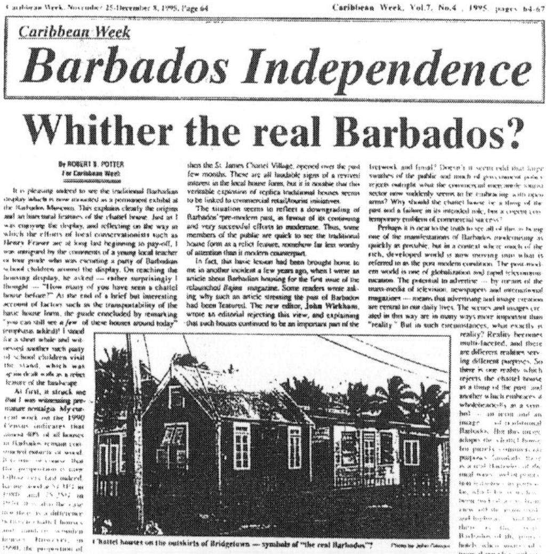
Caribbean Week

Barbados Independence

Whither the real Barbados?

By ROBERT B. POTTER

Chattel houses on the outskirts of Bridgetown — symbols of "the real Barbados"?

It is pleasing indeed to see the traditional Barbadian housing display which is now mounted as a permanent exhibit at the Barbados Museum. This explains clearly the origins and architectural features of the chattel house. Just as I was enjoying the display, and reflecting on the way in which the efforts of local conservationists such as Henry Fraser are at long last beginning to pay-off, I was intrigued by the comments of a young local teacher or tour guide who was escorting a party of Barbadian school children around the display. On reaching the housing display, he asked – rather surprisingly I thought – 'How many of you have seen a chattel house before?' At the end of a brief but interesting account of factors such as the transportability of the basic house form, the guide concluded by remarking 'You can still see *a few* of these houses around today!' (emphasis added). I stood for a short while and witnessed another such party of school children visit the stand, which was again dealt with as a relict feature of the landscape.

At first, it struck me that I was witnessing premature nostalgia. My current work on the 1990 Census indicates that almost exactly 40% of all houses in Barbados remain constructed entirely of wood. It is true, of course, that this proportion is now falling very fast indeed, having stood at 57.31% in 1980, and 75.25% in 1970. It is also the case that there is a difference between chattel houses and modern

Box 4.4 continued

wooden houses. However, in 1990, the proportion of houses built exclusively of wood was 45% of the total housing stock in respect of six out of the eleven parishes – the predominantly rural ones of – St John, St Thomas, St Joseph, St Andrew, St Peter, and St Lucy. Change may be occurring rapidly, but to describe chattel houses as a thing of the past does seem somewhat premature. More importantly, it seems to suggest the continued operation of a sharply divided nation: the modern and the not so modern. One might assume that the school parties were from the urbanised-suburbanised-touristised coastal belt of St Michael, Christ Church and St James. But the 1990 census data show that even in these parishes, wooden houses account for 41.69%, 29.91% and 28.54% of the total housing stocks respectively. More to the point perhaps, in 1991 some 559 house move permits were issued to enable chattel-type houses to be moved from one part of the country to another.

But on reflection, I found it more surprising that the children had not answered the teacher's original question with the reply, 'Sure, we've seen chattel houses before – why, I visited a tourist area only last week!' First there was the Chattel House shopping village in St Lawrence Gap and the chattel house and rum shop used to serve buffet meals at Sam Lords (a leading hotel). Then there was the Chattel Plaza, then Sandy Bank, and then the St James Chattel Village, opened over the past few months. These are all laudable signs of a revived interest in the local house form, but it is notable that this veritable explosion of replica traditional houses seems to be linked to commercial retail/tourist initiatives.

The situation seems to reflect a downgrading of Barbados' premodern past, in favour of its continuing and very successful efforts to modernise. Thus, some members of the public are quick to see the traditional house form as a relict feature, somehow far less worthy of attention than its modern counterpart.

In fact, this basic lesson had been brought home to me in another incident a few years ago, when I wrote an article about Barbadian housing for the first issue of the relaunched *Bajan* magazine. Some readers wrote asking why such an article stressing the past of Barbados had been featured. The new editor, John Wickham, wrote an editorial rejecting this view, and explaining that such houses continued to be an important part of the local cultural landscape. So much so, in fact, that he drew attention to the fact that the front cover logo of the magazine actually featured a drawing of a chattel house!

In a manner which parallels such downgrading, housing policy in Barbados has largely rejected the chattel house as part of a policy for providing houses for those on low incomes. The chattel house is an excellent example of self-help on the part of Barbadian people. Such houses can, of course, be put on a plot of land with basic services, together with a toilet (wet core), as part of a 'site-and services' scheme. Slowly, the inhabitants can upgrade, converting to a walled structure when finances and time permit. In a sense, this idea formed a component of the tenantries programme launched in 1980. But the idea of using the chattel house as a plank of housing policy has never been fully pursued by government. Instead, concrete starter homes, costing thousands of dollars were piloted, just as barracks-style and terraced row concrete houses were a feature of earlier government housing schemes.

Once again, it seems that the need to modernise is placing the premodern local system in a subordinate position. Even if bungalow-type houses are now the predominant feature of today's Barbadian housing scene, wouldn't it be nice for some of the architectural features of the chattel house to be retained, such as box-pelmets, fretwork and finial? Doesn't it seem odd that large swathes of the public and much of government policy rejects outright what the commercial-mercantile-tourist sector now suddenly seems to be embracing with open arms? Why should the chattel house be a thing of the past and a failure in its intended role, but a cogent contemporary emblem of commercial success?

Perhaps it is near to the truth to see all of this as being one of the manifestations of Barbados modernising as quickly as possible, but in a context where much of the rich, developed world is now moving into what is referred to as the post-modern condition. The post-modern world is one of globalization and rapid telecommunications. The potential to advertise – by means of the mass-media of television, newspapers and international magazines – means that advertising and image creation are central to our daily lives. The scenes and images created in this way are in many ways more important than 'reality.' But in such circumstances, what exactly is reality? Reality becomes multi-faceted, and there are different realities serving different purposes. So there is one real-

---

Box 4.4 continued

ity which rejects the chattel house as a thing of the past; and another which embraces it wholeheartedly as a symbol – an icon and an image – of traditional Barbados. But this image adopts the chattel house for purely commercial purposes. Similarly, there is a real Barbados of the rural zones, and of plantation tenantries in particular, which for years have been tucked away from view, off the main roads and highways. And then there is the 'real' Barbados of the tourist hotels, where images of a tropical paradise and rest and relaxation, replete with the chattel houses of a bygone era, are made 'authentic' for the visitor.

A final instance of the multiple realities which make up our modern society is also drawn from the realm of tourism: but this time in respect of the ways in which the realities of the history of Barbados are handled. Take the example of the floorshow that is offered to visitors several nights a week. Adverts appear here, there and everywhere, inviting the potential audience to come and 'see the cultures of the Caribbean as influenced by the Spanish, French and *African settlers*' (emphasis added). You perhaps wouldn't even notice if the advertising copy had been written as 'African, Spanish and French settlers.' But should a reality so central to the people, history and culture of Barbados be represented in this fashion? Should the greatest enforced movement of slaves be transmuted by verbal sorcery into an apparently voluntary migration? What reality is being served here, that one of the most salient aspects of the history of the island should be rewritten so blatantly for tourist consumption. A less cogent but interesting aspect of the same event is the exclusion of any reference to British history in the blurb. Presumably,

this aspect of reality might be construed as too near to reality, and distract from the Latin connotations of Spain and France. Reinterpretations of history – which are presumably intended to ease collective guilt about the realities of slavery – abound in the Barbadian mass-media aimed at the tourist. Hence, plantation house tours and dinners are advertised, stressing elegance, antiques, pine furniture and opulence, with no historical reference being made to the conditions for slaves that formed their counterpart. Elsewhere we are invited to trek across the original Barbados, as unspoilt now as it was 350 years ago, without any acknowledgement that the island would have been entirely wooded! It is hard not to reach the conclusion that all of these 'realities' are intended to increase the ability of the social and physical environments to handle tourists: to increase what the tourism expert would refer to as the carrying capacity of Barbados.

Whither the real Barbados? The truth is that it is a very successful semi-peripheral country, which is modernising very rapidly. In a number of situations modernising Barbados is clearly overthrowing premodern Barbados. But in other arenas, particularly those relating to tourism and cultural change, postmodern influences in Barbados are busily reviving aspects of the premodern Barbados. The net result is highly interesting and vibrant, but is in a number of respects highly contradictory, ambivalent and uneven. But that's the post-modern world for you and there seems to be no doubting the fact that Barbados is a part of it.

Source: This article first appeared in Caribbean Week, November 25–December 8 1995.

Already in Barbados there is considerable evidence that a growing number of concerns have begun to capitalise on the post-modern ethos of their guests. Mock villages consisting of traditional houses have sprung up as commercial outlets at the very same time as the state refuses to see such houses as representing a cogent force in the future housing equation of the nation. 'Pirate cruises' are offered to holiday-makers and traditional chattel houses appear on hotel premises as the locales for serving buffet dinners. Meanwhile, plantation floorshows and spectaculars even invite historically ignorant, or at best ahistorical, vacationers to see the cultures of 'Spanish, French and African

settlers', completely ignoring the condition of slavery and grossly misrepresenting the inhumanity of the system of slavery, presumably in the interests of increasing the enjoyment of the tourist product (Box 4.4). Recognition of the potential for this fragmentation of history in the so-called developing world by inhabitants of the core can have enormous financial rewards for the latter.

Although it is possible to put forward the argument that such 'staged authenticity' (McCannell, 1976) means the private lives of Barbadians are shielded from the tourist gaze (Urry, 1990), and that such developments may be making a real contribution to promoting

Plate 4.8 Barbados: 'just beyond your imagination' (source: Barbados Tourist Board)

long-term sustainability, it is again possible to take up an entirely different line of argument. Thus, there is the real fear that by rewriting the history, and indeed reinventing the physical environment, of the Caribbean region (Potter and Dann, 1996), the carrying capacity of tourist destinations may be exceeded, both in sociocultural and environmental terms; the only ones to profit are the developers.

We clearly live in a globalised world, but the technological changes that have allowed the development of mass tourism are also the means by which simplified, inaccurate or downright misleading images and representations can be propagated. It seems hard to refute the suggestion that the popular misconception of the entire Caribbean as a 'beach replete with swaying coconut palms' is the direct outcome of tourist advertising and promotion campaigns (Potter and Lloyd-Evans, 1998). At the same time, the daily realities of urban concentration, poverty and poor housing in the Caribbean are selectively weeded out from the stereotype. Thus, in their 1996 advertisement the Barbados Board of Tourism comments on the island as one 'with 340 days of sunshine a year, imagine how that reflects on you', and describes Barbados as 'just beyond your imagination', complete with the time-honoured view of sand, sea and coconut palm (Plate 4.8).

In a similar vein, the local population is also represented as a group of smiling, servile natives ready to respond to the bidding of predominantly White tourists. Such images and perceptions have unfortunate connotations in a context where pride in African origins and negritude (Lowenthal, 1960) are slowly increasing in the post-colonial era. Images can now be spread widely by means of colour brochures, television programmes, promotional video recordings, and almost instantaneously by websites and web pages.

The complexities of the development implications of tourism promotion in developing countries are thus considerable in a globalising post-modern world. The case of Barbados seems to illustrate the veracity of the argument that neither modernity nor post-modernity can exist without the other, and that rather than being different conditions, they are closely related. In an evolving world-order characterised by globalisation, fluidity and change, it is not altogether surprising that the notion of post-modernity represents a 'handy category to employ in the struggle for emancipation and a virtual synonym for postcolonialism' (Jones, Natter and Schatzki, 1993: 18).

Globalisation and unequal development

Throughout this chapter it has been argued that the notion of a basic sameness in respect of global culture is clearly a distortion and a gross oversimplification. Clearly, we live in a more globalised world, in which multinationals especially are coming to dominate world patterns of consumption and production, but there are many reasons why it is wrong to regard the outcome as increasing uniformity.

Firstly, strong resistance is sometimes shown by local and national cultures, especially in response to the influences of North America. The idea of a single global culture is clearly misplaced. As an example,

the opening of McDonald's was fiercely contested for some time in Barbados, despite its status as a leading tourist destination for North Americans and Europeans. When those in power relented, the fast-food chain only lasted six months, largely because Bajans are happier eating chicken rather than red meat. This is a simple and direct example illustrating that local customs and tastes can run directly across, and indeed against, apparently hegemonic global trends.

Secondly, rather than serving to erode local differences, global culture often works alongside them; and sometimes it even works via them. Particular groups within society may be targeted for the sale of certain products. In this manner, local differences may be explored and exploited wherever possible (Robins, 1995). Furthermore, increasingly within the global economy, cultural products are being assembled from all over the world and are being turned into commodities for an emerging cosmopolitan market-place. This is particularly true of music and tourism (Robins, 1995). Thus, from reggae to soca to African indigenous music, and in respect to Rastafarianism, the flow is not a one-way movement, and Third World products are being promoted and sold in First World market-places. Thus, globalisation has brought the possibility of the colony 'invading' the colonial power, and the periphery taking on, and 'winning against' the centre (Robins, 1995). Hence, in the new globalising system, we are encountering many incidences of what viewed historically is a reverse or counter flow. Again, such conflations can be interpreted as typical of the post-modern condition.

There is also the important argument that increasing globalisation and time–space compression in the end make us value more strongly than hitherto the notion of place as secure and stable. Thus, it can be posited that globalisation may well serve to engender localisation. This is sometimes described rather inelegantly as 'glocalisation', where there are multiple global–local relations through which locality becomes more salient than hitherto within the world system.

Furthermore, we have witnessed all too clearly that culture has always been characterised by hybridisation, difference, rupture and clashes, so it is possible to argue that nothing very new, strange or different is currently happening. Western European nation states may be seen as masters of modernity, whereas hybridised forms of culture are characterising the post-modern world. This, of course, reflects the fact that culture is never settled, finalised, complete and internally coherent (Hall, 1995). It has to be appreciated that cultures (systems of shared meanings), products

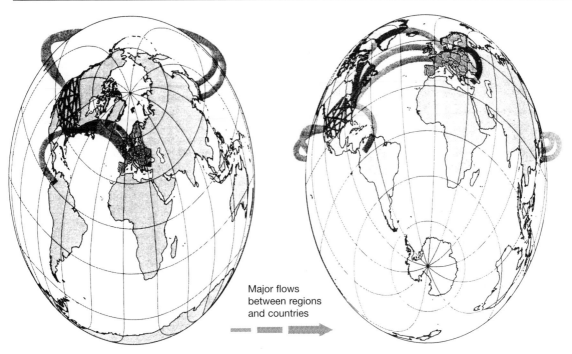

Figure 4.10 On and off the map: contemporary globalisation and the heterogeneity of global change. Adapted from Allen and Hamnett (1995).

and lifestyles will inevitably spread, and sometimes contract, in a highly heterogeneous manner. But aspects of production and ownership are far from evenly spread, due to the process of divergence and differentiation reviewed earlier in this chapter. Finally, the economic competition between places is now intense, given that major corporations can select between them, and this is leading to the possibility of ever sharper differences between areas, regions and places.

For all these reasons, we can conclude that uneven and unequal development are still characteristic of the global capitalist system (Figure 4.10). Globalisation is not all-encompassing and there is much that remains uneven about global relationships and global processes. All of these aspects of dynamic change are strongly skewed toward the developed North (Allen, 1995). The world may effectively be getting smaller, but the majority of its population does not yet have access to a telephone. For example, de Albuquerque (1996) states that over half the world's population have never made a telephone call, and there are more telephones in the New York–New Jersey metropolitan area alone than in all of Africa combined.

In discussing further 'net fever' in Haiti, the same author stresses that in a country where the average wage is US$3 per day and the per capita gross national product was US$220 in 1994, very few can afford the US$2,000–3,000 which was required in 1996 for a personal computer. Thus, access to cyberspace in Haiti is restricted to a well-heeled elite minority living in a state of the art, 21st century world, far removed from the impoverished majority (de Albuqueque, 1996). The same author summarises that Internet access is paralleling class systems of stratification, and threatens to splinter the globe into haves and have nots based on access to information/communications technologies (de Albuqueque, 1996).

In considering development, it has to be recognised that places in the globalising world system are not linked together in a uniform way. They are interrelated in very unequal ways, and such basic inequality would seem to be poised to increase rather than decrease in the near future. Competition between places for global capital is making the world more uneven, reflecting the trends of global divergence (Cochrane, 1995; Potter, 1993c; Armstrong and McGee, 1989). In the words of Cochrane (1995: 276), 'Globalisation and localisation are not the polar opposites which one might expect them to be', because 'globalisation is underpinned by the realities of uneven development' (Cochrane, 1995: 277).

PART II

Development in practice:
components of development

CHAPTER 5

People in the development process

People-centred development

People are, or certainly should be, central to the development process and an integral element in all development strategies. However, all too often in the past the needs of people have been ignored and there has been a failure to consider the possible implications of development policies on individuals, households and communities. As we have seen in Part I, there are many different and often conflicting views on the meanings of development, and the most appropriate strategies to be followed at different points in time and space. However, for one writer, Dudley Seers, development is unequivocally about improving the quality of people's livelihoods, and he argues that the reduction of three key variables – poverty, unemployment and inequality – should be central to the development process. As Seers observed: 'The questions to ask about a country's development are therefore: What has been happening to poverty? What has been happening to unemployment? What has been happening to inequality? If all three of these have become less severe, then beyond doubt this has been a period of development for the country concerned' (Seers, 1969).

Seers also emphasised the need for the true fulfilment of human potential and improvements in the quality of life, and in a later paper, entitled 'The New Meaning of Development', written after the oil crisis of the 1970s, he suggested that 'self-reliance' should be another important goal of development plans (Seers, 1979). In order to reduce poverty, unemployment and inequality, Seers and others have argued that development strategies are needed which fulfil basic human needs such as nutrition, water and sanitation, health and education. It is also important to recognise that these basic needs are inextricably linked, and policies must adopt a holistic approach towards improving human welfare. Too often in the past, development strategies have been driven by economic goals, whereas basic needs fulfilment has received less priority, commonly assuming that economic growth will somehow 'trickle down' spontaneously to the most marginal elements of society and space, as reviewed in Chapter 3. In fact, the Third World is littered with so-called development projects which, far from empowering people, supplying their basic needs and raising living standards, have on the contrary produced greater inequality, poverty and unemployment.

A further problem with many development strategies is that 'people' and 'communities' have all too frequently been perceived by developers as homogeneous and passive, rather than as diverse and dynamic entities. The peculiar needs, knowledge and skills of different individuals and groups within communities have often been ignored in favour of a broad and less sensitive approach. As a result, although development projects might have benefited certain sections of the population, other elements have lost out. For example, in The Gambia, where rice is a woman's crop and women possess the detailed knowledge and understanding of its production and processing, a series of overseas-funded irrigated rice development projects in the 1960s and 1970s achieved poor results, precisely because women's considerable expertise was ignored by the development teams. As Dey comments, 'By failing to take into account the complexities of the existing farming system and concentrating on men to the exclusion of women, the irrigated rice projects have lost in the technical sense that valuable available female expertise was wasted' (Dey, 1981: 122). It is essential, therefore, that future development strategies are built upon a detailed understanding of the individuals or communities which are the target of such policies, rather than being based upon the assumption that people and societies are stereotypically homogeneous.

This chapter will investigate the diversity of people and their role as a key resource in the development process. Firstly, a number of important demographic features will be considered, and this will be followed by an evaluation of some broad issues affecting the quality of life.

Population and resources: a demographic time bomb?

The question of the rate of population growth and its relationship to the availability of food and vital natural resources has exercised the minds of many scholars for centuries. Although some commentators see population growth as the 'big issue' in world development, painting a 'gloom and doom' scenario of population growth outstripping food supply, others are much less pessimistic and view population growth more as an 'engine of development' playing an important role in the development process.

A frequently cited starting point in the population and resources debate is Thomas Malthus's *An Essay on the Principle of Population*, published in 1798. Malthus described a highly pessimistic scenario of population growing more rapidly than food supply, and he advocated the need for 'preventative' and 'positive' checks on population growth. He further argued that the tension between population and resources was a fundamental cause of misery for much of humanity (N. Crook, 1997). More recently, Paul Ehrlich in his book *The Population Bomb* (1968) has commented: 'Americans are beginning to realize that the undeveloped countries of the world face an inevitable population-food crisis. Each year food production in undeveloped countries falls a bit further behind burgeoning population growth, and people go to bed a little bit hungrier. While there are temporary or local reversals of this trend, it now seems inevitable that it will continue to its logical conclusion: mass starvation' (Ehrlich, 1968: 17). The Club of Rome's *Project on the Predicament of Mankind* in the early 1970s further echoed Malthus's and Ehrlich's warnings, suggesting that, 'demographic pressure in the world has already attained such a high level, and is moreover so unequally distributed, that this alone must compel mankind to seek a state of equilibrium on our planet. Under-populated areas still exist, but, considering the world as a whole, the critical point in population growth is approaching, if it has not already been reached' (Meadows *et al.*, 1972: 191).

An alternative and much more positive perspective on the relationship between people, environment and resources was provided by economist Ester Boserup in her important book *The Conditions of Agricultural Growth* (Boserup, 1965, 1993). Boserup presented a convincing argument to show that population growth and increasing population density can in fact be key factors in generating innovation and intensification in traditional food production systems. She suggested that, provided the rate of population growth is not too rapid, populations will over time adapt their environment and cultivation strategies, such that increased yields can be obtained without any significant degradation of the resource base. This viewpoint has gained greater popularity in recent years, as detailed empirical research has revealed the considerable capacity of indigenous peoples to raise the productivity of their farming systems in the face of increasing population numbers (Tiffen, Mortimore and Gichuki, 1994). The continuing credibility of Boserup's thesis is indicated by the recent republication of her book, with a foreword by Robert Chambers, himself a key figure in development research. Chambers comments: 'The Boserupian thesis will continue to stimulate argument and inspire research on the links between population change, agricultural technology, and now sustainability. It has stood the test of time: it is repeatedly referred to by those who have read this book and by many who have not' (Boserup, 1993: 8).

The relationship between population and resources also featured strongly in the Brundtland Report of 1987 (WCED, 1987) and at the United Nations Conference on Environment and Development (the so-called Earth Summit), held in Rio de Janeiro, Brazil, in June 1992. This much publicised event, attended by many world leaders and non-governmental organisations, was concerned with key global resources and a number of major environmental issues, such as global warming and climatic change. It was at Rio, and in subsequent publications, that the concept of sustainable development was popularised. The essence of sustainable development is the need to achieve an equilibrium between the world's basic resources and their continuing exploitation by a growing world population, so as not to jeopardise these resources for future generations. Agenda 21, the comprehensive programme of action adopted by governments at the Earth Summit, continually emphasises the links between environment, population and development (United Nations, 1993).

The many different viewpoints expressed in the long-running debate on the relations between population and resources, themselves constitute distinctive 'geographies of development'.

Population distribution

In 1994 the world's population was estimated to be 5,601 million, but these people are by no means distributed evenly across the earth's surface and population density varies widely (Figure 5.1, overleaf). With the massive populations of China and India, the dominance of the Asia-Pacific region in the world distribution of population is particularly striking, and well over half the total population (2,935 million) live in this region, significantly more than the combined populations of the Middle East, Africa and Latin America. (World Bank, 1996b). Apart from some small and densely settled island states such as Barbados (with 593 persons per square kilometre), the Maldives (717) and Bermuda (1,160), and small enclaves such as Hong Kong (5,599) and Singapore (4,407), the most densely settled countries are to be found in Asia and also in Europe. India, with its massive population of 914 million, has a density of 259 persons per square kilometre, whereas Japan has 327, South Korea 432, Taiwan 556 and Bangladesh a staggering 803. Although European population densities nowhere reach the magnitude of Bangladesh, apart from the tiny enclaves of Monaco (14,737), Gibraltar (4,615) and Vatican City (2,273), Europe does have some relatively high population densities such as the United Kingdom with 234 persons per square kilometre, Belgium with 323, and the Netherlands 366. With the exception of the polar wastes of Greenland and Antarctica, the world's most sparsely settled areas include Australia and Canada, with only 2 and 3 persons per square kilometre, and large parts of Africa. In fact, Africa is still the world's least densely settled continent, with the desert states of Western Sahara, Mauritania and Libya having only 1, 2 and 3 persons per square kilometre, respectively. In southern Africa, Botswana and Namibia each have 2 persons per square kilometre. There are, however, certain countries in Africa with unusually high densities of population, notably the tiny countries of Rwanda and Burundi in central Africa, with densities of 275 and 197 persons per square kilometre, whereas Nigeria, the continent's most populous country with 108 million people has a density of 117 per square kilometre.

A word of caution is warranted concerning the reliability of national, and therefore regional and global, population statistics. Censuses are costly, time-consuming and require considerable expertise to administer and analyse. Consequently, some poor countries, and/or those which are politically unstable, have often been unable to conduct regular censuses, so the available data are often unreliable. Furthermore, national planning has been hampered by the lack of reliable and up-to-date figures. Nigeria, for example, undisputedly Africa's most populous country, had its first census for almost thirty years in November 1991. The previous census with any reliability (though this was questioned) was held in 1963, after which there were three further attempts to hold censuses, on each occasion failing because regional leaders submitted exaggerated head counts in an effort to show that their people were more numerous and thus entitled to 'a bigger share of the national cake'. The Nigerian federal government made great efforts to conduct a successful census in 1991, with three years of careful preparation, an investment of some £75 million and enlisting the help of half a million enumerators, including women who worked in Muslim areas. Estimates before 1991 had put Nigeria's population at well over 100 million and so many Nigerians seemed genuinely shocked when the latest census revealed a total population of only 88.5 million!

Population dynamics

For most of human history, population growth, averaged over long periods, has remained near zero. In fact, the modern expansion of the world's population only started in the eighteenth century with the slow decline of the death rate in Europe and North America. Population growth accelerated steadily in the twentieth century and has been particularly rapid in the developing countries since 1950 (Table 5.1, page 111).

Merrick (1986) points out that more people have been added to the world since 1950 than in all of human history before the middle of this century, a sobering thought. Bongaarts comments: 'The acceleration in growth is well demonstrated by the shortening of the time intervals needed to add successive billions to the world population. The first billion was reached early in the nineteenth century, the second billion took 120 years, the third 33 years, the fourth 14 years and the fifth (between 1974 and 1987) just 13 years. If projections turn out to be accurate, the next 3 billion will be added at an even faster pace, each taking just over a decade, to reach 8 billion before the year 2020' (Bongaarts, 1995: 8). Such projections suggest that it is likely the world's population will reach around 11.7 billion in 2100.

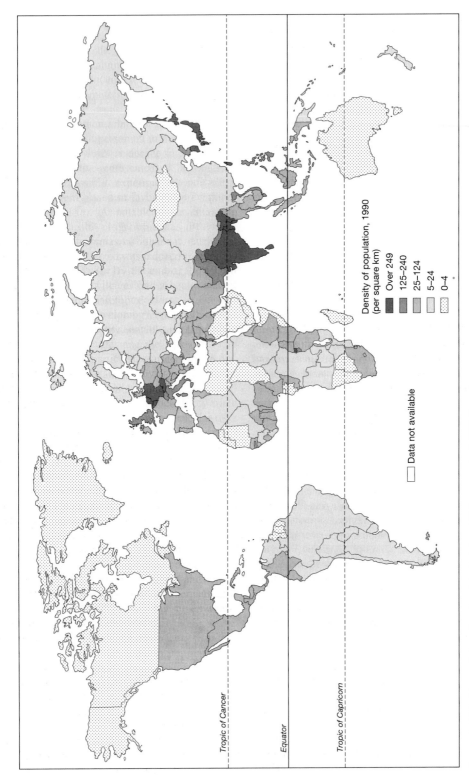

Figure 5.1 World population density. Adapted from Dickenson *et al.* (1996).

Table 5.1 World population growth, 1900–2100.

	Population (billions)			% increase	Estimated population (billions)		% increase
	1900	1950	1990	1950–90	2025	2100	1990–2100
Developing countries	1.07	1.68	4.08	143	7.07	10.20	150
Developed countries	0.56	0.84	1.21	44	1.40	1.50	24
World	1.63	2.52	5.30	110	8.47	11.70	121

Source: Adapted from Bongaarts (1995: 7–8).

Where and how these people will live in the future is of much interest and concern, since they will be by no means evenly distributed across the globe. In September 1994 the World Conference on Population and Development, held in Cairo, had the aim of drawing up a twenty-year programme (up to 2015) to combat overpopulation in the world, but reached deadlock on a number of issues, most notably the question of decriminalising abortion, a move which is rejected by the religious authorities of both the Catholic Church and Islam. The *1996 Human Development Report* estimated that the total world population in 2000 will be 6,120 million, of which no less than 4,880 million (79.7 per cent) would be in the developing countries (UNDP, 1996: 179).

The phenomenal growth of population in the developing countries is revealed in Table 5.1, which shows that, whereas population in the developed countries increased by 44 per cent between 1950 and 1990, the developing countries experienced a massive growth rate of 143 per cent in the same period. Looking towards the next millenium, projected growth rates for the period 1990–2100 indicate a significant decline in the rate of population growth for the developed countries (24 per cent) compared with the earlier period, but there is a marked acceleration in growth in the developing countries to 150 per cent. These countries have growth rates more than four times higher than developed countries, and the sizes of their populations are also generally much larger than those of the developed countries. Such growth is going to place even greater pressure on resources, which in many poor countries are already stretched. According to the World Bank, in 1996 Bangladesh had the world's thirteenth lowest per capita GNP (US$220), and in 1994 it had a total population of 117.9 million living on a land area of only 144,000 square kilometres, with

an average population density of no less than 819 people per square kilometre (World Bank, 1996b).

Projecting the size of national and global populations is fraught with difficulty, since it is affected by such unpredictable events as natural disasters, wars and medical advances. For example, the production and wide availability of an effective and cheap vaccine to combat malaria, endemic in so many tropical countries, could have a massive impact on reducing death rates, which in turn would affect population growth rates nationally and globally. The introduction of government policies designed to increase or reduce population, will also affect growth rates, as in China, where its one-child policy has had a marked effect on national population growth rates (Plate 5.1, overleaf) (Box 5.1).

A further issue which makes the future prediction of the size and growth of population in specific countries particularly difficult is the question of population redistribution. The movement of people within countries, perhaps from rural to urban areas, and between countries as voluntary migrants or maybe as refugees escaping from drought or civil war, has had a significant impact on population dynamics in certain countries and regions. Chapter 8 explores this in more detail within the context of movements and flows.

Changes in population growth rates over time are affected by a wide range of factors, but are essentially controlled by the changing relationship between birth rates and death rates. Table 5.2 (page 114) presents population statistics for a number of low-income, middle-income and high-income countries.

The crude birth rate is the most common index of fertility and is a ratio of the number of live births to the total population, usually expressed as so many per 1,000. Although the United Kingdom and the United States had birth rates averaging 14 between 1990 and 1995, Mali and Sierra Leone in West Africa had rates

Plate 5.1 'One-child' poster, Guangzhou, south-eastern China
(photo: Tony Binns)

BOX 5.1   China's one-child population policy

The population of the world's most populous country, China, passed the 1 billion mark in 1981 and by 1994 it had reached 1,196 million. However, with an estimated average annual growth rate of 1.0 per cent between 1993 and 2000, this represents a considerable slowing down of population growth, from the 1.8 per cent per annum experienced between 1960 and 1993, and is well below the 2.2 per cent average growth rate for all developing countries over the same period (UNDP, 1996). This significant decline in China's population growth rate is due to the implementation of a rigid birth control policy.

Following a marked decline in China's population during the late 1950s and 1960s, due to loss of life from disasters such as typhoons and flooding and the severe famines which followed, the country then experienced a 'baby boom' from 1963. The government increasingly questioned the value of a rapidly growing population in relation to resources under increasing pressure, and during the 1970s various attempts were made to encourage both family planning and delaying marriage. In Shifang County of Sichuan Province, the local authorities responded in 1971 by raising the age at which a person could get married; it went from 18 to 23 for women and from 20 to 25 for men. Additionally, the county

government made great efforts to spread birth control information, and to provide free services for contraception. When contraceptive methods failed, induced abortions became more common (Endicott, 1988). Reducing the fertility rate became a key priority and the slogan 'one is not too few, two will do and three are too many for you' was publicised nationally. Communities were encouraged to recommend which women should be able to have a baby and the practice of 'giving birth in turn' became widespread.

In 1980, as the baby boom of 1963 was approaching the age of marriage and childbearing, the policy of one child per family was officially adopted by the National People's Congress and was incorporated into the country's new 1982 Consitution (Jowett, 1990: 117). The State Council deemed it 'necessary to launch a crash programme over the coming twenty or thirty years calling on each couple, except those in minority nationality areas, to have a single child. . . . Our aim is to strive to limit the population to a maximum of 1200 million by the end of the century' (quoted in Jowett, 1990: 117). A number of relaxations to the policy were permitted, particularly where the first child was a girl, also among minority peoples mainly in western China, and in the rural areas of

Box 5.1 continued

certain provinces. It is probably in the urban areas where the policy has been most strictly enforced (Leeming, 1993: 61). To enforce the one-child policy, a system of rewards and penalties was introduced, such as parents being offered a 5–10 per cent salary bonus for limiting their families to one child and a 10 per cent salary deduction for those who produced more than two children (Jowett, 1990: 119).

In 1988 the woman director of Number 2 Neighbourhood Committee, Hua Long Chao Sub-district in the Sichuan city of Chongqing, commented that the main job of the neighbourhood committee is to 'educate the people to realise the importance of the one-child family' (personal communication). Permits to have children were given annually to about a hundred women in the neighbourhood, with priority being given to older women. The local factory, where most neighbourhood dwellers worked, evidently played a key role in awarding points to female workers, points that affected their relative position in the queue to receive a permit. Strong sanctions were imposed if a woman became pregnant without a permit, and the committee would notify the government authorities and the factory. An abortion was usually required, a fine imposed, wage increases frozen and a permit to become pregnant again would be further delayed.

Although undoubtedly slowing down the population growth rate, the one-child policy has had widespread social implications. The policy fundamentally conflicts with traditional Chinese family values, in which children are seen as a source of happiness and fulfilment as well as guaranteeing the continuation of the family line. The policy is undoubtedly unpopular, but much as they would like another child, many women accept official arguments that they must go without. The policy has also led to even greater prestige being given to the birth of a son, particularly in the rural areas where there are no state pensions. Much concern has been expressed in China about the long-term social effects of the creation of a generation of pampered 'little emperors' with no sisters and cousins. Furthermore, the underregistration of female births, abandonment or neglect of girl babies, prenatal sex testing followed by selective abortion, and instances of female infanticide were frequently reported in the Chinese press in the 1980s. Some have also criticised the one-child policy on the grounds that it represents an infringement of a woman's control of her reproduction process 'since many women, whilst perhaps not wanting to have to go on until they have one or more sons, do wish to have at least two children' (Endicott, 1988: 179).

A further concern is that the policy will completely transform the country's age structure, and in time will result in a declining workforce and a very aged population. The baby boom of the 1960s will eventually result in a large number of retired people in the 2030s. As Jowett comments: 'By then, over-65s could constitute more than 25 per cent of the population and thus, within a lifetime, the number of retired will have increased from one in twenty to one in four. Such a high level of old-age dependency is unprecedented even in today's developed countries where the over-65's generally constitute 10–15 per cent of a country's population' (Jowett, 1990: 121). In spite of reducing China's population growth rate, the policy has received much criticism, not least because it represents a massive 'experiment' in social engineering.

---

of 51 and 48 respectively. The crude death rate is the number of deaths per 1,000 of the population. Considering the same four countries, the United Kingdom and the United States had death rates of 12 and 9 respectively between 1990 and 1995, whereas Mali and Sierra Leone recorded death rates of 19 and 22 over the same period.

Two other important indicators of the quality of life and levels of development are infant mortality and life expectancy. The infant mortality rate measures the number of deaths of infants under one year old per 1,000 live births. This variable reflects general living conditions and the health and nutritional status of pregnant and lactating mothers. Infant mortality rates in 1994 varied from 4 in Japan and Sweden (6 in the United Kingdom), to 163 in Sierra Leone, one of the world's poorest countries. However, there is not always a direct relationship between wealth and infant mortality. For example, Vietnam, ranked 12 by the World Bank according to GNP per capita, has an infant mortality rate of 42 and per capita GNP of only US$200, whereas Gabon in central Africa, ranked 100 in 1994 with a per capita GNP of US$3,880, had an infant mortality rate of 89 (World Bank, 1996b).

Life expectancy at birth also reflects general living standards, nutrition and health care, and it reveals

Table 5.2 Population statistics for selected low-income, middle-income and high-income countries.

Country	GNP (US$) per capita (1994)[b]	Birth rate (1990–95)[a]	Death rate (1990–95)[a]	Infant mortality (1994)[b]	Life expectancy at birth (1994)[b]	Mean percentage annual population growth rate (1990–94)[b]
Bangladesh	220	41	14	81	57	1.7
Brazil	2 970	26	8	56	67	1.7
China	530	21	7	30	69	1.2
India	320	31	10	70	62	1.8
Jamaica	1 540	22	6	13	74	0.9
Japan	34 630	12	8	4	79	0.3
Mali	250	51	19	125	49	3.0
Sierra Leone	160	48	22	163	40	2.4
Sweden	23 530	13	12	4	78	0.6
United Kingdom	18 340	14	12	6	76	0.4
United States	25 880	14	9	8	77	1.0
Zimbabwe	500	40	9	54	58	2.5

[a] Data from Heinemann–Philip Atlases, *Philip's Geographical Digest 1994–95*, Heinemann, Oxford.
[b] Data from World Bank, *World Development Report 1996*, Oxford University Press, Oxford.

tremendous inequalities between developed and developing countries. A child born in Japan can expect to live for 79 years, but in Sierra Leone an average life span of only 40 years is all that a child can expect.

The demographic transition

The changing relationship over time between fertility and mortality rates is clearly demonstrated through the demographic transition model, which usually identifies four or five key stages (Figure 5.2).

Stage 1: high stationary or pretransition phase

This stage is characterised by high birth rates and high death rates, such that population growth is static or negligible. This situation applied to pre-eighteenth-century Europe and North America, but in many developing countries this remained the position up to the Second World War.

Stage 2: early expanding or early transition phase

In the next phase, the death rate (mortality) begins to decline with better living standards, largely due to im-

provements in nutrition and public health. The incidence of famines and epidemics also falls. However, the birth rate (fertility) remains high, so population growth accelerates. In much of Africa, Asia and Latin America, mortality did not begin to decline until the first half of the twentieth century. By the late 1960s, the annual average death rate had dropped to 15 per 1,000, whereas the birth rate remained high at 40 per 1,000, resulting in an annual growth rate of 25 per 1,000 people or 2.5 per cent. In 1979 Kenya recorded an annual population growth rate of 4.1 per cent, one of the highest in the world. Some sub-Saharan countries today still have the highest birth rates in the world. Kenya's crude birth rate in 1993 was 44.4, whereas Uganda and Niger had even higher rates of 51.9 and 52.3 respectively (UNDP, 1996). These growth rates are well above those observed in European populations when they were at the same stage in the transition.

Stage 3: late expanding or mid transition phase

Further change in the population growth rate is associated with improved technology in agriculture and industry, together with better education systems and legislation controlling child employment, such that the economic and social value of children typically declines. Furthermore, the breakdown of the extended family system places greater physical, emotional and

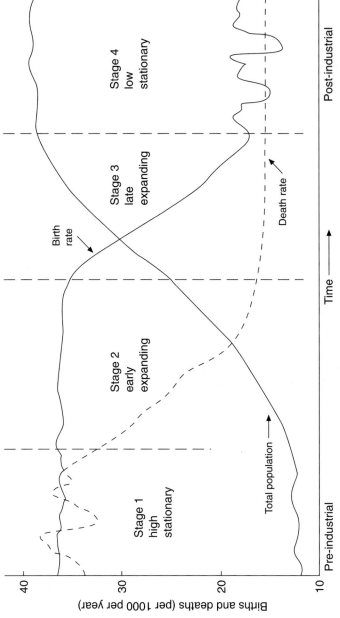

Figure 5.2 Demographic transition model. Adapted from Mayhew (1997) and G. Jones and Hollier (1997).

financial costs on the parents. Couples start to use contraception to limit family size and the birth rate begins to fall. The death rate continues to decline and population growth at this point reaches its maximum. This stage occurred in the developed countries in the late nineteenth and early twentieth centuries. Elsewhere, in the last two decades there have been marked reductions in birth rates in East and Southeast Asia, due to a combination of family planning programmes and socioeconomic development. Thailand had an annual population growth rate of 1.0 per cent for the period 1993 and 2000 and South Korea 0.9 per cent (UNDP, 1996).

Stage 4: low stationary or late transition phase

In this phase of the transition process, the death rate reaches its lowest level and as fertility steadily declines, the rate of population growth begins to fall. Most developing countries are currently in the mid and late transition stages. Hong Kong and Singapore, however, have nearly completed the transition and their respective population growth rates of 0.4 per cent and 0.9 per cent are similar to those of the United States (0.9 per cent) and Canada (1.1 per cent) (UNDP, 1996).

Some writers, such as Bongaarts (1994), add a fifth stage to the model, known as the 'declining or post-transition phase'. This stage is characterised by a new equilibrium being achieved between births and deaths, such that population growth is close to zero; in some cases there is even negative growth. It is probably true to say that most of western Europe and North America has now reached this stage. For example, average annual population growth rates for 1993–2000 are as low as zero (0.0 per cent) in Italy and Portugal, 0.9 per cent in the United States, 0.3 per cent in the United Kingdom and 0.1 per cent in Germany. The Russian Federation actually registers a negative figure of −0.2 per cent for the same period, whereas in Hungary and Latvia the figures are −0.4 per cent and −0.8 per cent respectively. The average figure for all industrial countries over this period is 0.4 per cent (UNDP, 1996).

Population policies

The reasons underlying the various changes reflected in the demographic transition are highly complex. It is unwise for commentators in developed countries to blindly advocate the strict control of population growth in poor countries. As O'Connor argues in the case of Africa: 'there is no evidence to suggest that rapid population growth is the main cause of poverty . . . which was just as widespread when the growth rate was much slower. In so far as the rapid growth results from high fertility it can be seen alternatively as a consequence of poverty' (O'Connor, 1991: 54). Population growth is undoubtedly a problem in some countries, but it is invariably a symptom of other problems such as poverty and lack of security. It should be appreciated that having more children among poor families is an insurance strategy to ensure household survival in places where infant and child mortality rates are high. Furthermore, children frequently play a vital role in generating household income, as well as providing social security in the absence of any state provision. The most successful initiatives to reduce family size start by tackling underlying causes and finding ways of ameliorating poverty and insecurity, helping to improve the health of mothers and children. Experience has shown that, with such strategies, both fertility and mortality rates should begin to fall.

Some governments have, however, taken a strong interventionist line to control population growth. In some cases, these policies have been aimed at reducing population; whereas in others, pro-natalist policies have been adopted to increase population. The Nazi regime in Germany during the 1930s and early 1940s was strongly pro-natalist, generating much propaganda on the need to create a master Aryan race. A wide range of measures was introduced, such as generous family allowances and tax concessions for large families, taxes on unmarried adults and prosecution for induced abortions. Rumania under the Ceausescu regime also adopted a strong pro-natalist line, with abortion becoming illegal in 1966 and tight controls placed on the availability of contraceptives. In the following year, 1967, the crude birth rate almost doubled to 27 per 1,000, causing considerable pressures on education, employment and housing as the 1967–70 bulge moved through the age groups. Elsewhere in the world, Israel and Saudi Arabia have also encouraged population growth, primarily to strengthen their political power. Israel's average annual population growth rate in the period 1990–1994 was 3.7 per cent, whereas Saudi Arabia's rate was 3.2 per cent, but reached as high as 5.2 per cent per annum during the period 1980–1990 (World Bank, 1996b). The upsurge of Islamic fundamentalism has resulted in some states becoming pro-natalist. By introducing its New Population Policy in 1984, Malaysia reversed a long-standing policy of promoting family planning to encourage women to 'go for five', to catch up with more populous neighbouring states and to prevent ethnic Malays from being dominated by the Chinese population. In response to a shortage of labour, Singapore from 1987 adopted a more selective pro-natalist stance, encouraging educated and professional couples to 'have three, or more if you can afford it' rather than 'stop at two' (Drakakis-Smith *et al.*, 1993).

However, with growing concern about the population–resource balance, anti-natalist policies are now rather more common than pro-natalist measures. India launched a family planning programme as early as 1952, and during the 1960s many other countries followed in response to current development thinking and the production of the contraceptive pill and intra-uterine devices. In 1965 US President Lyndon Johnson, addressing a United Nations audience argued, 'Let us act on the fact that less than five dollars invested in population control is worth a hundred dollars invested in economic growth' (Stycos, 1971: 115). The United Nations began to provide advisory services to family planning programmes in 1965, and by 1976 some sixty-three developing countries had initiated such programmes. It was probably the 1974 World Population Conference in Bucharest which brought population issues and family planning initiatives under the political spotlight. At this meeting, Western countries were strongly criticised by developing countries for placing too much emphasis on the population issue in development strategies, to the neglect of promoting social and economic progress.

Indonesia, the world's fourth most populous nation, adopted family planning in 1970 as a key element in the drive for economic growth. Considerable success was achieved between 1970 and 1995, with the average number of children per woman falling from 5.6 to 2.9. China's one-child policy is probably one of the most well-known and most rigid anti-natalist policies, affecting one-fifth of the world's population and enforced through tight community control and a series of rewards and sanctions. As a result, the 1970 crude birth rate of 33.43 per 1,000 fell to 17.82 in 1979 and 16.98 in 1996 (Box 5.1). Other family planning schemes have commonly rewarded individuals with various tax concessions, with free or preferential medical treatment (South Korea), priority schooling (Singapore) or even a new sari (Bangladesh). It is in Asia that family planning programmes have been most widely implemented, whereas progress has been slower in Latin America, largely due to the proscription of 'artificial' methods of contraception by the Roman Catholic Church. Although evidence suggests that individual families frequently disregard the Church's views on birth control, government policy makers are likely to be more strongly influenced.

Another important factor affecting the size of families is the status of women in society. Muslim societies are generally strongly patriarchal; men can have up to four wives and female education is a low priority, since women are expected to stay in the home. The fundamentalist revolution in Iran in 1979 swiftly overturned the Western-style policies introduced by the Shah, which had given a considerable degree of emancipation to women. Under the new regime, a woman was required to gain her husband's consent to work; access to sterilisation and abortion was tightly controlled and the minimum age of marriage for women was reduced from 18 to 15 years. As a result, from 1980 to 1990, Iran had an average annual population growth rate of 3.5 per cent (World Bank, 1996b).

As a result of such population policies leading to shrinking family size in the world's most populous countries, one British newspaper in January 1998 claimed that the population bomb had been defused (Figure 5.3, pages 117–18), although many of the world's poorest nations, and particularly those in Africa, continue to have rapidly growing populations. According to Figure 5.3, it is projected that in 2020, for the first time, Africa's most populous country, Nigeria, will be one of the world's six largest countries (*The Independent*, 1998: 11).

Population structure

Key elements in the demographic transition process, together with the impact of disasters such as war, famine and disease epidemics are manifested in changes to the population structure. Typically, developing countries with declining death rates, high birth rates and low life expectancy have youthful populations (Plate 5.2, page 120). Table 5.3 (page 120) clearly shows how some of the poorer countries with high population growth rates have a large proportion of their populations under the age of 18. In Mali and Sierra Leone, respectively 61.1 per cent and 52.3 per cent of their populations are under 18, whereas in more developed countries these figures are much lower (Japan 20.2 per cent, Sweden 22.7 per cent). These statistics have implications for child welfare, especially educational provision.

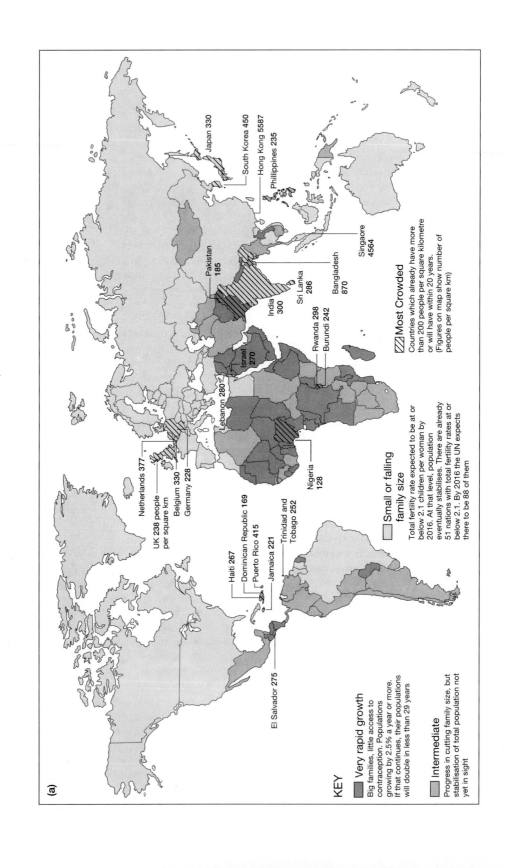

(a)

KEY

Very rapid growth
Big families, little access to contraception. Populations growing by 2.5% a year or more. If that continues, their populations will double in less than 29 years

Intermediate
Progress in cutting family size, but stabilisation of total population not yet in sight

Small or falling family size
Total fertility rate expected to be at or below 2.1 children per woman by 2016. At that level, population eventually stabilises. There are already 51 nations with total fertility rates at or below 2.1. By 2016 the UN expects there to be 88 of them

Most Crowded
Countries which already have more than 200 people per square kilometre or will have within 20 years. (Figures on map show number of people per square km)

Japan 330
South Korea 450
Hong Kong 5587
Phillipines 235
Singaore 4564
Pakistan 185
Sri Lanka 286
Bangladesh 870
India 300
Israel 270
Lebanon 280
Rwanda 298
Burundi 242
Nigeria 128
Netherlands 377
UK 238 people per square km
Belgium 330
Germany 228
Haiti 267
Dominican Republic 169
Puerto Rico 415
Jamaica 221
Trinidad and Tobago 252
El Salvador 275

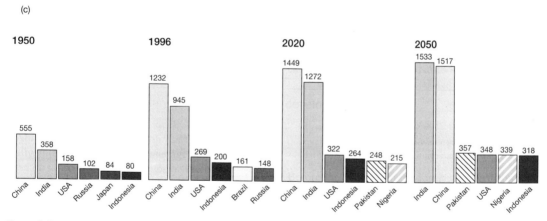

Figure 5.3 (a) Family sizes are shrinking rapidly, although human numbers are still soaring in many of the world's poorest nations. (b) European countries where population has already begun to fall. (c) The six biggest countries (in millions of people). The data are from the United Nations Population Division and the projections to 2050 are from the 'medium variant'. Adapted from *The Independent,* 12 January 1998.

The structure of a nation's population is clearly revealed in an age–sex diagram, commonly known as a population pyramid. The population pyramids for Ghana and the United Kingdom in 1990 reveal quite different population structures (Figure 5.4, page 121). In fact, only the diagram for Ghana actually resembles a pyramid, with a wide base indicating a youthful population and a steeply tapering top; there are fewer people in the older age groups due to an average life expectancy at birth of only 55 years. In sharp contrast, the diagram for the United Kingdom is hardly a pyramid, and is typical of a more stable

Plate 5.2 Children in The Gambia – a youthful population (photo: Tony Binns)

Table 5.3 Proportion of the population under 18 years old, for selected countries.

	Total population in 1994 (millions)[a]	Population under 18 years in 1995 (millions)[b]	Proportion of total population under 18 years (%)
Bangladesh	117.9	55.9	47.4
Brazil	159.1	62.1	39.0
China	1190.9	379.3	31.8
India	913.6	384.9	42.1
Jamaica	2.5	0.9	36.0
Japan	125.0	25.3	20.2
Mali	9.5	5.8	61.1
Sierra Leone	4.4	2.3	52.3
Sweden	8.8	2.0	22.7
United Kingdom	58.4	13.5	23.1
United States	260.6	68.6	26.3
Zimbabwe	10.8	5.7	52.8

[a] Data from World Bank, *World Development Report 1996*, Oxford University Press, Oxford.
[b] Data from UNICEF, *The State of the World's Children 1997*, Oxford University Press, Oxford.

population situation, with a smaller proportion of the population in the lower age groups and a greater proportion in the upper age groups, reflecting an average life expectancy that is 20 years greater than for Ghana. Interestingly, the UK figure also shows a higher proportion of women over 70 years, reflecting the differential in life expectancy between males and females, a common feature in a number of developed countries.

In some 'newly industrialising countries', such as Singapore and South Korea, the population has been steadily 'greying', with an increasing proportion in the higher age groups; this has implications for the size of the workforce and the provision of social welfare.

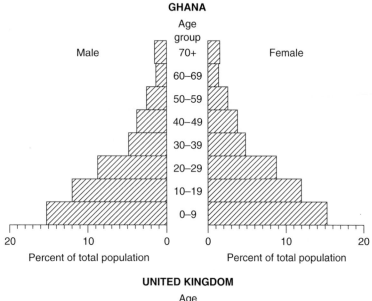

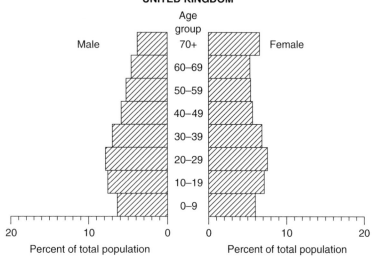

Figure 5.4 Population pyramids for Ghana and the United Kingdom. Adapted from Binns (1994a).

Quality of life

Household structure and development

Sadly, the world is a very unequal place and not all people enjoy equal access to basic needs and a satisfactory quality of life. As we have already seen, variables such as life expectancy and infant mortality vary greatly from one country to another. National statistics produced by the World Bank, the United Nations Development Programme (UNDP) and other agencies, also conceal marked variations which exist within and between different regions in specific countries, within and between rural and urban areas and also between different communities and households. The household is still the key living unit in most developing countries, which also usually controls production, consumption and decision making. Appropriately, households have become an increasingly important focus of study in recent years, since it has been recognised that 'the very success of development policy is likely to be undermined by a failure to view the household and family in a holistic manner' (Haddad, 1992: 1). Households in developing countries are typically larger than those in developed countries, and rather than merely comprising parents and children, they often also include grandparents, unmarried aunts and uncles, as well as more distant relatives. The term *extended family* is often used to describe such households. Urban-based households frequently retain strong links with their rural relatives, particularly those who live in the ancestral village, to which regular visits are common and where urban dwellers may have farmland.

All households pass through a developmental cycle, during which their size and composition may change and the all-important ratio between workers and dependents changes over time. At certain stages in this developmental cycle, households may be under considerable pressure and this may directly impact on the quality of life, perhaps in terms of nutritional intake. For example, households with a high proportion of very young or old members may encounter difficulties, since relatively few workers may have to support a large number of dependents. From a study in rural Malaysia, Datta and Meerman (1980) have suggested that most households experience a four-stage per capita income cycle, which is linked closely with 'dietary stress' in the following way:

1 Immediately following marriage, both husband and wife tend to earn income; per capita income is quite high, reducing the risk of dietary energy stress.
2 With the arrival of small children, there are more mouths to feed, but no additional income earners. Indeed, the mother must devote time to child care, which reduces her income-earning capabilities. The household suffers increased risk of energy stress.
3 As the children grow older, they are able to contribute to income earning and there are decreased demands on mother's time for child care. The risk of energy stress falls.
4 Finally, the income-generating children leave home and parents are at a greater risk of being ill, so there is increased risk of energy stress caused by a decline in command over food production.

As a result of this 'cycle', per capita household income in Malaysia typically fell as the age of the household head rose from 25.0 to 37.5 years and again from 50.0 to 62.5 years (Datta and Meerman, 1980).

In most developing countries there are also marked inequalities between households. More powerful households may be ethnically distinctive and better educated; they are key elements in the local, and possibly national, power structures, having good links with government officials, police, large landowners and traders; and they are frequently well endowed with assets and income. Poor households, however, are frequently less well educated, have poorer nutrition, are often ignorant of the law and have few assets apart from their labour. Poor households are often vulnerable and susceptible to exploitation as they become locked into cycles of debt, and they depend upon the assistance of one or more richer patrons. Recently, analysts have referred to the concept of social capital to denote the levels of loyalty and response that a household can tap during vulnerable or hard times.

Different pressures and responsibilities also exist within households. In some rural African societies, husbands and wives may live in separate houses in a village compound, have very different household responsibilities and quite separate incomes. Age and gender differentiation in household labour inputs and expenditure patterns has long been appreciated, with women appearing to display greater altruism, particularly in relation to children (Barrett and Browne, 1995). The gender of the household member who benefits from policy interventions may be relevant to intrahousehold distribution of wealth and resources, and there is a suggestion that 'enhancing and securing female earnings appears to make sound policy sense' Kabeer, 1992: 51). In fact, women's ability to maximise their own welfare and the welfare of their dependents may be severely constrained by the power relations of the household. The contributions of women to household income and welfare need to be much better understood. It is estimated that 70 per cent of the food of tropical Africa is produced by women, and yet far too frequently it is assumed that all farmers are men: 'Many of the most demanding farm jobs, such as hoeing, weeding and harvesting are done by women, in addition to taking care of young children, collecting firewood and water and processing and cooking food' (Plate 5.3) (Binns, 1994b: 88).

Many development schemes have failed and tensions within households and communities generated through misunderstanding the different roles and responsibilities of women and men. Household analysis should take note of relationships between households (interhousehold), as well as the composition of, and relationships within, individual households (intrahousehold). For example, in eastern Sierra Leone

Plate 5.3 Women farmers harvesting rice in The Gambia
(photo: Tony Binns)

among the Mende, Leach (1991) found that in addition to working on the household farm, women and men undertake separately a range of productive activities. Whereas men gained independent incomes from selling bushmeat, palm wine tapping and undertaking day labour, many women made individual rice swamps, cassava and groundnut farms and vegetable gardens. Some women also traded commodities, such as salt and dried fish, from their homes or were active in local markets (Leach, 1991). In polygamous households, co-wives usually had separate individual enterprises and their consumption unit consisted of themselves and their own children. Agricultural change has altered the division of labour between women and men, such that with the increasing production and sale of coffee and cocoa, 'wives are expected to assist with harvesting and processing, receiving only a discretionary gift of cloth and the right to glean fallen produce in return' (Leach, 1991: 47). Women have also increased their work on the family rice farm, whereas work on tree crops was regarded as a woman's duty for her husband. These duties could take significant time away from a woman's independent vegetable gardening or trading. However, Leach found that the life cycle of individuals and households was significant in that middle-aged men, who had built up resources through farming or trade, might relieve their wives' workloads by recruiting labour to help with the tree crop harvest or to plant rice, or might possibly help with clearing the swamp and the provision of trading capital. As Leach comments, 'Such wives tend to

complain less of overburden, although they are just as concerned to maintain their own income streams' (Leach, 1991: 48).

Another important, but frequently ignored, factor in many household budgets and survival strategies concerns the cost and benefits of having children, especially the important work they do (Box 5.2). Pryer's study of some ultrapoor households in an urban slum in Khulna, Bangladesh, reveals how, faced with low, unreliable and seasonal incomes, households attempted to achieve a diverse employment profile with as many family members working as possible. The extreme poverty of the community is reflected by the fact that in 1984 no fewer than 67 per cent of children under 5 in the slum were second- or third-degree malnourished; that is, they were under 75 per cent of the expected weight for their age (Pryer, 1987: 133). All seven severely undernourished households surveyed were deeply indebted, usually to landlords, employers, shops and neighbours. Since the only productive asset of these households is labour, all able-bodied adults and many of the children needed to do some form of paid work. Employment opportunities were invariably poorly paid in the informal sector, with men engaging in rickshaw pulling, petty trade, hawking and labouring on a daily basis. But during the monsoon months (June to September), less work was available. In two of the households, women were major earners, typically engaged in domestic work or home-based piece-rate work; this involved long hours and was poorly paid both absolutely and in relation to male wages.

BOX 5.2   Children: a neglected piece in the development jigsaw

Whereas children in developed countries can usually expect to be well fed, have good clean clothes, attend school until they are 16 or older, be well protected against ill-health and have a life span of well over 70 years, their counterparts in poor countries have few, if any, of these assurances. Furthermore, the welfare of children is a constant focus of attention in developed countries and a prime concern of many households, but children in developing countries are themselves expected to help in maintaining the household in so many ways. Fetching water and firewood, scaring birds from the fields, processing food and selling items in the urban informal sector are just some of the myriad tasks which children commonly perform. Yet, all too often, the vital contribution which children make to the welfare of households in poor countries receives remarkably little attention. As Robson comments, 'By the age of 10–12 years, some children may contribute as much to household sustenance as adults' (Robson, 1996: 43). Whereas in the last two decades there has been, quite appropriately, much debate and writing about the position and role of women, children still remain a rather neglected element in the development process. A recent report from Actionaid (1995) clearly shows that children and children's work have been neglected and empirical research reveals the importance of this in different social and environmental contexts, 'The inclusion of children and their participation in the development process is as much their basic human right as is their right to health provision and protection from hazardous work and living conditions' (Actionaid, 1995: 5). Children need to be viewed as 'social actors' and the conventional model of childhood must be modified to support the notion of a 'plurality of pathways to maturity' (SCF, 1995).

Research has shown that gender is also important in determining children's roles and responsibilities. A study in Jamalpur, a rural area in Bangladesh, revealed that boys are mainly involved in cultivating family plots of land and occasionally they go to neighbouring farms to work as wage labourers. They enjoyed their work and in earning cash from casual labour they felt they were an asset to their families. Girls in the same area, however, were somewhat resentful that their work, which involved mainly household duties and looking after younger children, was

invisible. The girls saw education as a way of earning greater respect from their families and communities. Both girls and boys agreed that they should work hard to help their parents and to contribute to family income and/or welfare (Actionaid, 1995: 49).

For children in many developing countries, attending school is regarded as a privilege rather than a right. Commonly, far fewer girls than boys in developing countries go to school and this is reflected later in life in a significant differential in adult literacy rates between males and females. In Sierra Leone, where the male adult literacy rate in 1993 was 43.3 per cent, the rate for females was only 16.7 per cent. A similar pattern is seen in India, where the female adult literacy rate is 36.0 per cent compared with the figure for males of 64.3 per cent. This differential is often much greater in Moslem countries, where girls are more likely to stay at home to assist their mothers with household chores, rather than go to school. In Libya, whereas the male adult literacy rate is 86.3 per cent, the female rate is only 59.3 per cent. Saudi Arabia similarly has a large male–female literacy differential, with a male literacy rate of 70.4 per cent and a female rate of 47.6 per cent (UNDP, 1996). Whereas girls in poor countries may often attend the first few years of primary school, the proportion of girls in school classes frequently falls sharply among higher age groups. In Uganda, whereas 83 per cent of girls were enrolled in primary schools in 1993, only 10 per cent were enrolled in secondary schools. In Saudi Arabia the female enrolment falls from 73 per cent in primary schools to 43 per cent in secondary schools (World Bank, 1996b).

A study in the Kyuso area of Kitui district in eastern Kenya found that the major reason for school non-enrolment and drop-out was poverty and the high costs of school materials and uniforms. During drought conditions, parents frequently withdraw their children from school to assist with water collection and to look after younger children (Actionaid, 1994). In many developing countries, more and more children, particularly girls, are being withdrawn from school, as investment in their education is seen as being lost when they marry.

Gender and the roles of children of different ages need to be much better understood. Development agencies should seek to identify the effects of

Box 5.2 continued

development projects on children and to reduce the need for children to be sent out to work as part of household survival strategies. Appropriate forms of education should be introduced, and, if necessary, this should be vocationally oriented and undertaken in the workplace. Some progress has been made, notably through The International Year of the Child in 1979, the 1989 UN Convention on the Rights of the Child, and the 1990 UNICEF-sponsored World Summit on Children. These were important landmarks in attempting to raise the profile of children in all communities, but in some countries there is still a very long way to go in improving the lives of children.

In the seven households, an average of 68 per cent of income was spent on food; the rest went towards fuel, rent and repayment of debt. Food intake was well below the recommended daily allowance. One mother, whose husband died of tuberculosis, went out to work from 7 am to 4 pm and again from 6 pm to 11 pm every day, while her 10-year-old daughter assumed responsibility for child care and domestic work. In another family, the wife, with a chronically sick husband who was unable to work, sold saris illegally on the black market, assisted by her 17-year-old niece, and her 12-year-old daughter was a servant in the main market. Given the family's circumstances, the wife was clearly the economic and social household head. In both families, illness of the chief earners had led to sale of assets, indebtedness and deepening poverty, leading to inadequate, unreliable and seasonal flows of food. One of the two women household heads was forced into heavy dependence on a patron who was her employer and landlord, while the other was engaged in illegal trading, both highly precarious strategies (Pryer, 1987). This study clearly illustrates both the different roles of members in attempting to ensure the household's survival, and the cycle of poverty and malnutrition from which it is extremely difficult to escape.

Health and health care

The health status of a population, or elements of a population, can be crucial in the development process. It might be argued that a healthy population is more able to contribute to development efforts and will also be better placed to benefit from the fruits of these efforts. We have already seen that many developing countries have high rates of infant mortality and their life expectancy levels are well below those of richer, more developed countries. Variables such as these are often a good reflection of the health status of a population and the quality of health care. In many poor developing countries health facilities are inaccessible to a large proportion of the population, especially those living in remoter rural areas. Where hospitals and clinics do exist, there are frequently shortages of trained health workers, drugs and basic equipment.

Providing good health care for everyone is an expensive undertaking for the governments of poor countries. In many parts of the world, but notably tropical Africa, missions play an absolutely crucial role in providing health care in certain areas, together with a variety of non-governmental organisations (NGOs). Many developing countries still cling to a top-down style of health care inherited from the colonial period, with a considerable proportion of health expenditure being allocated to a few key hospitals, particularly in the main towns and capital city, whereas the rural areas remain relatively neglected.

In China under Mao Zedong, various efforts to improve public health were introduced, particularly from 1968 onwards, when a central government document called the 'June 26th Directive on Public Health' demanded that the focus of medical and public health work should be transferred to the countryside. Mao suggested that although city hospitals should keep some doctors, a greater proportion should be sent to work in the villages (where 85 per cent of the population lived) to teach medical knowledge to the peasant youth (Endicott, 1988: 157). Although these 'barefoot doctors', as they were popularly called, had a lot to learn, Mao believed they were better than 'fake doctors' and 'witch doctors'. Furthermore, he argued that villages could afford them as medical funding was redistributed such that the bulk was now directed to prevention and cure of the common, frequent and widespread diseases. Village commune hospitals received much more finance, which could be used to purchase equipment such as X-ray machines, as well as to develop both Chinese and Western medicine. There was also a campaign to extend the recruitment of barefoot doctors, midwives and medical orderlies. The impact of these policies was clearly seen in rural areas such as Shifang County in Sichuan Province: 'In three-month, sometimes six-month, courses, qualified doctors, sent down to the countryside by rotation during the Cultural Revolution, trained 658 barefoot doctors in basic first aid, Chinese medicine, acupuncture, the use of thermometers, the dispensing of vaccines by injection and drugs for influenza, stomach upsets and other common ailments. [As a result] the total number of medical personnel in the county rose from 592 in 1965 to 3,420 a decade later. . . . [The barefoot doctor initiative represented] . . . a good start on creating an accessible, experimental, non-elitist public health system biased in favour of prevention' (Endicott, 1988: 158). A notable achievement in Shifang County's health programme was the virtual eradication of schistosomiasis (bilharzia), such that the number of people affected was reduced from 5,700 in 1959 to only 4 in 1982. By the end of the 1970s, 85 per cent of Chinese villages had a health station staffed by one or more barefoot doctors.

Health problems in developing countries are frequently the result of poverty, notably inadequate or poor quality food and water and the lack of proper sanitation. Children are particularly susceptible to diarrhoea, and measles also claims many victims, whereas vitamin A deficiency may lead to blindness and infection, particularly after measles. Iron deficiency often leads to anaemia, causing weakness and particular risks for newborn children. Children are also very susceptible to a number of nutrition-related diseases such as pellagra, beriberi, rickets and kwashiorkor. Diseases associated with water are a major problem in many developing countries, and there is some evidence that the expansion of irrigated agriculture has encouraged

their spread with large areas of slow-moving water in dams and reservoirs. Schistosomiasis is transmitted by snails in slow-moving water, whereas onchocerciasis (river blindness), which is endemic to large parts of tropical Africa, is also associated with water and is transmitted by the black fly. Possibly the most serious threat to health in tropical regions is malaria, transmitted by mosquitoes which breed close to stagnant or slow-moving water. It is estimated that in tropical Africa alone as many as 200 million people are affected by malaria, which weakens the victim and lowers their resistance to a wide range of other possible infections. Although draining swamps and spraying pools with insecticide might help, mosquitoes are becoming resistant to certain chemicals and antimalarial drugs. Mosquitoes are also responsible for transmitting dengue fever and yellow fever.

The health and nutritional status of households and household members may vary over time. We have already seen how a particular stage in a household's life cycle may affect income and nutrition. This longer-term variation may be compounded by marked seasonal pressures in tropical countries, where the period of hardest work commonly coincides with the rainy season, which is also the time when the occurrence of many diseases, such as malaria, increases. Furthermore, in many communities, towards the end of the rainy season, but before the harvest, food stocks are often getting low and the quantity and quality of food intake declines. This time of year, sometimes called the 'hungry season', is associated with greater susceptibility to infection, though people still have to work hard in the fields. Certain elements of communities and households, such as pregnant and breast-feeding

---

BOX 5.3 AIDS: the scourge of Africa

AIDS (acquired immune deficiency syndrome) is a disease in which the body's natural protection or immune system is damaged. The first AIDS case was reported in the United States in 1981 and the first recorded African case was in 1983. The extensive spread of human immunodeficiency virus (HIV), the aetiologic agent that causes AIDS probably began as early as the 1960s, but spread rapidly during the mid to late 1970s and early 1980s. Two strains of HIV have been identified (HIV1 and HIV2) and infection occurs when blood from an infected person passes directly into another's bloodstream, and also through sexual intercourse. By July 1994 the World

Health Organisation estimated that there were some 16 million adults and more than 1 million children infected with HIV, the majority of whom lived in sub-Saharan Africa (WHO, 1994: 1). In almost all cases, those infected by HIV develop AIDS, which is inevitably fatal. The global spread of HIV continues apace, such that Asia in 1993 had only 1 per cent of the world's AIDS cases, but just a year later it accounted for 6 per cent of the total, due to the rapid growth of AIDS in South Asia and East Asia.

But it is in sub-Saharan Africa that the problem is greatest. On top of Africa's general impoverishment and lack of any meaningful 'development' in

Box 5.3 continued

many countries over the last two decades, the AIDS problem adds further to the challenges which African governments, and indeed the world community face today. In Botswana, Malawi, Uganda, Zambia and Zimbabwe, AIDS is now considered to be the leading cause of death between the ages of 15 and 39. HIV in sub-Saharan Africa is mainly spread through heterosexual intercourse and perinatal transmission, which can occur *in utero*, during delivery or after birth through breast milk.

Although the problem is greatest in urban areas, it is not insignificant in rural areas. It was estimated in 1994 that in Nairobi (Kenya) and Abidjan (Côte d'Ivoire) the prevalence of HIV1 among prostitutes was well over 80 per cent (US Bureau of the Census, 1994). But given the fact that Africa is still predominantly rural, in absolute numbers AIDS cases in rural areas predominate, though accurate data are difficult to obtain. The table shows that the average incidence of AIDS in sub-Saharan Africa is 12.4 per 100,000 people, well above the figures for all developing countries and the world as a whole. Indeed, certain African countries have staggeringly high incidences of AIDS. For example, the figure of 96.7 for Zimbabwe is, according to UNDP (1996), the second highest in the world after the Bahamas (131.4). A number of other African countries also have among the world's highest rates of AIDS, notably Botswana (65.6), Congo (58.4), Côte d'Ivoire (44.6) and Malawi (49.2).

Prothero (1996) has considered the possible effects of population migration on the transmission and diffusion of AIDS in West Africa and concludes that there is a need for more research on the complex interactions of socioeconomic, cultural and biomedical mechanisms. AIDS has now been added to the list of other child-killers in sub-Saharan Africa: diarrhoea, malaria and measles. Browne and Barrett (1995) specifically examine the impact of the African AIDS epidemic on children and suggest that although more children still die from malaria, diarrhoea and acute respiratory infections, the long-term effect on economic and human development at national, community and household levels gives much concern. HIV-infected children have a short life expectancy, with 80 per cent dying from AIDS-related causes by the age of 5. In Zambia, the under-5 mortality rate has increased from 125 per 1,000 live births in 1989 to 203 in 1993, due at least in part to AIDS-related causes. A further problem concerns the number of

The incidence of AIDS in selected countries.

Country	AIDS cases per 100 000 people (1994)
Bangladesh	nd
Brazil	7.3
China	nd
India	nd
Jamaica	13.5
Japan	0.2
Mali	5.8
Sierra Leone	0.5
Sweden	2.2
United Kingdom	2.7
United States	22.7
Zimbabwe	96.7
All developing countries	6.7
Sub-Saharan Africa	12.4
World	7.7

Note: nd = no available data.
Source: UNDP (1996).

children orphaned as a result of AIDS, which it is estimated in ten Central and East African countries could reach between 5 and 6 million children by the year 2000, or about 11 per cent of the total 10–15 year old child population (Browne and Barrett, 1995).

AIDS could also have a wider impact on population growth rates, which are likely to fall, having a Malthusian effect that could lead to an improvement in the ability of the world to sustain and feed itself. However, the spread of AIDS will increase the demand for curative health care, placing considerable pressure on poor African countries and perhaps meaning less health care for the rest of the population. For example, treating the estimated number of AIDS cases could represent 23 per cent of 1990 public health spending in Kenya and as much as 65 per cent in Rwanda. However, as Brown (1996b: 17) suggests, 'The most critical impact of AIDS is likely to be in the damage inflicted on the productive capacity of an economy and its potential to achieve food security through domestic production or economic access in world markets.' What, sadly, does seem clear is that sub-Saharan Africa already has the dubious distinction of being the world's poorest region, and AIDS is likely to further exacerbate poverty as the poor lose access to what is often their only resource – their own labour.

Table 5.4 Primary school enrolment and literacy for selected countries.

Country	Gross primary school enrolment ratio (1990–95)[a]	Male adult literacy rate (1995)	Female adult literacy rate (1995)
Bangladesh	79	49	26
Brazil	111	83	83
China	118	90	73
India	102	66	38
Jamaica	109	81	89
Japan	102	100	100
Mali	31	39	23
Sierra Leone	51	45	18
Sweden	100	99	99
United Kingdom	112	99	99
United States	107	99	99
Zimbabwe	119	90	80

[a] The gross enrolment ratio is the total number of children enrolled in a schooling level – whether or not they belong in the relevant age group for that level – expressed as a percentage of the total number of children in the relevant age group for that level.
Source: UNICEF, *The State of the World's Children 1997*, Oxford University Press, Oxford.

women and young children may suffer disproportionately at such times.

If individuals and households are to fulfil their true potential, in addition to the provision of adequate nutrition and health care, it might be argued that an entitlement to education and freedom of expression should be key elements in the development process.

Education

Like health care, education is an expensive item for poor countries and its quality and availability show considerable variations between and within countries, as does student attainment. And also like health care, education systems are frequently a legacy of the colonial period, often totally inappropriate for the present-day needs of individuals, communities and nations. Indeed, there has been much debate on what is the most appropriate form and structure of educational provision in poorer countries. For example, what proportion of the budget should be allocated to the different sectors (primary, secondary and tertiary), and should more attention be given to non-formal education, such as farmer training and the acquisition of craft skills, rather than formal classroom tuition? Most commentators would probably agree, however, that providing everyone with basic primary education, especially literacy, should be the first priority of all

countries. Table 5.4 shows how primary school enrolment and adult literacy rates in two of the world's poorest countries, Mali and Sierra Leone, are well below those of other countries. In many countries there is also a marked difference in the number of boys who attend school compared with the number of girls. Generally fewer girls attend school, particularly in Muslim countries, and this is reflected in later years in male and female adult literacy rates (Box 5.2).

The 1990 World Conference on Education for All, held in Jomtien, Thailand, proclaimed the need for diverse, flexible approaches within a unified national system of education (UNICEF, 1997). The conference agreed a number of objectives for primary education:

Teach useful skills. Courses should be relevant and linked to community life.
Be more flexible. Use child-centred approaches; adjust to the daily routine and the seasonal farming calendar.
Get girls into school. Be sensitive to social, economic and cultural barriers to ensure equal participation.
Raise the quality and status of teachers. Improve pay and conditions and retrain teachers with negative and stereotypical ideas.
Cut the family's school bill. School fees and equipment charges deter participation; basic education that deters child labour must be free of such costs for poor families.

The difference in enrolment ratios between developed and developing countries becomes significantly greater at secondary and tertiary levels. As W.T.S. Gould (1993) has shown for secondary enrolment, Africa lags very far behind other regions, whereas for South Asia the relatively higher proportion of the secondary age group enrolled in India is pulled down by much lower proportions in Bangladesh and Pakistan, both with populations of well over 100 million. Both Pakistan and Bangladesh are Muslim countries with low female participation in secondary education. At the tertiary level, although 40 per cent of the age group is enrolled in education in the high-income countries, an average figure of under 10 per cent is common in low- and middle-income countries, ranging from 2 per cent in sub-Saharan Africa to 17 per cent in Latin America and the Caribbean (W.T.S. Gould, 1993).

Some governments have introduced wide-ranging reforms to their education systems in an effort to make them more appropriate for national development needs. Whereas the small West African state of The Gambia is setting up its own university with Canadian assistance, rather than sending students overseas for higher education, Africa's most populous state, Nigeria, is debating whether it should rationalise its university system to reduce expenditure. Meanwhile, Ghana has also undertaken major educational reforms. In the 1950s, the country probably had the highest proportion of its children in school in Africa, and this expanded further after independence. However, as the national economy deteriorated, so did the schools and the quality of education, such that education spending fell from 6.5 per cent of GDP in 1976 to only 1 per cent in 1983. Since 1985, under the Economic Recovery Programme of Jerry Rawlings' government, schools have been encouraged to make a more positive contribution to economy and society. A single Ministry of Education and Culture was created along with decentralised planning in 110 districts. The school system was reduced from 17 to 12 years, the cost of boarding was passed on to pupils and student loans were introduced in the tertiary sector. The curriculum was restructured to emphasise practical and life skills rather than academic subjects. A strong emphasis was placed on expanding school enrolments, especially for girls (Binns, 1994b). Investment in female education must receive top priority, not least because studies have revealed strong links between education and health, notably a strong correlation between high levels of infant and child mortality and low levels of maternal education, particularly basic literacy.

Human rights

The Vienna Conference on Human Rights in 1993 revealed significant differences in opinion on the nature of human rights and related policies. For example, some Asian countries questioned external criticism of their human rights records; in particular, they showed their resentment at having imposed on them a set of values based upon Western traditions (Drakakis-Smith, 1997b). However, many would agree that an important issue affecting the quality of life is the ability of all people to voice their opinions freely and without fear of retribution. In some countries it is apparent that certain elements of the population, such as women, are denied complete freedom of speech because of religious and/or cultural attitudes. Across the world, there are many examples of repressive regimes, both military and civilian, which have clamped down with varying degrees of severity on any opposition.

China's continuing 'occupation' of Tibet is a continuing source of much controversy. Although the extent and nature of Chinese influence and control over Tibet through history is disputed, in 1950 the People's Liberation Army clashed with Tibetan troops as the new Chinese government sought to integrate Tibet into the Chinese state. This policy was given a significant boost when, in 1954, China and India reached an agreement, whereby India recognised Tibet to be an integral part of China in return for China undertaking to respect religious and cultural traditions. However, in 1959 there was a rebellion in Tibet against Chinese control and the Tibetan Buddhist spiritual leader, the Dalai Lama, and many of his followers fled to India, where they still remain. Meanwhile, China has steadily strengthened its control over Tibet, increasing the number of troops, introducing inappropriate policies and suppressing religious and cultural activity. China has even installed a young boy as its own 'puppet' spiritual leader in place of the Dalai Lama. In 1987 Beijing reasserted that 'Tibet is an inalienable part of Chinese territory' (*Beijing Review*, 19 October, p. 14).

In other countries, freedom of speech has been denied to specific racial groups, and nowhere was this more entrenched than under the apartheid regime in South Africa. Racial discrimination and separation in South Africa can be traced back long before 1948, when the National Party took power and formally introduced a policy of 'apartheid' or separateness. Apartheid was a policy based on fear, notably fear of the minority White population being dominated by the majority Black population. But the White regime was also well aware that economic survival was completely and unavoidably dependent on the plentiful supply of cheap non-White labour. The National Party government argued that different racial groups should be allowed to live and develop separately, each at its own pace and in accordance with its own cultural heritage, resources and abilities.

In reality, however, the regime was harsh and introduced a wide range of oppressive legislation to control the lives of non-White groups, which together comprised over 85 per cent of South Africa's population. Black political parties such as the African National Congress (ANC) were banned in 1960, police powers increased in 1962, Black newspapers suppressed in 1976 and censorship of political pamphlets introduced. Meanwhile, the Group Areas Act of 1950 extended the principle of separate racial residential areas on a comprehensive and compulsory basis. Its application was felt most strongly in cities such as Pretoria, where Indian traders were moved out of the city centre, and Cape Town, where coloured inhabitants in the suburbs were relocated in segregated areas despite local council objections. Under the 1955 Natives (Urban Areas) Amendment Act, the rights of Blacks to live in a town were confined to those who had either been born there or who had worked there for fifteen years, or ten years with a single employer. All other Blacks required a permit to stay for longer than three days.

A catalogue of legislation enforced what was known as 'petty apartheid', which took such forms as segregated transport, public toilets and even beaches. The broader national development strategy known as 'grand apartheid' was manifested in the creation of homelands called Bantustans. Through the 1959 Promotion of Bantu Self-Government Act, eight (later extended to ten) distinct 'Bantu homelands' were created, each with a degree of self-government and based largely on the historic homelands of different Black tribal groups. Some of these homelands were geographically nonsensical, such as KwaZulu with 48 large and 157 small isolated tracts, and Bophuthatswana with one of its six segments located 320 kilometres from the others. All Black South Africans were given the citizenship of a particular homeland in 1970, but then some subsequently lost their South African citizenship when four homelands were given 'independence': Transkei (1976), Bophuthatswana (1977), Venda (1979) and Ciskei (1981). Although these homelands had all the trappings of independent states, they were not recognised as independent by any country other than South Africa.

Many of the homelands were located on poor quality marginal land away from major urban areas, yet the South African economy, and particularly the mines and industries, were totally dependent on a plentiful supply of cheap Black labour. Consequently, male family members were frequently absent from the homelands, spending much of their time living in crowded hostels close to the mines and on the edge of the large cities. Meanwhile, the women and children were largely left to fend for themselves in the rural areas, but heavily dependent financially upon remittances sent from their absent male relatives.

Growing internal and international pressure, however, gradually forced the minority government to consider dismantling certain elements of apartheid, and this process was accelerated after F.W. de Klerk took power in 1989. In February 1990 an important and historic signal of intent was given to the world community, when Nelson Mandela and several other ANC leaders were released from prison. The country's first democratic elections were held in April 1994 and the charismatic Mandela was proclaimed as first president of the 'new' South Africa. Mandela and the new ANC government then set to work on implementing a range of policies through its Reconstruction and Development Programme, designed to dismantle the strucures of apartheid and address the practical problems facing one of the world's most 'unequal' nations.

Conclusion

Returning to a point made at the beginning of this chapter, people are central to the development process. Unfortunately, in recent years, people have too often been a secondary consideration after the quest for wealth and profit. There is a need to reformulate development strategies so they place people at the heart of development. There have been some brave calls to the world community to make this a reality, but sadly their impact has been small. For example, the Independent Commission on Population and Quality of Life (ICPQL, 1996) has argued that the world faces a linked crisis of environment, quality of life and population, and proposes a number of guiding principles in relation to population growth and improving the quality of life: equity, caring, sharing, sustainability and human security. The commission takes issue with the prevailing concept of development, describing it as 'exclusively economic and obsessed with deregulation

[it] . . . inevitably produces massive exclusion, inside every society, among nations, on all continents. This requires a shift in the way policies and measures are shaped and in how political decisions are made' (ICPQL, 1996: 4). The commission argues that a number of issues must be tackled urgently, including 'making life more liveable through improved individual and collective health and security; dealing with the scourges of poverty and exclusion; raising the levels of literacy, education and access to needed information; rationalizing production and consumption in terms of what the planet's resources can continue to provide and bringing fairness and equity to all through better-balanced exploitation and use of these resources [such as keeping more profits from raw materials 'at home'; utilizing them in a sustainable manner]; more effective policies of aid and assistance; and finding new funding mechanisms between North and South. And, last, but hardly least, caring for ourselves, our neighbours, and the environment by observing the rights pertain-ing to all of humankind' (ICPQL, 1996: 286). The commission stresses the importance of not only environmental sustainability but also social sustainability, and it emphasises the synergy between the two. Importantly, the commission places much emphasis on improving the quality of life, which 'should become the chief focus of governments north and south.' It suggests: 'We urgently need a new synthesis, a new balance between market, society and environment, between efficiency and equity, between wealth and welfare. A new balance between economic growth on the one hand, and social harmony and sustainability on the other' (ICPQL, 1996: 16).

These are admirable sentiments, but given the lamentable record of national government and international community action following earlier well-intentioned initiatives, such as the commissions of Brandt (1980) and Brundtland (1987), there is inevitably some scepticism about progress in the future. But we live in hope.

CHAPTER 6

Resources and the environment

Introduction

There is no doubt that human societal development depends on the physical resource base of the globe. We live ultimately in a closed system in which matter and energy cannot be created or destroyed. Part I has shown something of the diversity of opinion on normative questions of how wealth and well-being should be created or redistributed. Similarly, there has been much debate as to the precise relationship between prospective development achievements and environmental resources at all scales of analysis. At the international scale, resource inadequacies have been forwarded as the cause of underdevelopment (Huntington, 1945) and in direct contrast, development has been modelled to have put 'mankind on the brink of extinction' (Ehrlich, 1968). In recent decades, however, there has also been a more widespread understanding of the links between poverty and degradation (Elliott, 1994).

The continuing debate over the meaning and practice of sustainable development is fundamentally about integrating these apparently contradictory opinions and reconciling development and the environmental resources on which society depends. As noted in Chapter 3, although sustainable development has become the development paradigm of the 1990s, there is no longer universal meaning ascribed to the notion of development. Indeed, it has been suggested that the concept of sustainable development can mean everything or anything you want (O'Riordan, 1995). Distinct views of sustainable development can certainly be identified according to different perspectives on the relationship between human societies and the environment. Ecocentrists view humankind as part of nature and a global ecosystem in which nature must be respected regardless of its value to society; technocentrists see nature as separate from society and as an instrument for exploitation and material gain for

human benefit (Pepper, 1996). It is therefore not surprising that there are also quite varied orientations to environmental management and action, emerging from these fundamentally different philosophical entry points to the sustainable development debate. Deep Ecologists favour radical changes in political and economic structures, in contrast to the 'self regulation through enlightened conscience' promoted by the Dry Greens (O'Riordan, 1995: 13).

Despite this evident diversity of views surrounding sustainable development in theory and in practice, it is apparent that sustainable development is considered to be inherently desirable and a policy objective which should be striven for (Elliott, 1994). 'Like motherhood, and God, it is difficult not to approve of it' (Redclift, 1997: 438). Indeed, the attractiveness of the concept may lie in precisely the way in which it can be used to support varied political and social agendas. This chapter is a detailed investigation of the interrelationships between the key resource sectors and their historical and contemporary development patterns and processes. In so doing, it illustrates the key challenges of sustainable development in alleviating environmental degradation and in finding new forms of development, particularly in overcoming poverty.

The importance of resources in development

The predominant view of resources is that they are given value by society in respect to the functions they can perform and according to the levels of development and aspirations of society. As Zimmerman (1951: 7) has stated, 'Resources are not, they become.' However, there is also the view of environmental resources as stocks of substances or materials found in nature. The substantial debate concerning the adequacy of resources to support the demands of modern society, particularly over the last two hundred years, flows from these conceptual differences:

> If environmental resources are simply stocks of substances found in nature, then they are inevitably fixed and limited in quantity. Limits to resource use must inevitably exist. If, on the other hand, resources reflect human appraisal, then the conclusion is quite different. In this case, their limits are not imposed by the non-human environment, but rather by human ingenuity in perceiving usefulness or value. (Mather and Chapman, 1995: 3)

In Chapter 5 the doomsday predictions of Thomas Malthus were noted in which he considered that there were definite environmental limits (a fixed amount of land) to human development. Subsequently, influential publications such as the *Limits to Growth* (Meadows *et al.*, 1972) have promoted similar views of an ultimate limit on economic development presented by the availability of resources. In contrast, authors such as Ester Boserup (1965) and Julian Simon (1981) have been central in promoting the functional view of resources encapsulated in Zimmerman's definition, and in argu-ing against the inevitability of societal collapse. They point rather to the social, institutional and technological factors which serve to extend the boundaries of development. In contrast to Malthus's emphasis on the physical limits of resources, these authors focus on the stimulus to, and opportunities for, innovations and developments in resource use which population growth brings.

There is no denying, however, the importance of resources in development. All forms of productive activity make demands on the resource base; as raw materials and as energy sources in industrial and agricultural processes and in terms of the varied 'sink' functions which the environment provides in absorbing, dissipating and transporting the by-products of these activities. Resources are also consumed through human social activities associated with fulfilling the basic need for shelter, sustaining urban lifestyles, and so on, confirming a close relationship between resources and development. Intrinsically, the environment supports life itself through the regulation of the earth's temperature and atmosphere.

Water resources

Water resource management is fundamental to human existence and economic development; Ohlsson (1995: 3–4) put it like this:

> Life on this planet was conceived in water. . . . The dawn of civilisation was ushered in by people in arid areas who were able to create gardens from the desert surrounding them by intricate irrigation systems. . . . Industrialisation was heralded, only a few hundred years ago, by the noise of a myriad of water-driven mechanical hammers reverberating through European forests.

Global water withdrawals have accelerated sharply over the twentieth century, as shown in Figure 6.1 (overleaf), and they have increased at rates in excess of population growth (Mather and Chapman, 1995). Total and per capita water consumption are generally higher in the developed world, and with economic development, relatively more water is consumed in the industrial and domestic sectors. At a global level, agriculture is currently the largest user of water, accounting for approximately 70 per cent of water withdrawals and reflecting the importance of agricultural production in developing countries. Ohlsson (1995) suggests that increases in agricultural output since the 1950s and the era of the green revolution in regions of the developing world (Chapter 10), have been 'roughly proportional to the amount of water supplied' (p. 7). Certainly, the investment in the construction of dams worldwide was most rapid at this time, as shown in Figure 6.2 (page 135). Globally, over 36,000 dams of at least 15 metres high now impound more than 5,000 cubic kilometres of freshwater for use in agricultural and industrial development (World Resources Institute, 1994).

Water resources, however, are fundamentally limited; the amount of water which can be made available to various groups of people is determined by the precipitation which falls. Augmenting water supply in one place over a particular period therefore necessarily impacts on supplies elsewhere. Although there is some indication that historical increases in per capita water consumption may be slowing with the more conservation-oriented approach in some developed nations, access to quality water supplies remains a primary factor in realising future development aspirations in many regions. This is highlighted in the example provided in Box 6.1. Indeed, water has been called 'the oil of the 1990s', referring to the centrality of the resource for future world development and regional political and economic stability. It has also been suggested that many more people will have their lives adversely affected by issues of water in the near future than was the case with respect to oil in the 1970s and 1980s (Biswas, 1993). There are no alternatives to replace water as there are with oil.

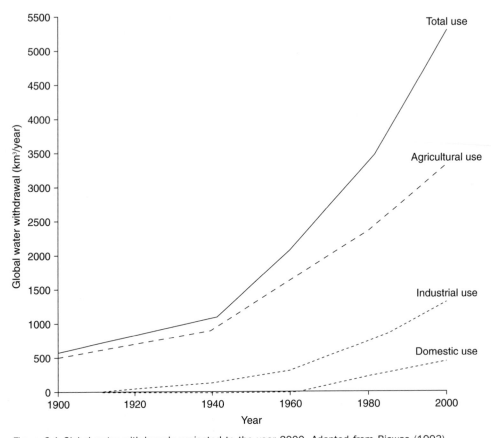

Figure 6.1 Global water withdrawals projected to the year 2000. Adapted from Biswas (1993).

Energy resources

On the global scale, there is a strong relationship between GNP per capita and energy use (Figure 6.3, page 137). In 1994 over 60 per cent of all commercial energy sources were consumed in the high-income economies as defined by the World Bank (World Bank, 1997: 228). In agricultural production specifically, the consumption of commercial energies has been fundamental to enabling increases in yields, partly through the mechanisation of production; it has also fuelled the manufacture of fertilisers and pesticides. There are strong regional variations in the use of such energy sources, however, as can be seen in Figure 6.4 (page 138).

Energy consumption is influenced by many factors including climate and the size of a country (Canada in Figure 6.3), and reliable time-series data are often hard to obtain. There is evidence, however, to suggest that the transformation of society from a predominantly agricultural and rural base to an urban and industrial base is associated with a rise in total energy consumed and an increased reliance on commercial fuels rather than traditional fuels such as wood, charcoal and other biomass sources. This pattern has led to the classification of 'high- and low-energy societies' on the basis of the level and type of energy consumed (Mather and Chapman, 1995). Despite the unexploited potential of renewable sources of energy such as hydroelectric power and nuclear energy, the newly industrialising economies rely heavily on fossil fuels and there is wider evidence in the developing world that urbanisation is accompanied by an increased dependence on imported oil. Oil prices in relation to non-oil commodity prices rose rapidly in the early 1980s, leaving many developing nations with over 25 per cent of their export earnings being used solely to finance the importation of energy needs (Barke and O'Hare, 1991). The high costs of oil imports were a major contribution to the debt problems experienced by

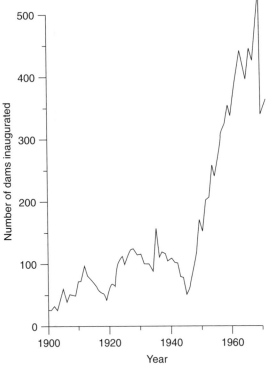

Figure 6.2 The expansion of dam construction in the twentieth century. Adapted from Mather and Chapman (1995).

many developing countries in the 1980s, leading to the current situation shown in Figure 6.5 (page 138), where debt servicing and energy imports account for the bulk of all foreign trade receipts in many areas of the developing world.

Despite some relationship between the level of economic development (as measured by GNP) and the types and quantity of energy use, this relationship is far from perfect. Economic growth need not be accompanied by increased energy usage, as shown for many developed countries by the slowing in the amount of energy consumed per unit growth in GNP (World Resources Institute, 1994). In fact, Kats (1992) argues that a failure to use energy efficiently helps explain the relatively weak economic performance of the former Eastern bloc countries in the 1980s: 'In every country where energy use is wasteful, continued high levels of investment in energy production without adequate investment in improving efficiency of energy use will slow rather than enhance development' (Kats, 1992: 266).

Mineral resources

The presence of diverse mineral supplies within a nation is a potential source of comparative advantage in economic development, as shown by the historical experience of Europe and North America. Currently,

---

BOX 6.1   Transboundary water resources in development

Although there is much progress to be made globally in the more efficient management of existing water sources to enhance supply, increasingly (and particularly in the developing world), the prospects of securing the water resources required to facilitate future development depend on the use of transboundary river and lake basins. This is exemplified by Biswas (1993: 167):

> There is no question that it is going to be an increasingly complex task to provide an adequate quantity and quality of water for various human needs. Difficult though it is going to be to institute more rational and efficient management policies and practices for water sources that are contained wholly within the geographical boundaries of individual sovereign states for a variety of interrelated technical, economical, social, institutional and political reasons, the problem is likely to be intensified by several orders of magnitude when the management and

development processes for water sources that are shared by two or more countries are considered.

Before the changes in the former Soviet Union and Eastern Europe, approximately 47 per cent of the land area of the globe lay within 214 transboundary river or lake basins, in that they were shared by two or more countries (World Resources Institute, 1992). Table 1 highlights the predominance of such systems within the developing world. Although the majority of international water bodies are shared by only two countries, there are nine river and lake basins which cut across more than six countries. Except for the Danube (twelve countries) and the Rhine (eight countries), all of these systems – the Niger, Nile, Zaire, Zambezi, Amazon, Lake Chad and the Mekong, are in the developing world.

The particular and challenging standing of Africa as host to many of the largest river systems of the

Box 6.1 continued

Table 1 Distribution of transboundary river and lake basins by region.

Region	Number of rivers and lakes extending into two or more countries	Percentage of total transboundary river and lake basins
Asia	40	19
Europe	48	22
North and Central America	33	15
South America	36	17
Africa	57	27
Total	214	100

Source: Biswas (1993).

globe is clearly evident. However, part of the explanation of the large number of international basins in Africa is due to the characteristics of the political boundaries within the continent, which were 'drawn by the European powers with scant regard even for the physical geography of Africa, let alone the Africans' (Griffiths, 1993: 66). Despite the potential for resource disputes presented by the geography of the African continent, conflicts over water have surfaced more predominantly in the more developed and faster developing regions of the world, as shown in Table 2.

There are over 2,000 treaties relating to common rivers and water basins aimed at ensuring the water needs of all parties are met (World Resources Institute, 1994). However, it is thought that in many cases these regulations are either disregarded or inadequate (Biswas, 1992). In the Middle East, the Nile Water Agreements signed in 1959 between Sudan and Egypt are regularly ignored, with Egypt exceeding her quota for water extractions on an annual basis (Lee and Bulloch, 1990). The agreements are also insufficient in that they do not accommodate the upstream needs of Ethiopia (Biswas, 1992).

The resultant struggles over limited water resources could threaten already fragile ties between states in this region. Considering the case of the Euphrates alone, Syria and Iraq nearly went to war in 1975 after Syria and Turkey filled reservoirs behind two new dams, causing a sharp drop in the level of the river (Vesiland, 1993). Iraq depends on the Euphrates and the Tigris for its water, most of which originates in Turkey and Syria. In recent years, Turkey has once again begun to harness the waters of these rivers within the Greater Anatolia Project. The project includes the great Ataturk Dam, a further twenty-two dams, and the world's two largest irrigation tunnels at 25 feet in diameter, to divert water to the Harran Plain (the land between the two great rivers which hosted the ancient agricultural civilisations of Mesopotamia). In 1990, when Turkey started

Table 2 The location of major international water disputes.

River	Countries in dispute	Issues
Nile	Egypt, Ethiopia, Sudan	Siltation, flooding, water flow and diversion
Euphrates, Tigris	Iraq, Syria, Turkey	Reduced water flow, salinisation
Jordan, Yarmuk, Litani, West Bank Aquifer	Israel, Jordan, Syria, Lebanon	Water flow diversion
Indus, Sutlei	India, Pakistan	Irrigation
Ganges-Brahmaputra	Bangladesh, India	Siltation, flooding, water flow
Salween	Burma, China	Siltation, flooding
Mekong	Cambodia, Laos, Thailand, Vietnam	Water flow, flooding
Parana	Argentina, Brazil	Dam, land inundation
Lauca	Bolivia, Chile	Dam, salinisation
Rio Grande, Colorado	Mexico, United States	Salinisation, water flow, agrochemical pollution
Rhine	France, Netherlands, Switzerland, Germany	Industrial pollution
Maas, Scheide	Belgium, Netherlands	Salinisation, industrial pollution
Elbe	Czechoslovakia, Germany	Industrial pollution
Szamos	Hungary, Romania	Industrial pollution

Source: Middleton, O'Keefe and Moyo (1993).

Box 6.1 continued

to fill the Ataturk Dam, it stopped the flow of the Euphrates and 'suddenly achieved what years of diplomacy had failed to do – the bringing together of Iraq and Syria' (Lee and Bulloch, 1990: 13). In the longer term, the Anatolia project is expected to reduce Iraq's receipt of water from the Euphrates by 90 per cent and Syria's to 60 per cent of normal flow (World Resources Institute, 1994: 183).

The problems associated with individual transboundary water systems are very country-specific. They accommodate factors including fears over national sovereignty, political sensitivities, historical grievances and national self-interest. Thus, developing international principles for the management and

control of such resources remains problematic: 'To a great extent, international organisations such as the UN system have deliberately stayed away from the issue of international rivers and lakes primarily because they have considered such issues to be politically sensitive' (Biswas, 1992: 7). Despite the lack of direct international governance of water resources, international institutions and bilateral organisations do play a role. For example, few developing countries have adequate funds for capital-intensive water development projects such that they have no alternative but to follow donor conditions which may include the adoption of international treaties and the management of needs across the whole system.

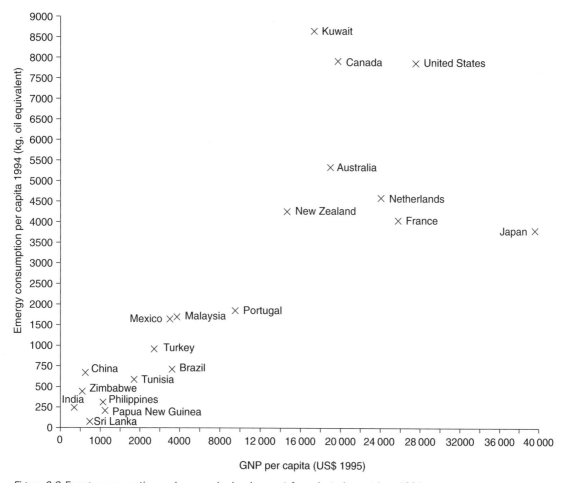

Figure 6.3 Energy consumption and economic development for selected countries, 1994.

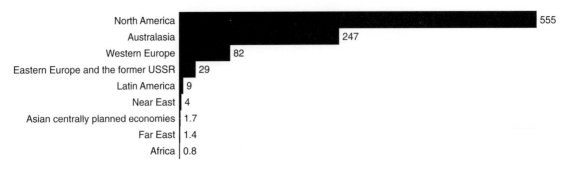

Figure 6.4 Agricultural production: the use of commercial energy per agricultural worker by major region, 10^9 joules. Adapted from Grigg (1995).

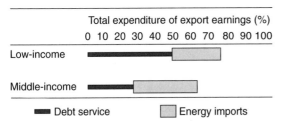

Figure 6.5 Energy imports, debt service and export earnings in developing countries. Adapted from Middleton, O'Keefe and Moyo (1993).

mineral production is extremely important in the economies of many developing nations, as shown in Figure 6.6. As well as contributing to the export earnings of a country, mineral development can assist in attracting foreign capital, in creating jobs and stimulating demand for local goods and services, in raising taxes, in prompting infrastructure developments and in providing options in a country's route to industrialisation. Yet recent evidence suggests that, since the 1960s, the 'mineral economies' of the developing world, defined as having over 40 per cent of their export earnings from the mineral sector, have not performed as well on conventional economic development indicators as was predicted or in relation to the less well-endowed countries, as shown in Table 6.1 (page 140).

Auty (1993) has proposed a 'resource-curse' thesis to explain this pattern. The suggestion is that, in practice, mineral production is highly capital-intensive and is therefore controlled by a few multinational companies that are able to raise the necessary capital. In turn, there may be little impact on national employment, production is often heavily mechanised with expatriate labour being used to provide specialist skills,

and mineral exploitation may yield only modest local production linkages due to factors including the often remote location of mineral resources and the importation of specialist technologies.

Resource constraints and the development process

The assertion that resources are fundamental to economic development has the corollary that limitations in the quantity or quality of resources will hamper these processes. This section investigates the concept of resource scarcity and, through the detailed example of water, it shows that resource constraints do operate in development. However, the nature of the 'constraint' is highly dynamic; it needs to be considered in relation to particular groups of people and specific locations; and it may owe less to the characteristics of the resource base than to issues of control, use and management of resources.

The works of Malthus, Ehrlich and Meadows continue to attract support and to be applied in contrasting socioeconomic and historical contexts, although their envisaged limits to human development due to insufficient quantities of resources to meet demands, have not materialised. 'Absolute resource scarcities' have not generally arisen as predicted; 'the fears of imminent resource exhaustion that were widely held 20 years ago are now considered to have been unfounded' (World Energy Council cited in World Resources Institute, 1994: 169). It is now recognised that 'threats to the sustainable use of resources comes as much from inequalities in peoples' access to resources and from the ways in which they use them as from the sheer numbers of people' (WCED, 1987: 95). Rather than absolute resource scarcity, it is evident

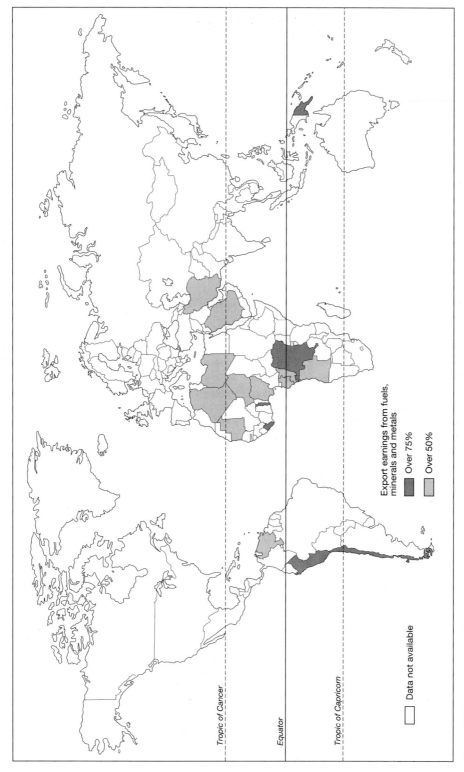

Figure 6.6 Export earnings from minerals, 1990. Adapted from Dickenson *et al.* (1995).

Table 6.1 Economic growth rates in mineral-rich and other developing countries.

	Hard-mineral exporters		Oil exporters		Other middle-income countries		Other low-income countries	
	1960–71	1971–83	1960–71	1971–83	1960–71	1971–83	1960–71	1971–83
Mean annual growth of per capita GDP (%)	2.5	−1.0	2.9	1.9	3.7	2.0	1.3	0.7
Number of countries	10	10	10	10	29	29	20	20

Source: Mather and Chapman (1995).

that particular groups of people and locations have suffered (and continue to experience) resource constraints in development relative to others. As Mather and Chapman (1995: 17) suggest, the 'habitual human condition' is that there has 'never been sufficient abundance to satisfy all human wants.'

The concept of resource scarcity becomes much more complex when applied beyond the 'stock resources' such as non-renewable minerals and fossil fuels. Indeed, the physical existence itself of a resource does not ensure its availability for development; access may be difficult or there may be a lack of sufficient capital or appropriate skills and technology to bring the location into production. In contrast, a resource which exists only in small quantities in a few locations may not be scarce if demand is low (perhaps due to issues of income levels and distribution). Furthermore, 'geopolitical' resource scarcity (Rees, 1990) may also be produced in circumstances where a resource is heavily localised and producing countries are able to restrict output and/or exports, such as prompted by the Organisation of Petroleum Exporting Countries (OPEC) during the Arab-Israeli conflict in 1973.

Rees (1990) forwards a further category of resource scarcity referring to scarce 'qualities' of resources such as with respect to attractive landscapes, wildlife or clean air. Here quality may refer to aesthetics (in the case of valued landscapes) or physical characteristics (such as the ability of the atmosphere to absorb pollutants). For these renewable resources, market forces (to date) rarely operate as they have with commodity resources, such as oil, to influence demand or to generate alternatives (Mather and Chapman, 1995). Furthermore, as seen later in this chapter, scarcity in terms of quality is in many instances increasing under the impacts of development based on other categories of resources. And resource scarcity may be shaped by factors not only of supply, but also of

demand, the economics of the market and political decision making. These ideas are now developed in relation to water resources.

Water resource 'constraints' in development

Despite the evident complexity of the factors shaping the relationship between resources and development, it is possible to identify resource constraints that are common across a number of locations and groups of people, particularly in the developing world. Within agricultural production in the tropics, it is precipitation not temperature that controls plant growth, in direct contrast to the 'boundary conditions' (Biswas, 1992) within temperate agriculture. For the majority of farmers dependent on rainfed production (as discussed in Chapter 10), precipitation is highly variable (particularly where rainfall totals are low), often localised in distribution and having a high intensity. Challenges for the development of tropical agriculture therefore include managing the variable extent and timing of rainfall and the potential implications for soil erosion and fertility presented by the high kinetic energy of rainfall.

In the arid zones of the developing world, two distinct strategies for survival evolved in response to the varied pattern of water resources over space. One is based on the low-intensity use of dispersed resources (water being of fundamental importance) such as within pastoral societies; the other is based on the intensive use of more favoured locations, such as where plentiful water resources enable permanent cultivation and higher densities of population. Although distinct, they may be operated in conjunction. Box 6.2 looks at the characteristics of the irrigation technologies and agro-ecosystems which the Marakwet peoples in Kenya developed in order to sustain occupation of the arid Kerio valley.

BOX 6.2 Indigenous irrigation and the extension of ecological margins for development: the Marakwet people of the Kerio Valley, Kenya

Adams and Anderson (1988) refer to the system of stream diversion and canal networks along the east wall of the Rift Valley in Kenya, the 'Kerio cluster', as the most complex and extensive indigenous water management system in Africa south of the Sahara. Since the nineteenth century at least, the Marakwet peoples have applied considerable engineering skills to exploit the challenging ecologies of the Rift Valley, to enable permanent settlement and the cultivation of cereals and vegetables across the valley floor.

The Kerio Valley itself is an arid area lying around 3,000 feet above sea level, receiving less than 600 mm annual rainfall and with scrub-like natural vegetation. However, less than 10 miles away, but separated by the precipitous Rift Valley escarpment, is the substantially different natural ecology of the Cherangani Plateau at 9,000 feet above sea level. This is a well-watered zone, receiving around 1500 mm of rainfall annually, supporting evergreen forests, and crossed by two major and several minor rivers and streams. Over the years, the Marakwet people have constructed dams, furrows, channels and terraces to divert and carry the water from the minor streams of the plateau to the fields of the valley floor. Earth and stone channels with simple brushwood and stone dams are used to modify the variable flow of the natural streams

and to carry water to the majority of fields between the foot of the escarpment and the east bank of the Kerio River. Irrigation furrows are also used to irrigate kitchen gardens and fields among the villages of the hillsides (Adams and Anderson, 1988). Complex systems of water allocation for the maintenance of structures have developed within the Marakwet communities (Adams, 1996).

By the 1930s, the system of irrigation and livelihood was the focus of much commentary by outsiders, including colonial officials such as the district officer in the area at that time:

> The plan of using streams at the top of the escarpment to water parched fields 3,000 feet below shows a practical imagination which has sometimes been supposed not to exist in the African. (Henning, 1941: 270)

Subsequently, there have been many different views concerning the environmental sustainability of the system and projects to modify and extend it. It is considered, however, that the current extent of cultivation and settlement along approximately 50 kilometres of the Rift Valley above the Kerio River, is at least as extensive as it was in the early nineteenth century (Adams, 1996).

In 1987 the Brundtland Commission identified a large group of the world's agricultural producers who could be considered 'resource-poor'. An estimated one-quarter of the global population live in very poor rural environments in terms of the inherent characteristics of water resources, among other things, and the acquired features of those environments (acquired through development interventions). These farmers, because they are poor, lack the financial resources to invest in the capital equipment and necessary inputs to raise production or manage the resources in these areas appropriately, yet they live in areas of the world which require precisely such levels of investment if they are not to be degraded further. In addition to what could be considered their economic and ecological marginality, these farmers are often also marginal in a political sense. For example, they may have little political power in the sense of their participation

in, and control over, institutions which structure their lives.

Similarly, in the urban sector, an estimated 600 million urban residents in the developing world live in 'life and health threatening homes and neighbourhoods' (Hardoy, Cairncross and Satterthwaite, 1990: 4). Many are classified as such due to factors of inadequate water supply and sanitation. The centrality of these issues in human health and development is confirmed in Table 6.2 (overleaf), which shows the potential impact of resource improvements on mortality from the most common causes of death. Such diseases are a function not only of the presence of disease vectors (i.e. water quality), but also the quantity of water a household can command (through public piped supply, purchase or collection) and the provisions for removal of water once used. Furthermore, water resource constraints operate in conjunction with other

Table 6.2 The potential health benefits of water supply and sanitation improvements.

A. Effects of improved water and sanitation on sickness

Disease	Millions of people affected by illness	Median reduction attributable to improvement (%)
Diarrhoea	900[a]	22
Roundworm	900	28
Guinea worm	4	76
Schistosomiasis	200	73

B. Effects of water supply and sanitation improvements on morbidity from diarrhoea

Type of improvement	Median reduction in morbidity (%)
Quality of water	16
Availability of water	25
Quality and availability of water	37
Disposal of excreta	22

[a] Refers to number of cases per year.
Source: World Bank (1992).

features of the urban environment – such as the cramped housing conditions which make for the rapid transmission of disease – to aggravate ill-health.

The close interrelationship between water resource constraints, human health and development can be illustrated through the example of the estimated 20–30 per cent of the urban population of the developing world who depend on water vendors for their supply (Hardoy, Cairncross and Satterthwaite, 1990). Those who have to buy water in this way tend to use smaller quantities and therefore may suffer compromised standards of hygiene in food preparation and washing. In the metropolitan areas of southern Brazil, members of those households without piped water were found to be almost five times as likely to die from diarrhoea as those who had such access (Hardoy, Cairncross and Satterthwaite, 1990). Water purchases may account for in excess of one-fifth of a typical household's budget, leaving less income for other basic necessities such as food, again with nutritional and health implications. Malnutrition and susceptibility to disease are strongly linked. Typically, households dependent on water vendors pay many times in excess of those who enjoy water piped by the municipality. In referring to a study in Lima, where

the price of water bought from tanker trucks was sixteen times the metered rate, Hardoy, Cairncross and Satterthwaite, (1990: 114) comment that 'it is expensive to be poor'. In this way, many urban households remain in poverty and individuals continue to lack the basic right of good health and the ability to participate in their own development.

No simple models

Change is an inherent characteristic of all livelihood systems. The nature of resource constraints in development also varies over time. Processes of change do not impact on all individuals or groups of people equally. Resource scarcity can therefore constitute a variable constraint at this scale. Continuing the example of water resource constraints, consider some recent research into the way households respond to periods of drought and the consequent prospect of food insecurity. Working in Zimbabwe, Campbell, Zinyama and Matiza, (1991) identified significant differences in the nature and sequencing of household responses to drought according to factors including socioeconomic status and length of residence in the areas under study. The poorest groups in all areas under study liquidated their assets earlier in response to food shortages. As the food deficit persisted, these groups also relied increasingly on the sale of their labour. In contrast, the wealthier groups took on alternative options to raise cash, such as the brewing of beer for sale. Irrespective of socioeconomic status, those households resident in the long-settled communal villages were able to draw upon established social networks to facilitate the redistribution of food in the early stage of a deficit situation, whereas those newly arrived in resettlement areas were more reliant on the sale of their labour and crafts. It is evident in this research that water shortage impacts most detrimentally on the poorer groups and the newly settled households who become more vulnerable to food insecurity. Further research is needed into the capacity of those households that have responded via the diversion of labour away from subsistence production to cope in the future and, critically, to become more secure in the future.

The Zimbabwean example confirms that whether or not resource conditions constitute a 'constraint' on development is defined at the level of the individual and is place-specific. This is in response to varied and dynamic political, economic and social forces in addition to factors of the environment such as meteorological drought. All shape the nature of the resource 'constraint' and the ability of certain groups in society to respond and take action to overcome such limitations. On a wider scale, and in relation to resource sectors other than water, national governments have been able to import food to overcome internal limitations of land and agricultural production. Since 1970, food imports to developing countries by volume have increased by 250 per cent (World Resources Institute, 1994). Similarly, innovations related to the mobility of energy supplies have enabled the transformation from a low-energy to high-energy society in countries such as Japan, despite the relative paucity of indigenous energy resources.

In summary, it is evident that resource constraints on development do exist; many people in the developing world experience the harsh reality of hunger, disease, pollution and hazard, and the persistence of poverty is testimony to the unresolved challenge of finding the means for overcoming continued resource constraints in development at the local level. However, it is increasingly acknowledged that the persistence of these development challenges owes less to the geographical characteristics of the natural resource base than to issues of control, use and management of resources. These factors also underpin much of the discussion of resource degradation in the following sections. However, the fundamental challenge for sustainable development in the future is in overcoming the resource constraints of the poorest groups in society, where options for development are most restricted. The examples given so far show how these groups have a very close relationship with the physical resource base, live in some of the most impoverished environments of the world, lack precisely the means required to prosper in those areas, and through their poverty may contribute to the further degradation of those resources.

Environmental impacts of development

Just as the characteristics of the resource base evidently shape the challenges and opportunities for development, development processes themselves impact on the environment and the varied functions it performs, often to their detriment. Indeed, before the late 1970s, development was portrayed by environmentalists as incompatible with conservation. Particularly in America, the undesirable side-effects of industrial development such as air and water pollution

were being experienced and an emerging environmental movement campaigning on these issues was fuelled by the generally anti-establishment middle-class sentiments which prevailed at that time (Biswas and Biswas, 1985).

The dissonance between development and the environment was further reinforced by the primacy in development thinking during the 1960s given to economic growth (Chapters 1 and 3), within which the limiting factors were considered to be finance and technology, not natural resources. Although it is now widely acknowledged that a lack of development can also be a prime factor in resource degradation, this section emphasises the ways in which the characteristics of past development patterns and processes have contributed to some of the major contemporary global environmental issues.

The environment provides a number of interrelated resource functions for the development of society: as inputs into the economic system; as a sump for the waste products of human activities, including economic production; and in terms of 'services' such as the maintenance of the gaseous composition of the atmosphere or for aesthetic pleasure and recreation. Subsequent sections have confirmed that such functions are neither discrete nor time-bound; resources may serve multiple ends and the significance of particular functions may change with use, political interest or economic developments.

Rarely do resources cease to exist in absolute terms as a result of development. Instead, they become 'degraded' in relation to the actual or possible future functions they can perform. For example, as land becomes 'reduced to a lower rank' (Blaikie and Brookfield, 1987: 1), agricultural productivity declines, requiring capital and labour inputs to rectify the situation and prevent further losses. Although absolute exhaustion of the resource may be avoided, in this case through the application of chemical fertilisers, the costs of resource degradation in terms of required remedial actions may be substantial. Furthermore, environmental impacts are not necessarily felt solely by a particular land user at that point in time. In the case of chemical fertilisers, these may contribute to the nitrification of water courses and impact on downstream users over a period of time. Indeed, Blaikie and Brookfield (1987) refer to land degradation as the 'quiet crisis'; over longer periods of time, processes of degradation make land users more vulnerable to adverse conditions such as drought. Critically, any discussion of environmental degradation needs to be not only in relation to particular resource

Plate 6.1 Deforestation in Surinam – rainforest along the Tapanahony river is cut down to make room for agriculture (photo: Ron Giling, Panos Pictures)

functions at specific times and places, but also in relation to identified interest groups:

> To a hunter or herder, the replacement of forest by savanna with a greater capacity to carry ruminants would not be perceived as degradation. Nor would forest replacement by agricultural land be seen as degradation by a colonising farmer. Usually there are a number of perceptions of physical changes of the biome on the part of actual or potential land-users. Usually too, there is conflict over the use of land. (Blaikie and Brookfield, 1987: 4)

Deforestation

The removal of forests and woodlands through cutting or deliberate fire at rates in excess of natural regeneration processes is perhaps the most obvious global pattern of resource degradation (Plate 6.1). In 1982 the Food and Agriculture Organisation (FAO) estimated the rate of tropical forest loss to be around 114,000 square kilometres per year. In 1990 the World Resources Institute reported a substantially higher figure of 204,000 square kilometres lost annually throughout the 1980s. Note that the problems of data accuracy and comparability concerning deforestation are substantial, as detailed in Figure 6.7. Grainger (1993) concludes that all estimates of tropical deforestation rates are unsatisfactory in some sense, although some

1 Lack of uniformity in the definitions of *forest* and *forest land* between different countries
2 Variations in the definition of *forest destruction* (it may include selective felling or it may be complete clearance only)
3 Differing techniques in undertaking forest inventories
4 A common concentration on commercial timbers to the exclusion of non-commercial trees and other woody plants
5 Data is often for individual countries rather than subdivided by forest type
6 Data may be withheld by governments or companies for strategic reasons

Figure 6.7 Some problems with data on 'deforestation'. Adapted from Reading, Thompson and Millington (1995).

can be considered more unsatisfactory than others. In considering the particular case of Nepal, Thompson and Warburton (1985) argue that the 'uncertainty' surrounding forest degradation must be made explicit and part of future planning and development:

> If the most pessimistic estimates are correct, the Himalaya will become as bald as a coot overnight . . . if the most optimistic estimates are correct, they will shortly sink beneath the greatest accumulation of biomass the world has ever seen. (Thompson and Warburton, 1985: 116)

Deforestation itself is not a new phenomenon. Much of Europe was cleared largely of forest in the Middle Ages, and the forest cover of North America has been reduced from an estimated 170 million hectares to 10 million hectares (1 hectare = 10,000 square metres = 0.01 square kilometres) over the period since colonisation (Goudie, 1990: 38). However, it is the contemporary rates and extent of forest removal across the humid tropics which are unprecedented and the focus of much global environmental concern at the present time.

The degree of popular and scientific interest in deforestation reflects the varied functions woodland resources currently fulfil, and are expected to perform in the future, for diverse interest groups across all spatial scales. Indeed, forests and woodlands could be considered the archetypal multiple resource in that they provide a range of raw materials, including fuelwood, fruits and medicines; they also play a vital role in the maintenance of plant and animal life support systems and global biodiversity; they modulate climate; and they perform a host of less obvious resource functions which are now acknowledged to be essential for societal well-being, including opportunities for recreational enjoyment.

There is a close association at the global level between shrinking forests and expanding areas of cropland, as can be seen in Figure 6.8. The driving factors, however, behind both deforestation and reforestation are quite poorly understood (Grainger, 1993). In particular, there is a need to distinguish between the proximate and fundamental causes of deforestation. For example, although 'farmers are probably the most significant anthropogenic influence' (Adams, 1990: 124) particularly in Africa and Latin America, the ultimate cause of deforestation may lie in a complex mix of factors that operate in any particular locality. In the Rondonia region of Brazil, where impoverished farmers have colonised areas bordering newly constructed highways and are accepted to be the principal agents in the exponential increase of deforestation rates since the mid 1970s, Colchester and Lohmann (1993) stress the underlying failure of economic development planning as fundamental. They argue that agrarian reforms

> have failed to achieve their targets, they have failed to alleviate rural poverty, they have failed to secure peasant tenure, they have failed to effect adequate redistribution of land, they have failed to stem the rising tide of landlessness and, above all, they have failed to respond to the needs and demands of the peasants themselves. (p. 15)

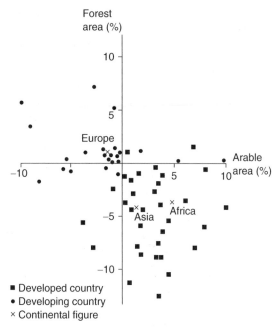

Figure 6.8 Rates of change in forest and cropland areas, 1989–90. Adapted from Mather and Chapman (1995).

In summary, the inadequacies of past development are prevalent throughout the most regularly proffered explanations of tropical deforestation, whether they are stated as the inappropriateness of government policies, the lack of agricultural skills on behalf of farmers or the penetration of capitalism and the modern debt crisis.

Soil erosion and desertification

Soil erosion is a further and related global environmental issue that arises fundamentally from resource use at rates in excess of regeneration through natural processes. Soil-forming processes occur at slow rates (a few millimetres per century) and are controlled at the local scale by highly specific factors of geology, climate and topography. Processes of soil erosion may in contrast be very rapid, perhaps several centimetres a year. Soil degradation may occur naturally as a result of the action of wind and water over time, but accelerated rates of erosion occur through the interaction of human activities with such biotic agents. The removal of tree cover exposes soil surfaces to the direct impacts of rainfall and also removes the effect of root binding on soil stability. Soil may subsequently be degraded 'quantitatively' through its physical removal from one location to another or 'qualitatively' referring to losses in fertility, in moisture and nutrient content or changes in chemical composition, in soil flora and fauna (C.J. Barrow, 1995).

Stocking (1995) argues that land degradation is the single most pressing current global problem. It is estimated that since 1945, an area roughly the size of China and India combined, has been eroded 'at least to the point where the original biotic functions are impaired' (Stocking, 1995: 223). Estimates of land damaged or lost for agricultural use through soil degradation range from 'moderate to apocalyptic' (World Bank, 1992: 55). In continuity with processes of deforestation, assessments of soil erosion also suffer from problems of data quality and comparability of measurements. It is also extremely difficult to attribute the complex processes of soil erosion to a single cause. However, there is evidence to suggest that soil degradation is getting worse on the global scale (Stocking, 1987) and human impacts on soil degradation could be expected to widen in the future through processes of acid deposition, radioactivity and pesticide pollution (C.J. Barrow, 1995).

In large measure, the physical processes of soil erosion are well understood and technical packages for conservation are available. However, soil erosion clearly persists. Blaikie (1985: 50) suggests that the most promising direction in terms of the explanation of continued soil degradation is that 'conservation is as much about social processes as physical ones and the major constraints are not technical, but social'. This work has done much to raise understanding of the individual land user and the wider political economic context in which resource management decisions have to be taken. For example, many colonial interventions in African agriculture were justified on the basis of the conservation of soil. Measures such as compulsory construction of contour banks and limitations on the number of stock owned, however, may have done more to protect White settler interests in agriculture in southern Africa than they did to prevent soil erosion (Elliott, 1990). Many soil conservation programmes in developing countries are now delivered through foreign aid institutions within which the implicit assumptions of this 'colonial model' may persist (Blaikie, 1985).

Soil erosion is one component of the wider processes of desertification; desertification occurs wherever

land is periodically deprived of adequate moisture, where soils are infertile, or poor drainage leads to saline conditions, crusts or pans, if vegetation is

present it will probably be easy to disturb and slow to re-establish. Because the end product often resembles desert the process has been termed desertification. (Barrow, 1995: 105)

In continuity with soil erosion, desertification may be triggered through natural processes, such as drought or the actions of wild animals in destroying vegetation, e.g. rabbits, termites or locusts. However, human actions most regularly introduce or speed up these processes, fundamentally through resource use in drylands at rates in excess of natural processes of soil, moisture and vegetation regeneration. Population pressure, overgrazing and deforestation are the most widely proposed explanations of desertification, but as already seen, these processes themselves may reflect deeper underlying forces of change.

In 1977 the United Nations Environment Programme (UNEP) organised a World Conference on Desertification (UNCOD) in Nairobi, which was instrumental in forwarding desertification as a major contemporary environmental issue. It was suggested that 35 per cent of the world's land surface (D.S.G. Thomas, 1993) and one-sixth of the world's population (C.J. Barrow, 1995), were at risk from desertification. Since this landmark conference, there have been substantial efforts to assess and address the problem of desertification (Plate 6.2). For example, there is now a consensus on the definition of desertification, an agreed assessment database and the Convention to Combat Desertification was signed in 1995 by 115 countries (UNEP, 1997).

Between UNCOD and UNCED (United Nations Conference on Environment and Development), images of advancing deserts have been refined, particularly as improved understanding of dryland ecologies has forced a distinction between natural fluctuations and long-term degradation. It has also been recognised that many actions to combat desertification had been costly, focused on technical interventions, were largely initiated by the international aid community, and were rarely sustained beyond the initial donor input stage (E. Dowdeswell, undated). At UNCED, often known as the Earth Summit, and within the subsequent convention, the social dimensions of desertification (which had been almost totally lacking in the plan of action which emerged from UNCOD), are now considered to be of paramount importance. Although the centrality of effective actions on behalf of the international community remains 'equal to that demanded by global warming, the destruction of the ozone layer and the loss of biodiversity' (A. Hentati, undated: 7),

Plate 6.2 Anti-desertification scheme – building a digue (barrage) to hold water in the river bed during the rainy season to irrigate nearby crops, near Timbuktu, Mali. (photo: Jeremy Hartley, Panos Pictures)

the convention also contains innovative plans for the promotion of participatory, decentralised policies and projects to assist the most marginalised groups in sustaining themselves in dryland areas.

Global warming

Past development processes and patterns have also impacted negatively on the ability of the environment to absorb the waste products of those activities. In so doing, a further function of the environment – maintaining key life support systems – is often impaired. Up to a certain threshold, natural sinks such as the atmosphere, oceans, vegetation and soils of the earth may be able to absorb gases, particulate matter and chemicals, which are created as waste in production processes. Beyond that threshold, however, pollution effects may occur, whereby the presence of those substances may compromise that sink function in the future and concurrently impact negatively on human health and the operation of plant and animal systems.

The issue   One of the most fundamental sink functions of the natural environment is in respect to the carbon cycle, within which oxygen is emitted through processes of photosynthesis and respiration as carbon

is circulated between the atmosphere, oceans, soils and land vegetation. The amount of carbon dioxide in the atmosphere is a critical determinant of global surface temperature. The reradiation of solar radiation back into space (i.e. the amount of heat lost from the earth's surface) is controlled by the insulating effect of 'greenhouse gases' in the atmosphere, in the presence of water vapour. These gases include carbon dioxide and also methane, nitrous oxide and a range of chlorofluorocarbons (CFCs). If there were no such insulation or 'greenhouse effect' from the atmosphere, the average temperature of the earth's surface would be at least 30 °C cooler than at present (Kelly and Granich, 1995: 77). However, the average temperature of the earth's surface has risen by approximately 0.5 °C over the last century (Reading, Thompson and Millington, 1995) as concentrations of these warming agents in the atmosphere have increased. This 'global warming' has been referred to as 'the major environmental challenge facing humanity over the next 100 years' (Hill *et al.*, 1995: 83). Figure 6.9 (overleaf) refers to the substantial debate which persists regarding the directions of climate change. However, the 'warming consensus' is for global mean temperature to rise by 3 °C in the next century (Reading, Thompson and Millington, 1995: 353).

Figure 6.9 Debate continues over the direction of global climate change. Cartoon by David Hughes.

Global warming is a 'supranational' environmental issue in that the causes and impacts extend across all national boundaries to include all peoples and environments: 'A molecule of greenhouse gas emitted anywhere becomes everyone's business' (Clayton, 1995: 110). The explanation of global warming illustrates clearly how the sink function of the environment has been overloaded via the destruction of vegetation, the production of pollutants at rates in excess of those which can be rendered harmless by natural processes and through the creation of artificial substances which cannot be so absorbed. Furthermore, an increase in the earth's surface temperature threatens many of the life-support functions which the environment serves through processes such as the disruption of sea levels and ocean currents and alterations of the world's major biomes and agricultural potentials, as discussed below.

Causes   Carbon dioxide currently accommodates approximately 55 per cent of the warming or 'forcing' effect (C.J. Barrow, 1995). Natural 'sinks' such as the oceans and green plants are thought to absorb approximately half of the carbon dioxide produced by human activities (Foley, 1991). Increasingly, however, additional carbon is transferred into the atmosphere more quickly than these natural processes of diffusion and photosynthesis can remove it. In consequence, the proportion of carbon dioxide in the atmosphere rises. Although the proportion of carbon dioxide in the atmosphere has varied substantially over geologic time, there has been an unprecedented 25 per cent increase in carbon dioxide concentrations in the last hundred years (World Resources Institute, 1990).

The major source of carbon dioxide production to date has been the burning of fossil fuels which currently transfers approximately 5 billion tonnes of carbon every year into the atmosphere (Foley, 1991). Development processes globally have required a progressive increase in per capita consumption of coal, oil and natural gas, as noted earlier in this chapter. The United States is the largest gross energy consumer and among the highest in per capita terms (Figure 6.3, page 137). In contrast, in many developing countries, in excess of 90 per cent of total energy needs may be supplied via non-commercial sources, as shown in Table 6.3. And the United States currently has the highest greenhouse gas emissions from all sources (World Resources Institute, 1994: 201). However, the energy-related carbon dioxide emissions from the developed world increased by only 28 per cent over the period 1972–1990 in comparison to 82

Table 6.3 Percentage of total energy supplied via fuelwood sources in selected developing countries.

Ethiopia	93
Bangladesh	80
Burkina Faso	94
Nepal	97
Peru	34
India	36
Brazil	33
Nigeria	82
Kenya	70
Thailand	63

Note: Of the commercial energy used in the developing world, 45 per cent is consumed in only three countries, India, China and Brazil; in contrast, less than 3 per cent is used on the African continent.
Source: Soussan (1988).

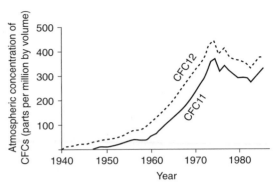

Figure 6.10 Atmospheric concentrations of CFCs. Adapted from R.D. Thompson (1992).

per cent over the same period in the developing world (United States Department of Energy, 1994). The OECD countries have accounted for a declining share of world energy consumption in recent decades, falling from 65 per cent in 1973 to less than 50 per cent in the 1990s (Mather and Chapman, 1995: 141). This is in part due to the rising consumption in the newly industrialising countries in particular, but also due to the increased efficiency in energy use in many developed countries, as already noted (Mather and Chapman, 1995).

Deforestation is estimated by the Intergovernmental Panel on Climatic Change to account for approximately 10–15 per cent of the recent increase in atmospheric carbon dioxide (IPCC, 1990). The loss of tree cover itself reduces the uptake of carbon dioxide from the atmosphere (i.e. compromises the sink function of the environment), and the associated burning of logs and biomass sources contributes further to the production of carbon through oxidation processes. The continued reliance in many developing countries on fuelwood and biomass sources of energy for national development and for daily survival of the majority of the people in these regions, is evidently in itself a major factor in deforestation and in global warming. Other aspects of past development patterns in the developing world further compound these processes. The felling and export of timber to secure foreign exchange to finance national development and service debts; regional development strategies which encourage large-scale woodland clearance such as to support ranching, often owned and operated by multinational companies; and development policies which have

failed to provide more sustainable livelihood options for small-scale cultivators; all three are factors in the rising contribution of the developing nations to global carbon emissions via deforestation.

In the future, the forcing contribution of carbon dioxide is likely to decline. For example, chlorofluro-carbons (CFCs) may be up to 100,000 times more effective than carbon dioxide in forcing global warming. CFCs came into widespread use after the Second World War and are primarily emitted by the industrial world (Foley, 1991). Atmospheric concentrations of the two most commonly used CFCs have risen from zero in 1940 to 320,000 tonnes in 1985, as shown in Figure 6.10. Annual rates of increase up to the Montreal Protocol were an estimated 5 per cent (Foley, 1991). CFCs are used for purposes including refrigeration, air-conditioning, aerosol propellants, insulation foams, the cleaning of electronic components and fire-fighting gases. Although global production of CFCs has been declining in the 1990s (Brown, 1996a), they remain active in the atmosphere for periods up to 75 years in the case of CFC 11 and 110 years for CFC 12 (R.D. Thompson, 1992: 70); eventually they are destroyed by ultraviolet radiation in the stratosphere.

CFCs are also the primary cause of ozone depletion in the upper atmosphere, which leads to the exposure of plant and animal systems to the damaging effects of increased ultraviolet radiation. Not only, therefore, have global industrial production processes been a large consumer of fossil fuels and a major producer of carbon dioxide, they have impacted on the environment through the development of synthetic chemicals such as CFCs. But there is also the prospect that industrial developments will produce less environmentally damaging substitutes for these chemicals, such as has occurred with aerosol propellants.

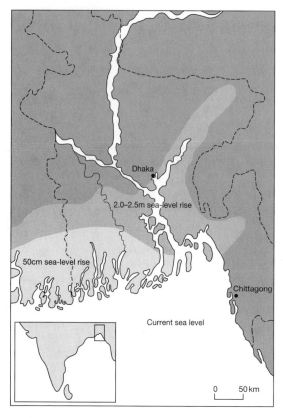

Dhaka

2.0–2.5m sea-level rise

50cm sea-level rise

Chittagong

Current sea level

0   50 km

Figure 6.11 The inundation of Bangladesh under proposed sea-level changes. Adapted from Reading, Thompson and Millington (1995).

Impacts   The major impacts of global warming are likely to relate to water resources. Sea level could rise by about 5 millimetres per year in future (compared to the current rate of 1 millimetre per year) as a result of the thermal expansion of the oceans, with the increased temperatures and the melting of mountain glaciers and ice sheets (Reading, Thompson and Millington, 1995). An estimated 10 million of the world's population live less than 3 metres above sea level (Hill *et al.*, 1995: 84). Countries such as Bangladesh and the Netherlands therefore will be particularly vulnerable to the inundation of lands and the physical loss of property resulting from global warming. The Dutch have already initiated a huge and costly programme of coastal defence construction to protect themselves from a possible 50-centimetre rise in sea level (C.J. Barrow, 1995). A 50-centimetre rise is also the 'best-case' scenario currently predicted by the Intergovernmental Panel on Climate Change for Bangladesh. A sea-level rise of this magnitude would lead to substantial inundation, displacement of people and the loss of farmland in Bangladesh. Figure 6.11 illustrates the spatial extent of such impacts under an even more serious scenario of a 250-centimetre rise.

Global warming will also impact on water availability for human use. On average, the world will be wetter in future by between 3 and 15 per cent (Clayton, 1995) due to greater evaporation from the warmer seas. However, there is substantial uncertainty as to how the enhanced rainfall will be distributed over time and space, or the resultant likely impact on agricultural potential. Two of the most likely outcomes are expected to be increased aridity in the deserts and desert margins, and drier summers in the temperate latitudes (Parry, 1990) where water availability is already a serious and costly issue. The contamination of groundwater as a result of sea-level rise is likely to compromise water supplies on small islands in particular.

Urban atmospheric pollution

Climate warming is a major atmospheric pollution problem which stems from the production of substances in quantities in excess of their absorption by natural processes with implications for the functioning of the earth as a whole. Additional atmospheric pollution problems such as urban air pollution may have quite local (largely industrial and transport-related) sources and localised impacts, but are global environmental issues in the sense that all cities tend to experience similar problems. For example, in 1993 every 'megacity' in the world surveyed as part of a UNEP and WHO programme had at least one major pollutant which exceeded the threshold, determined by the World Health Organisation, above which deleterious effects on human health could be expected (UNEP/WHO, 1993) (Plate 6.3).

Sulphur dioxide is probably the most well-known urban pollutant, with most industrial cities having been subject to sulphurous smogs at some point (Elsom, 1996). An estimated one billion urban residents worldwide are exposed currently to levels of sulphur dioxide in excess of WHO guidelines, many of them in developing nations (World Bank, 1992). The key health effects are the impairment of respiratory function (including the aggravation and/or causation of bronchitis, asthma and emphysema) and the subsequent strain on the heart, leading to premature death. The primary sources of sulphur dioxide production are the smelting of metallic ores, the burning of coal and oil

Plate 6.3 Smog enveloping Mexico City: air pollution has been estimated at 100 times above acceptable levels – equivalent to smoking 40 cigarettes a day. (photo: Mark Edwards, Still Pictures)

in power production and heating, and in transport. Increasingly, over the last two decades, cities in the developing world have larger concentrations of sulphur dioxide than in the more developed nations, where these emissions are falling (World Bank, 1992). The World Health Organisation recommends that sulphur dioxide exposures should not exceed an average of 40–60 micrograms per cubic metre over the course of a year. However, many cities far exceed this average, as shown in Table 6.4.

In an attempt to reduce the localised health impacts of sulphur dioxide pollution at the ground level, some of the measures taken have contributed to the wider pollution problem known as acid rain. This phenomenon includes the dry deposition of sulphur and nitrogen oxides onto damp surfaces, thereby converting to sulphuric and nitric acids, and the precipitation of acids formed in the atmosphere through similar processes of oxidation. As emissions of sulphur dioxide have increased and strategies such as the building of taller stacks have been implemented, pollutants have been placed into the larger atmospheric circulation system for longer periods, causing major problems in the natural environments often at substantial distances from the source of the pollutant. Drifting pollution from Europe was linked to acidification of Scandinavian water bodies in the 1960s.

Table 6.4 Sulphur dioxide pollution in selected cities.

City	Site years[a]	Mean (μg)[b]
Cairo	12	41.7
São Paulo	10	36.6
Caracas	2	24.3
Beijing	8	86.6
Guangzhou	8	45.5
Shanghai	8	63.6
Bangkok	2	1224.1

[a] Site years = the number of sites multiplied by the number of years in operation during recent years.
[b] Data are presented as the annual mean for each pollutant during the years of observation; μg = microgram.
Source: World Resources Institute (1996).

The average annual precipitation in Britain in the early 1980s was estimated to be between pH 4.5 and 4.2 in comparison to the 'normal level' for 'naturally acid' rainfall of pH 5.0. The impacts of acid rain extend from damage to fish species, through the disruption of soil nutrient cycles, and as a result of the liberation of heavy metals from soils and bedrock, to possible human health impacts.

Biodiversity loss

Prompted largely through Western environmental thinking, it is now appreciated that the environment performs new resource functions, including protecting the global commons and providing amenity services. There is evidence that development processes are also degrading these functions. For example, the first comprehensive global assessment of biodiversity undertaken by UNEP concluded that between 5 and 20 per cent of some groups of animal and plant species are threatened with extinction in the near future (World Resources Institute, 1996: 247). However, ascertaining the resultant loss of quality or value of these 'resources' is very problematic.

For example, diversity within the biosphere can be considered in terms of ecosystems, species and genetic material (Mather and Chapman, 1995). Development processes have certainly depended on the manipulation of the gene pool of species considered useful to societies. The particular case of high-yielding varieties in agricultural development is discussed in Chapter 10. Earlier sections of this chapter have also detailed the substantial modifications of habitat which have occurred with 'development'. The capacity of human society to impact on biodiversity has increased with economic development; current rates of species extinction are thought to exceed current rates of species emergence by over a million times (Mather and Chapman, 1995: 120). The impacts of such biodiversity losses are hard to determine since relatively little is known about ecosystems themselves, but also because the potential value of diversity in ecosystems, species and genetic materials, even solely within future agricultural production, is impossible to quantify.

The primary cause of loss of biodiversity has been the intensification of agriculture. This includes the manipulation of the gene pool of species, the increased use of pesticides, the overcropping of animal species such as through hunting, and the modification and loss of habitat with the encroachment of agriculture. For example, the food supply for over 85 per cent of the global population is based on only twenty plant species (Mather and Chapman, 1995: 119). Furthermore, the development of high-yielding cultivated varieties has led to a loss in genetic diversity and a vulnerability to predation by bacteria or aphids which, 'unlike their genetically uniform prey, are constantly evolving' (Murray, 1995: 22). As an illustration, 70 per cent of all rice grown in Indonesia is descended from a single maternal plant (Mather and Chapman,

1995: 119). Although pesticide use to combat such vulnerability to predation has increased rapidly in agriculture, particularly in the developing world (Mather and Chapman, 1995), crop losses have remained constant over the same period, as an ever increasing number of pest species develop resistance to known pesticides. Further threats to genetic and species diversity are caused by development processes which have commoditised species for purposes other than food production; for example, rhino horn is commoditised as an aphrodisiac and exotic species are commoditised as pets.

However, it is the modification and loss of habitat, especially forests and wetlands, resulting from the conversion of lands for agricultural production, that has been the major factor in the loss of species diversity. Tropical forests are particularly 'species diverse'; they occupy less than 7 per cent of the earth's surface but account for between 50 and 90 per cent of all known plant and animal species (Mather and Chapman, 1995: 122). Yet development, as earlier identified, has accelerated the processes of tropical deforestation during the twentieth century. And wetland habitats worldwide are increasingly under threat from agricultural intensification, but also through urban developments. 'Notwithstanding the wide gaps in knowledge, there are many indications that marine biodiversity is in real trouble' (World Resources Institute, 1996: 249).

Wetlands include areas of marsh, fen, peatlands, swamp forests, estuaries and coastal zones. They provide many resource functions, including direct use value in terms of supporting agriculture, fishing, tourism and transport. Wetlands also have indirect hydrological uses, including groundwater recharge and discharge, sediment trapping and flood protection (Hughes, 1992). Their biodiversity function is in the diverse habitats supported for aquatic species, waterfowl and other wildlife. It is estimated there are sixty species of trees and shrubs and over two thousand species of fish, invertebrates and epiphytic plants that depend for their survival on the wetland environments of mangrove swamps alone (Maltby, 1986). However, over half of the world's mangrove forests may already have been destroyed. Table 6.5 highlights the extent of this loss in selected countries.

Riparian and coastal zones have long been favoured areas for human populations and settlement, particularly in arid zones, as illustrated in Box 6.2. Distinct agricultural systems have evolved to exploit the productive habitats of wetlands such as in aquaculture, recessional cultivation and wet rice cropping. Many modern agricultural systems involve substantial modi-

Table 6.5 Mangrove extent and loss in selected countries.

Country	Current extent (000 hectares)	Approximate loss (%)	Period covered
India	100–700	50	1963–77
Peninsular Malaysia	98.3	17	1965–85
Philippines	140+	70	1920–90
Singapore	0.5–0.6	20–30+	Pre-agricultural period to present
Thailand	196.4–268.7	25	1979–87
Vietnam	200	50	1943–90
Puerto Rico	6.5	75	Precolonial to present
Ecuador	117+	30+	Pre-agricultural to present
Guatemala	16	30+	1965–90
Cameroon	306	40	Pre-agricultural to mid 1980s
Kenya	53.0–61.6	70	Pre-agricultural to mid 1980s
Guinea-Bissau	236.6	75+	Pre-agricultural to mid 1980s
Liberia	20	70	Pre-agricultural to mid 1980s

Source: World Resources Institute (1996).

fication of natural ecologies and investment in drainage and irrigation works, as illustrated in the case of the fenlands of eastern England. Forces of environmental disturbance to wetlands in developing countries, including the intensification of wet rice cultivation in Asia and wider engineering works, are discussed in Chapter 10.

One of the most powerful contemporary forces of wetland degradation globally is the encroachment of urban, industrial and infrastructural developments. Riverside and coastal locations are highly valued sites for housing and tourism developments throughout the world and are exhibiting some of the highest levels of urban growth (World Resources Institute, 1996). Pressure for land generally is an important factor. In Singapore an additional 6,000 hectares of land have been created by filling along the shoreline, converting habitats and altering water flows and causing siltation. In eastern Calcutta, 4,000 hectares of inland lagoons have been filled to provide homes for middle-class families at the expense of land and livelihood for tenant farmers. In São Paulo an estimated one million squatters now live in protected watershed areas (World Resources Institute, 1996).

Box 6.3 highlights the varied environmental impacts, including loss of habitat, of developments in the Far East based on golf tourism. Tourism is an example of a further new resource function currently ascribed to the environment and which is increasingly promoted as a route to development in developing countries (Chapter 4). In the case of other forms of tourism, such as wildlife tourism, rare species of plants and animals are given value by mainly inter-national visitors who may threaten, quite directly through their activities, the very resource on which such tourism development depends (Cater, 1992).

The search for sustainable resource management

It is quite evident on the basis of the processes and patterns discussed in this chapter that development in the future is intrinsically related to the environment. Although the concept of sustainable development remains contested, the central tenet of sustainability in practice is this interrelationship of environment and development: development challenges are environmental challenges and one end will not be achieved without the other. However, there are no blueprints for sustainable development in the future.

This chapter has also shown that the nature of the challenge to reconcile environment and development in the future is necessarily highly specific in time and space. For example, it was seen that conceptions of the environment, of resources and of their value to society are constantly being redefined as development occurs. It has also been clear that the significance and implications of resource constraints, scarcity and degradation are similarly highly conjunctural at the local level.

Although the nature of the challenge may be highly place-specific, these challenges and opportunities of sustainable development depend on actions throughout the hierarchy of societal organisation. The consideration of a number of environmental issues of

BOX 6.3   Golf-course development in the Far East: the environmental impacts of tourism development

From its origins as a game for idle amusement among King James and his courtiers in Scotland in the fifteenth century, golf has expanded to provide a source of enjoyment for an estimated 50 million people worldwide (Pleumarom, 1992). Golf is also big business, being worth approximately $27 billion in the United States alone (Wheat, 1993). Holidays centred on golf are now an important segment of the tourism industry (F. Pearce, 1993). In Asia, golf is the fastest-growing sport (Hiebert, 1993). In the early 1990s in Thailand, new golf-courses were being built at a rate of one every ten days (F. Pearce, 1993: 32) with an estimated 100,000 tourists playing on Thailand's courses every year (Wheat, 1993).

Much of the investment in golf in Asia has come from the four Tiger economies (investing beyond their own borders) and from Europe, Japan and the United States. Japan has an estimated 12 million golfers (twice as many as in the United Kingdom) and seventeen hundred courses (F. Pearce, 1993) (Plate 6.4). However, high membership fees, long waiting lists, the need to book rounds months in advance and the emerging opposition to proposed new courses in Japan by local environmental groups, has led to such developments overseas to cater for the Japanese demand for golf. The high costs of golf-course design, construction and maintenance, ensure that the majority of golf projects in Asia are owned by foreigners and the remainder by a local elite. In the early 1990s, half of Indonesia's golf courses were owned by the then-President Suharto and other member of his family (M. Williams, 1994).

The rapid expansion of golf has not gone unchallenged: 'Golf courses are emerging as one of the most environmentally rapacious and socially divisive forms of tourist and property development' (F. Pearce, 1993: 33). Golf-course developments in Thailand consume on average over 100 hectares of land compared to 64 hectares in Europe (Pleumarom, 1992). It is the associated hotel developments, residential plots, department stores, sports complexes and other recreational facilities which take up the space, but which are often essential to the economic viability of the development. Of the golf-courses now in operation or under construction in Thailand, Pleumarom (1992) suggests that 65–75 per cent are located on agricultural land and up to 25 per cent on classified forest land.

Not only do golf-course developments displace other land uses, but there is concern as to the manner in which such land-use changes are effected. Japanese golf-course construction methods have been described as being of the 'destroy and rebuild' (M. Williams, 1994) or 'scorched earth' (Pleumarom, 1992) variety, whereby landscapes are flattened and recontoured from scratch to provide wide fairways, limited rough and sculptured greens (very much in common with the American style). The large-scale removal of soil, the creation of artificial lakes, the flattening of slopes, the destruction of forests and wetlands and the removal of natural vegetation all change the way water and soil move within the overall system.

Golf-courses in Thailand use between 3 million and 6.5 million litres of water per day (Pleumarom, 1992; Wheat, 1993; Traisawasdichai, 1995). Such levels of consumption would satisfy the domestic needs of around 60,000 rural villagers over the same period (Traisawasdichai, 1995: 17). In 1994 Thailand experienced its worst ever drought year, in which farmers were prohibited by the government from growing a second rice crop to conserve water. Despite this, golf-courses were able to continue pumping water from the reservoirs (Traisawasdichai, 1995). It is also known that, when government restrictions are attempted, golf-courses illegally divert water from irrigation channels and draw off water at night. Not only, therefore, does golf-course development compete with other uses for scarce water supplies, it may also contribute to the degradation of sources, through the pumping of underground water and in other ways.

The environmental impacts of the heavy use of pesticides and fertilisers required to maintain this style of golf-course are now becoming known. Although there is no data for Thai courses, an average of 2 tonnes of pesticides annually are applied to each course in Japan, twice what is applied to agricultural lands (Pleumarom, 1992: 107) Only half of these pesticides remain in the soil, the rest lingering on grass and trees or being dispersed into the air. Caddies, greenkeepers and players are all

Box 6.3 continued

Plate 6.4 A Tokyo golf-driving range at night (photo: Simon Bruty, Allsport)

therefore exposed to this pollution as well as the pollution created by other chemicals used in golf-course maintenance, e.g. soil coagulants that are used to hold water in artificial lakes or preparations to make the turf look greener. A survey conducted by Japan's Medical and Dental Practitioners Organisation showed that 40 per cent of almost 1,500 agricultural poisoning cases treated nationwide involved amateur golfers, or people working at or living near golf-courses (Wheat, 1993: 12).

There are also gendered aspects to the social disruption caused by golf-course developments. In the Thai situation, the sex trade has a long association with tourism generally, and golf tourism does little to challenge the status quo. Most golf caddies are girls and the 'services' they provide are used to promote golf-courses to the visitor (Traisawasdichai, 1995). Thai caddies have been sent to Japan to learn the 'finer arts of etiquette' and the women themselves recognise that it is not solely advice and club carrying which the golfers value; according to one Thai caddie, 'beautiful women who are good at talking and pleasing their customers can make a lot of money' (Pleumarom, 1992: 107).

global concern has shown how ideas about resource use and development may be shaped to a greater extent by forces and events at substantial distances from a particular society and its local environment. Many resources are now increasingly mobile through technological developments in exploitation, transport and communications; and through similar processes of development, the potential of human actions to impact on environments currently extends far beyond the site of production or use.

The analyses presented in this chapter have confirmed a number of core challenges for resource management and alluded to some of the ways in which new processes and patterns of development need to be found. Addressing the welfare needs of the poor is essential; the persistence of poverty restricts the power of national governments and individuals to engage in development activities, to avoid degradation of resources and ill-health and to rectify existing environmental damage. Inequalities in access to resources, at all scales and across sectors such as land and energy, simultaneously inhibit the incentive to reuse, recycle or conserve on behalf of those who control such resources; and they restrict the options of marginalised people in avoiding actions which further degrade the resource base on which they depend for immediate survival. In such ways, these processes lead to highly contrasting and plural geographies of development.

Fundamentally, the burgeoning literature on sustainable development has been important in rejecting ideas of any simple, deterministic link between resources and development. Part III provides insight as to how these challenges are being variously taken up, rejected, moulded or compromised by the activities of different organisations involved in influencing development policy and action within particular places and spaces.

CHAPTER 7

Institutions, communities and development

Introduction

Decision making in development is undertaken by various individuals, agents, bodies and organisations. Matching the resources for development (Chapters 5 and 6) and the ideologies of development (Part I) is perhaps most overtly undertaken by governments in the definition and implementation of policy and planning at a national level. Indeed, the state has traditionally been a primary institution or organisation of development in the developing world. However, the outcomes of those government decisions are 'filtered' (Messkoub, 1992) through additional varied institutions, in that these policies depend on the actions of many more individuals operating, for example, as part of a wider society. Whilst governments 'propose', people and institutions 'dispose' (Messkoub, 1992: 186). Household and gender relations are two important forms of institution (although less tangible than the state) that structure the lifestyles and livelihoods of individuals; they are extremely varied over time and space (Women and Geography Study Group, 1997; Pearson, 1992). The more formal organisations at all scales, from international institutions such as the United Nations to grazing associations or borehole syndicates, are also of prime importance in that it is around these organisations that the 'cultural norms or rules, behaviour, values and expectations revolve' (Dooge, 1992: 283). Evidently, institutions across these scales are not neutral factors in the development process; 'they represent values, which in turn represent the interests of some political or social group' (Sharp, 1992: 55). This chapter considers key patterns and trends in the functioning of a range of institutions operating to influence the nature and impact of specific development interventions and the development process more widely.

In the post Cold War era of the 1990s, institutions are receiving substantial interest in many fields of research and development. At the international level,

the number of institutions has increased rapidly in the last fifteen years, particularly in the environmental area (World Resources Institute, 1994). Existing institutions are also becoming subject to critical evaluation, especially those of the state, in terms of their capacity to address social and economic needs. Furthermore, there is an emerging debate concerning the relevance of specific institutional forms in relation to others and what needs are best fulfilled by a particular organisation.

Within this chapter, three specific but interrelated forces are considered to be directing such institutional developments. Firstly, the global search for sustainable patterns and processes of development from the late 1970s onwards has brought environmental issues into both international relations and to the understanding of, and response to, the 'grass roots realities' of the most impoverished groups in developing societies (Chambers, 1983, 1993). The concept of sustainability is built on the fundamental interdependence of development and environmental conservation at all scales of analysis and the requirement for new norms of behaviour within institutions across all spheres of human activity (Elliott, 1994). Secondly, recession within the global economy in the 1980s has demanded a worldwide reassessment of the role of particular institutions and the relationships between various organisations in the search for ways to ease debt, to stimulate economic growth and to make the most effective use of financial and human resources in the delivery of products and services. Responses have included changes in the lending practices of major multilateral organisations such as the World Bank and within Official Development Assistance programmes of bilateral donors. The conditions for receipt of development finance now regularly include requirements for institutional change within the borrowing nations. New relationships are also being fostered between multilateral organisations and institutions such as voluntary organisations, in order to enhance the development impact of project and policy lending, and in turn to promote stability in the global monetary system.

Finally, although the 1990s have witnessed much progress towards democracy, as evidenced by the overthrow of dictators such as in the former Yugoslavia, the majority of the population of the developing world continue to live under non-elected governments, in the absence of a functioning judiciary or civilian control of the military, and without basic rights such as freedom of speech and information (Chapter 5). Furthermore, progress towards democracy in terms of making institutions more responsive to public needs and more responsible for their own actions (as well as in terms of securing citizens' access to decision making and power) are increasingly stated by institutions themselves as the means through which their objectives in development will be met. For example, the World Bank referred to the near impossibility of economic development in sub-Saharan Africa in the absence of 'good governance' (World Bank, 1989) referring largely to 'efficient' institutional mechanisms to formulate and implement policy (Sharp, 1992) but including the basic foundations of democratic freedoms and the rule of law (Righter, 1995).

Development assistance in the 1990s is increasingly being channelled not through national governments, but via diverse non-governmental organisations (NGOs) and it is within this sector that the most radical agendas and strategies for community participation and empowerment as 'the key to the poor's struggle for equity, human rights and democracy' (Craig and Mayo, 1995: 2) have been set. Fostering people's participation is now central to the agenda of the major development institutions, as noted in Chapter 1, and NGOs are seen as the critical facilitators of an alternative 'people-centred' development.

This chapter assesses the nature and activities of a number of institutions prominent in the development process in terms of these forces of international change. Although they are considered independently, the interdependence of these three forces for institutional development is highlighted throughout and it is perhaps at these junctures that the key challenges and uncertainties for future development lie. For example, there is concern as to whether democratic governments will be more or less able to commit themselves to long-term sustainable development strategies due to issues of short-term electoral opportunism.

The rise of global governance

International institutions are created by states, usually as a 'means of achieving collective objectives that could not be accomplished by acting individually' (Werksman, 1996: xii). At the end of the Second World War, the challenges of the resurrection of the global economy (of avoiding a return to the national protectionism which created the Great Depression of the 1930s) and of reconstruction in Europe in particular, prompted the creation of new forums such as the United Nations and the International Bank for Reconstruction and Development (World Bank) for the development and coordination of international efforts towards preserving peace, resolving conflicts and promoting social and economic development. Fifty years on, the reconstruction of Europe is evident and peace has been retained in that region. However, poverty and armed conflict persist in many areas of the developing world, and in the 1990s there are also new challenges for international institutions, including the need to resolve development aspirations across the globe within the limits of natural resources and the environment (Chapter 6).

International institutions are highly varied in terms of their shape and function, the rules and practices concerning their activities and their power to influence the behaviour of other organisations. Fundamentally, the pursuit of collective goods, including conservation of the global environment, demands some devolution of sovereign power, but the capacity of international institutions to take on characteristics and powers distinct from the states which created them depends on the willingness of those states to make such investments (Werksman, 1996). This provides the context for any critical analysis of the operation and development outcomes of these institutions. This chapter briefly considers the origins and current functioning of a number of institutions at this scale and the influence on these organisations of the three global forces of change identified above.

The United Nations system

Disillusion with the UN has become one of the clichés of our time. The reality is much more complex. (Ignatieff, 1995)

Everyone expects everything of it. (Evans, 1993: 24)

Few would question the need for reform of the UN, crafted in another era, haunted by recent failures in Somalia and former Yugoslavia and facing grave financial woes. (Littlejohns and Silber, 1997)

The Charter of the United Nations (UN) was signed in San Francisco in June 1945 by 51 countries as the successor organisation to the League of Nations created

Plate 7.1 Repatriation of Cambodians from refugee camps in Thailand under United Nations' protection (photo: Teit Hornbak, Still Pictures)

in the interwar years. The UN was a more complex and ambitious undertaking, intended to go beyond merely ensuring stability in international relations in order to 'systematise the promotion of change' (Righter, 1995: 25) (Plate 7.1). The UN today is a complex web of institutions, many with overlapping responsibilities, but in line with the original broad purposes including a commitment to equal rights of people of all nations, to free succeeding generations from the scourge of war and to promote social and economic progress. The UN system, as illustrated in Figure 7.1 (page 161) (Buckley, 1995), includes the United Nations organisation itself (which has six main organs, including the General Assembly and the Security Council), various special commissions and programmes (such as the United Nations Environment Programme), seven functional commissions (such as in human rights), five regional economic and social commissions, numerous standing and ad hoc committees, plus a number of specialised agencies (such as the International Labour Organisation and the World Bank). It also includes varied associated but autonomous bodies and institutions such as the World Trade Organisation, which liaises closely with the UN.

By 1995 membership of the UN had widened to 185 nations. Almost every country in the world is now a member of the UN, with notable exceptions such as Switzerland and Taiwan. Its headquarters are in New York and it is funded through a mixture of member state assessments and voluntary contributions. In 1995 its total budget was US$1.35 billion (including salaries for a staff of over 500,000) with a further US$3.1 billion for peacekeeping purposes. Currently 135 members are in arrears to the UN, which is itself US$3 billion in debt (Bogert, 1995: 14).

Each member state is represented on the General Assembly (GA) of the UN with equal voting power irrespective of the size, population or power of that country. The GA meets annually but serves largely as a 'forum for forging consensus and influencing state behaviour' (Werksman, 1995: 11) since decisions taken by the GA (unlike those of the councils) have no legally binding force for governments. Such decisions do, however, 'carry the weight of world opinion' (Buckley, 1995: 4) with many resolutions being subsequently incorporated into international treaties such as the Climate Change Convention of 1994. The GA controls the budget and staffing levels of the UN and deputes its work to six committees dealing with disarmament, economic and financial matters, social humanitarian and cultural matters, decolonisation, administrative and budgetary affairs and legal matters (I. Williams, 1995: 19).

In 1992 the largest assembly of heads of government in history occurred at Rio de Janeiro for the United Nations Conference on Environment and Development (UNCED) in the pursuit of sustainable development patterns for the future. Included within Agenda 21, the substantial framework document adopted at the conference, were statements of commitment for reform of the UN itself towards sustainable development. For example, Chapter 38 set out mechanisms for strengthening cooperation and coordination on environmental and developmental issues across the UN system. The United Nations Environment Programme (UNEP) is the UN's primary environmental policy coordinating body, with its headquarters in Nairobi. It was created in 1973 following the United Nations Conference on the Human Environment in Stockholm of the previous year. UNEP's constituent act stressed the need to assist developing countries to implement environmental policies and programmes that are compatible with their development plans. Its coordinating role in environmental matters was given impetus by the report of the Brundtland Commission (WCED, 1987) and reaffirmed at UNCED.

Preparations for the creation of a new international institution within the UN system, the Commission on Sustainable Development (CSD), were also initiated at Rio. The new commission was placed under the Economic and Social Council (rather than the GA itself) and met for the first time in 1993. This 'overarching international environmental organisation' has been called the 'UN Committee of the Whole' (Righter, 1995: 305) relating to its principal function in monitoring progress in the implementation of Agenda 21 through evaluation of all reports from all relevant organisations, programmes and institutions of the UN system. The CSD has also been specifically charged with monitoring commitments by UN member nations to provide financial resources and to the transfer of technology (World Resources Institute, 1994: 225). It has no legal or budgetary authority such that it is essentially a forum for review, exchanging information, building political consensus and for forging partnerships.

There are 53 elected members of the CSD: 13 from Western Europe and America, 13 from Africa, 11 from Asia, 10 from Latin America and the Caribbean, and 6 from Eastern Europe (World Resources Institute, 1994). It also has a mandate to foster the participation of NGOs, industry, scientific and business communities through encouragement of such groups to make written submissions or to address meetings of the commission at the request of the organisation itself or the commission. Various working groups have been established to consider certain cross-sectoral elements annually (e.g. poverty and financial resources); others consider clusters of sectoral issues every three years, such as health and human settlements.

The quest for sustainable development globally has also impacted on other UN organs concerned with the environment. For example, in 1993, the United Nations Development Programme (UNDP) launched Capacity 21 to assist developing countries to incorporate principles of sustainable development into national programmes and processes. This initiative aims to strengthen both national and local capacity to implement sustainable development strategies, and to strengthen related programmes through facilitating the involvement of all stakeholders in development planning and environmental management. UNDP and UNEP also share responsibility with the World Bank for managing and implementing the Global Environmental Facility (GEF). GEF was pioneered in 1991 and aims to create new funds for governments of low-income countries (defined as those with per capita GNP of less than US$4,000) to enable them to take on environmental actions which have clear global benefits in areas such as greenhouse gas emissions, biodiversity, ozone and international waters. The Security Council of the UN also recognised in 1992 that there were non-military sources of instability, including in the environmental field, which have become threats to national peace and security. The legal and political challenges of giving the Security Council a greater role in environmental disputes, such as over international waters (Chapter 6) and international refugees, are currently under urgent review (Werksman, 1996).

Through most of the UN's life, the bulk of money and effort has gone towards economic and social development, coordinated by the Economic and Social Council (ECOSOC) and effected through seven functional commissions and the five regional commissions shown in Figure 7.1. Buckley (1995) suggests that 75 per cent of the UN effort during the last decade (measured in terms of money spent and the people involved) has gone to these activities. There is substantial debate, however, as to the development impact of UN activities. For example, the high proportion of UN spending on economic and social development needs to be viewed in terms of the very limited overall budget of the UN system. In 1995 the budget of the New York Police Department was almost double that of the UN (*Newsweek*, 1995: 19). Only 14 per cent of Official Development Assistance globally goes through the UN (World Resources Institute, 1994: 226) and efforts post Rio to try to get more funds for development administered centrally have largely failed. Furthermore, levels of development assistance overall to the developing world have decreased from US$60 million in 1992 to US$50 million in 1994 (Sahnoun, 1994: 11).

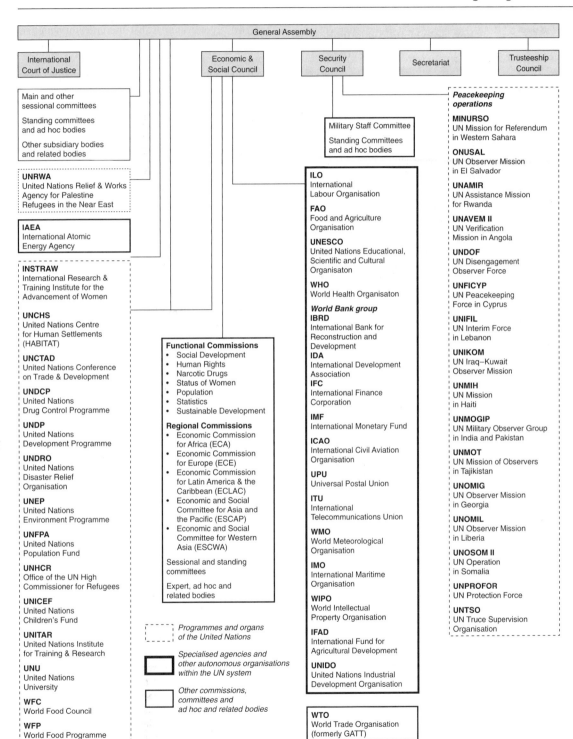

Figure 7.1 The UN system. Adapted from Buckley (1995).

Righter (1995) suggests that the development impacts of the UN have been limited by the number of agencies competing for the same monies and the subsequent 'sectoral distortions of focus and consequent special pleading' (p. 187). This has resulted in what has been termed the UN's 'butterfly' or 'laundry list' approach to development: 'thousands of mini projects working on mini objectives' (Righter, 1995: 59). Rather than UN agencies giving some kind of neutral advice (which was what the agencies claimed made their assistance programmes superior to those of the national aid departments of donor governments), intra-agency rivalry and 'agency salesmanship' have characterised UN development interventions. In 1985, UNDP money was being spent through twenty-nine 'executing agencies' of the UN and a further eight 'participating' UN organisations (Righter, 1995: 58). Inevitably, problems of duplication and a lack of coherence in UN programmes supported by excessive bureaucracies resulted. It has been estimated that there were over eleven thousand UN meetings and conferences annually in New York and Geneva alone (Righter, 1995: 212). However, such criticisms are not to deny the significant role played in directing development research and action by many individuals working within the UN system or as guided by the annual publications of UN organs such as the Human Development Report of the UNDP.

In terms of responding to global economic recession and the international debt crises, central UN organs have exhibited less change than associated institutions such as the World Bank (see next section). UN agencies have traditionally been concerned with projects and technical cooperation. Indeed, Righter (1995) suggests that even in terms of policy analysis in their own sectoral specialisations, UN agencies are being outperformed by the development banking system: 'The World Bank's expertise in education now exceeded that of UNESCO for example' (p. 268). However, programmes of UN organs such as the UNDP are now increasingly taking a multifaceted strategy in their activities, e.g. towards encouraging private sector growth in developing countries. In this regard, UNDP now collaborates with private enterprise as well as governments, is supporting measures to promote capacity building within the business sector and is working to coordinate its activities with other international and bilateral donor agencies.

Ensuring a level of democracy within member nations through the maintenance of international peace and security was the principal purpose of the UN as originally stated within its founding charter. Although it is generally considered that the latter part of this century has been an era of unprecedented world peace, 'there have been 120 conflicts in the Third World since the Second World War and 22 million people have died' (ul-Haq, 1994: 21), suggesting there is much work to be done from within the UN system to promote democratic processes in the world. Many of the inhabitants of the developing countries still do not enjoy basic freedom from conflict, and fewer than half of the 185 UN member states have elected governments (Buckley, 1995: 4).

Responsibility for effecting peace and security lies with the Security Council, of which there are five 'permanent members', the United States, the United Kingdom, China, France and Russia (the leading powers at the time of the establishment of the UN). Any UN operation, including peacekeeping, can only proceed with the agreement of the member states. Yet 85 per cent of global arms sales are made by these permanent members of the Security Council of the UN (Buckley, 1995: 9). The United Kingdom has 22 per cent of the global arms market, the second biggest market share after the United States. The United Kingdom supplies battle tanks to Turkey, CS gas and rubber bullets to Nigeria, extensive and varied arms to Saudi Arabia (the United Kingdom's largest customer) and Hawk jets to Indonesia (K. Booth, 1997: 18).

Each of the members of the Security Council has the 'power of veto', in that a no-vote by any one of them can stop a resolution being passed even if the other fourteen (co-opted) members vote yes. During the Cold War, the power of veto served to limit substantially the UN's peacekeeping role, since either the United States or the Soviet Union would veto any substantive proposal in the Security Council. Since the late 1980s, the power of veto has been used only rarely (Buckley, 1995). Concurrently, the number of peacekeeping operations undertaken and the costs of UN military responses, increased dramatically in the mid 1980s, as shown in Figure 7.2(a). Despite this, peacekeeping dues (calculated in the same manner as general contributions) totalled US$2.29 billion in 1995 (*Newsweek*, 1995: 20) with some of the world's richest countries (and also members of the Security Council) as the major debtors (Figure 7.2(b)).

The reactive rather than preventative actions taken by the Security Council to ensure peace around the world have been the source of much debate, particularly after the tragedies of Somalia and Rwanda. The independence of the UN from national strategic interests has also been questioned. For example, Buckley

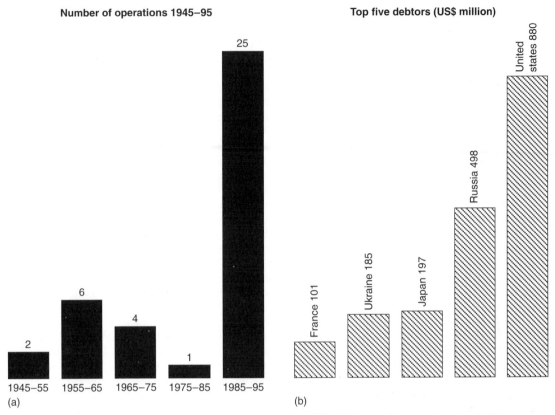

Number of operations 1945–95

Top five debtors (US$ million)

Figure 7.2 (a) The number of UN peacekeeping operations, 1945–95; (b) the top five peacekeeping debtors. Adapted from *Newsweek* (1995).

(1995) notes the extensive financial arrears to the UN, 'but when the chips were down, the big powers had no trouble coming up with the US$1.5 billion a *day* that was needed to support allied action against Iraq in the Gulf War' (p. 20). Brazier (1994) goes further to suggest that the military campaign of UN forces in the Gulf War was 'transparently a US action' involving largely US soldiers, commanded by US generals, and using 'state-of-the-art US hardware' (p. 5).

Within Agenda 21, the UN system has committed itself to internal reforms, along principles including democracy. This statement not only encapsulates the right of each state to participate as sovereign equals within the various constituent agencies of the UN system, but also refers to the need to involve non-state actors in the work of those institutions to ensure that decision-making is 'based on the full range of information, expertise and depth of conviction' (Werksman, 1996: xviii). Over one thousand NGOs attended the UNCED in Rio and over nine hundred such organisations now have consultative status with ECOSOC

(World Resources Institute, 1992). This gives NGOs the right to attend meetings of the central and subsidiary bodies, to submit written statements, to testify before ECOSOC and its committees and in some cases to propose agenda items for ECOSOC consideration.

In 1993 the UNDP made 'people's participation' the central theme of its Human Development Report. The stated vision of societies in the future was built around people's genuine needs, with profound implications 'embracing every aspect of development' (UNDP, 1993: 2). The report called for 'at least 5 new pillars of a people-centred world order: New concepts of human security; new models of sustainable human development; new partnerships between state and markets; new patterns of national and global governance; and new forms of international co-operation' (p. 2). While recognising the role of more decentralised governments and the emergence of NGOs as powerful processes for people-centred development, as discussed in subsequent sections, the UNDP report also stated clearly that these institutions could be effective only if the overall framework of national governance becomes genuinely democratic. Furthermore:

> Let us also recognise that the forces of democracy are not likely to be so obliging as to stop at national borders. *This has major implications for global governance.* States and people must have the opportunity to influence the global decisions that are going to affect them so profoundly. *This means making the institutions of global governance broader and more participatory.* There should, in particular, be a searching re-examination of the Bretton Woods organisations. *And the United Nations must acquire a much broader role in development issues.* (UNDP, 1993: 7; all emphasis added)

It is evident that the constituent institutions of the UN system are currently undergoing substantial change. It is suggested, however, that the UN is generally struggling to find a future. Bogert (1995) questions whether the UN is the 'global emergency number', a development agency or a 'floating cloakroom', where NGOs can meet and 'even begin to take on some functions from overburdened governments' (p. 15). She is confident that there is a future for the UN in addressing global issues beyond the reach of any one country, issues such as fighting terrorism, curbing nuclear proliferation and combating disease. However, the logic of the UN is predicated on the idea of members that are nation states. Yet, as discussed in Chapter 4, civil society today has few borders, states are becoming increasingly fractured and multinational corporations are bigger than some countries:

> It [the UN] is trying to re-invent itself for a world its founders never envisaged. It is no longer policing disputes between states, but within states. It no longer arbitrates between sovereignties, but struggles to keep sovereignties from disintegrating under the strain of civil war. It was intended as an organisation of states and yet it is now called upon, time after time, to protect people against their states. (Ignatieff, 1995)

The World Bank Group and the International Monetary Fund

The World Bank Group includes four key international institutions in development: the International Bank for Reconstruction and Development (IBRD), established in 1944 at the United Nation's Bretton Woods conference; the International Development Association (IDA), established in 1960; the International Finance Corporation (IFC), established in 1956; and the Multilateral Investment Guarantee Agency (MIGA),

established in 1988. All four are based in Washington. The International Monetary Fund (IMF) was also created at the Bretton Woods conference and is closely associated with the World Bank Group.

Although only 'peripherally conceived as a development agency' (Righter, 1995: 187), since 1950 the World Bank (WB) has increasingly lent monies to governments of the developing nations. It is currently the major source of finance for development in the Third World, financing over nineteen hundred projects through a portfolio of US$148 billion dollars (Werksman, 1996: 132). In addition, the World Bank influences national development policy in terms of directing research, technology transfer and other forms of institutional support. The country and sector reports drawn up by the World Bank are used by commercial lenders and aid agencies to plan their own activities. The annual *World Development Reports* of the World Bank and associated publications are also an important source of information and opinion for academics and practitioners (not all of whom take the full institutional line). Certainly, and particularly through structural adjustment programmes, the World Bank has moved far beyond its original function as a straightforward credit institution; it has become involved in policy determination and planning in developing countries to an unprecedented extent. As such, the World Bank is a major player in shaping development outcomes. C. Crook (1991) goes so far as to suggest that currently 'the fate of hundreds of millions of people turns on the decisions these institutions make' (p. 3).

The IBRD was founded on the principle that many countries in Europe after the Second World War would be short of foreign exchange for reconstruction and development activities, but would be insufficiently creditworthy to borrow all the necessary funds commercially. In contrast, IBRD as a multilateral institution, with share capital owned by its member countries, could borrow on world markets and lend more cheaply than commercial banks. The IBRD raises monies through selling bonds and other securities to individuals, other banks, corporations and pension funds around the world. It lends money over 15- to 20-year periods, and loans are subject to interest. IDA provides no-interest loans to the poorest countries, defined as those with per capita incomes of less than US$1,305. IDA loans constituted 25 per cent of World Bank lending in 1993. It cannot raise funds on capital markets as does the IBRD, so it depends entirely on wealthier nations for finance (supplemented by a portion of general World Bank profits). Loans are to be repaid within 35 to 40 years with a 10-year grace

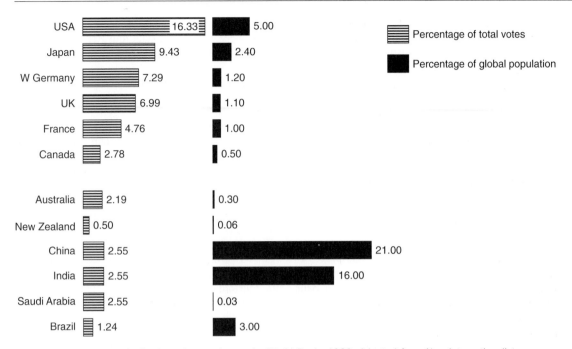

Figure 7.3 Voting powers of selected countries at the World Bank, 1989. Adapted from *New Internationalist*, December 1990.

period. The IFC lends directly to the private sector without the requirement of government guarantees, and MIGA also promotes private investment in developing countries through providing guarantees on investments against non-commercial risks such as war or nationalisation.

Like any other bank, the World Bank Group pursues profit and return on its investments and has been successful in this respect every year since its inception. World Bank securities are considered to be among the world's safest, enabling it to borrow on very favourable terms. Unlike commercial banks, however, the World Bank takes no financial risks since its loans are to governments with whom it has 'preferred creditor' status, in that debts are paid out of the borrowing country's general revenue or from new loans. Werksman (1996) estimates that for every US dollar the US government has invested in the World Bank, US companies have received back US$1.10:

> The other major shareholders are not faring badly either. Income accruing from procurement contracts when compared to every dollar contributed to the Bank leads to benefits amounting to US$1.90 for the United Kingdom, US$1.78 for France, US$1.47 for Germany. Only Japan with US$0.97 appears to receive a little less than its contributions. (Werksman, 1996: 131)

There are currently 176 member countries of the World Bank, each with a representative on the Board of Governors. A smaller group, the Board of Executive Directors, is responsible for general operations and policy making within the group, the chairman being the World Bank president (who to date has always been a US citizen). Votes are made by member countries on the allocation of World Bank funds according to the financial contribution of that nation to the Bank (one dollar one vote). On this basis, the G7 group of industrialised countries have 45 per cent of voting rights in the IBRD and 48 per cent in the IDA (US 17 and 16 per cent respectively) while the remaining 148 countries share the rest: 'only seven Third World countries have over 1 per cent of the votes and most of the poorest countries have less than 0.1 per cent' (Dickenson *et al.*, 1996: 264). Figure 7.3 shows the voting power of selected countries in the World Bank.

To gain membership to the World Bank Group, a country must first be a member of the IMF. Member countries of the IMF pay a subscription and agree to abide by a 'mutually advantageous code of economic conduct' (C. Crook, 1991: 3). Traditionally, the primary distinction between these two institutions is that the IMF is concerned with the health of the international monetary system and may lend (from subscriptions) to member countries briefly to over-come short-term financial instability. The World Bank is evidently more concerned with the financing of longer-term development in the poorer countries of the world. However, in the 1980s, insolvency of the middle-income debtors threatened the international financial system, such that the IMF moved increasingly into the area of lending to members. The distinctions between the roles of the World Bank and the IMF have subsequently become less clear: 'The Fund had begun to worry about longer-term development, and the Bank was taking a new interest in short-term macro-economic policy . . . in one developing country after another, the two institutions devised overlapping programmes of economic reform and backed them with cash' (C. Crook, 1991: 4).

The search for sustainable patterns and processes of development has prompted many changes in the structure, function and activities of the World Bank Group. In the 1970s, the World Bank itself recognised that in many instances, environmental degradation was compromising the impact of its project lending on economic development. In addition, the role of World Bank lending in causing environmental destruction was taken up at this time by environmentalists, particularly in the United States. Contributors to *Ecologist* magazine were particularly prominent in the condemnation of World Bank practices (e.g. vols 14–17, 1984–87). The impacts of the Carajas iron ore project in eastern Amazonia, which involved the clearance of tropical forest the size of England and France combined, were among those documented. Similarly, the Polonoroeste highway project in the Brazilian Northwest, was exposed in the same publication as requiring the resettlement of 30,000 families and the large-scale destruction of tropical forests.

Such media exposure and emerging public pressure, particularly from US environmentalists, have been important factors in prompting changes in the World Bank towards greater environmental sustainability within its operations. The United States contributes approximately one-fifth of all capital funds to the World Bank each year; environmentalists have extensively lobbied congressional subcommittees to press for reform of the World Bank and to withhold funds if changes are not forthcoming. In 1973 the World Bank established an Office of Environmental Affairs to review the prospective environmental impacts of its project lending. By 1985 that office had only five staff, suggesting its capacity to review all projects was severely limited (Rich, 1994). In 1987,

under President Conable, a central Environment Department was created with four Regional Environment Divisions to oversee and promote environmental activities.

In 1986 the first no-vote from the US executive director on environmental grounds was cast for a World Bank project in the Brazilian energy sector. Since 1989, US government representatives on the boards of directors of all multilateral development banks (MDBs) are required by US law to vote against loans for projects if specific environmental procedures are not followed. Furthermore, under the Pelosi amendment of 1992, all MDBs to which the US government gives finances are required to carry out an environmental impact assessment and to make it available to all affected groups and NGOs, unless preliminary investigations show no environmental impacts are associated with the project. In 1993 the World Bank articulated its 'fourfold environmental agenda' towards sustainability in development (Box 7.1).

The debt crisis of the early 1980s prompted profound changes within the institutions of the World Bank. Since 1979 an increasing proportion of World Bank lending has been not to projects, but to programmes of broadly based policy reforms. By 1994, 30 per cent of World Bank lending fell into this category (World Bank, 1994a). As developing world governments became increasingly cash-strapped at a time of world recession and declining terms of trade (particularly for non-oil commodities), the World Bank noted that any developmental progress from its traditional portfolio of project-based lending was 'being swamped by macroeconomic imbalances in most of its client countries' (Reed, 1996: 9). At the same time, the IMF was looking beyond 'crisis management' in the monetary and financial sectors of developing countries, towards assisting countries in building up productive capacity. Comprehensive solutions to the debt crisis were considered to be essential. From within both institutions in the 1980s, the economic crisis in developing countries was conceived as more than a temporary liquidity issue (as it had been viewed in the 1970s). Multilateral cooperation and significant changes were required in the mandate and practices of the institutions of the World Bank and the IMF themselves.

The term *structural adjustment programmes (SAPs)* is used generically to describe the activities of the World Bank and the IMF in the design and support of packages of broad-based policy reform within a country. The central aim is debt reduction via

---

BOX 7.1   The fourfold environmental agenda of the World Bank

1 Assisting member countries in setting priorities, building institutions and implementing programmes for sound environmental stewardship.
2 Ensuring that potential adverse environmental impacts from bank-financed projects are addressed.
3 Assisting member countries in building on the synergies among poverty reduction, economic efficiency and environmental protection.
4 Addressing global environmental challenges through participation in the Global Environment Facility (GEF).

The first dimension of the agenda encompasses a variety of measures, including financial resources for targeted (i.e. primarily environmental) projects such as pollution control, land conservation and natural habitat protection. In 1994 the World Bank supported one hundred environmental projects, representing a commitment of more than $13 billion (World Bank, 1994a). However, although 'many of these projects have been characterised as green because they involve natural resource management, whether they are actually environmentally beneficial is open to question' (Werksman, 1996: 135). In addition, the World Bank now supports national and regional environmental planning exercises, largely National Environmental Action Plans, which aim to integrate environmental considerations into a nation's overall economic and social development strategy over the longer term. The World Bank also disseminates knowledge on sustainable development through the preparation of papers and policy guidance based on best practice.

   Procedures for assessing and mitigating the potential adverse environmental impacts of World Bank project lending are set out in Operational Directive 4 of October 1989. Under this directive, at the preapproval stage of a proposed investment, projects are 'screened' for environmental impacts and assigned to one of four categories requiring varying levels of subsequent environmental analysis on the basis of the nature, magnitude and sensitivity of environmental issues. Staff of the Regional Environment Divisions then assist the prospective borrower in carrying out the necessary environmental analysis. In 1993, 14 per cent of projects were screened as

Category A and required a full environmental impact analysis in the light of their projected 'diverse and significant environmental impacts' (World Bank, 1994a). Some 50 per cent of all projects at the pre-approval stage in that year were subject to some degree of environmental analysis. The borrower is then notified as to the level of environmental assessment required and assisted in preparing it through the identification of Terms of Reference. A borrower may also be offered funds to finance the use of consultants in the process, wherever necessary or desired.

   An enhanced appreciation of the synergies between poverty alleviation and environmental conservation was a central contribution of the Brundtland Report (Elliott, 1994). It is now recognised that addressing the welfare needs of the poorest groups is essential in the move toward sustainable development, not only in terms of the moral imperative, but that positive environmental ends will be achieved in so doing. In 1990 the World Bank's Development Report had poverty as its central focus. Financial flows to the world's most vulnerable groups made up 21 per cent of its total lending in 1993. Such flows are monitored under the Programme of Targeted Interventions (PTI) and are defined as interventions where the proportion of poor among project beneficiaries exceeds the proportion of the poor in that population.

   The final element of the World Bank's environmental agenda is its contributions to the Global Environment Facility. In practice, critics suggest that the 'noble aspirations' of the GEF have not been achieved (Werksman, 1995) in large measure due to the domination of the World Bank in decision-making procedures. For example, membership of GEF is dominated by those developed countries wealthy enough to contribute the minimum $4 million required. The chair of GEF is appointed and employed by the World Bank and as many as 80 per cent of GEF projects (up to 1994) were linked in some way to larger World Bank projects. There remains a lack of consensus regarding how NGOs can and should participate in GEF projects.

Source: World Bank (1994a) World Bank and the environment. *Fiscal 1993*.

Table 7.1 The principal instruments of structural adjustment.

Currency devaluation
Monetary discipline
Reduction of public spending
Price reforms
Trade liberalisation
Reduction and/or removal of subsidies
Privatisation of public enterprises
Wage restraints
Institutional reforms

> programs of policy and institutional change necessary to modify the structure of an economy so that it can maintain both its growth rate and the viability of its balance of payments in the medium term. (Reed, 1996: 41)

The first SAP was initiated in Turkey in 1980 and by the end of the decade, 187 SAPs had been negotiated for 64 developing countries (Dickenson *et al.*, 1996: 265). It is argued that the impact of structural adjustment has now gone far beyond the original national context for which they were designed, to become an instrument for global economic policy and the re-structuring of the world economy (Reed, 1996). The specific instruments of structural reform are varied but may include those in Table 7.1.

Although there has been some 'nuancing' (Reed, 1996) of the basic objective of SAPs such as with the addition in the early 1990s of the qualifier of 'sustained' to the original growth objective, there is evidence to suggest that improvements in the macroeconomy may be at the cost of other desirable ends including sustainability. Ghana is often heralded by the World Bank as one of its success stories in terms of progress with economic adjustment, with average annual growth in GDP of 3.8 per cent throughout the 1980s (Rich, 1994). However, timber exports increased from US$16 million in 1983 to US$99 million in 1988 under the trade liberalisation required by structural adjustment. As a result, the forest area within Ghana has reduced to 25 per cent of its original size (Rich, 1994). In Java, a dramatic increase in the terms of trade for horticultural goods through liberalisation of markets has encouraged upland farmers to move from less profitable basic starch staples to more profitable fruits and vegetables (Conway and Barbier, 1995). Although this has created an incentive to invest in soil conservation measures, the increased profitability of these crops is also encouraging farmers to extend the cultivation into steeply sloped volcanic soils where run-off and soil erosion

are enhanced. Bryant and Bailey (1997: 79) conclude that 'multi-lateral institutions have rarely, if ever, been on the side of poor and marginal grassroots actors – rhetoric notwithstanding,' pointing to the widespread enclosure of land and other environmental resources, used by these local actors, which has occurred under policies pursued by these institutions.

The desire for democracy and the growing power of civil society in developing nations is evidenced by the rising number of diverse non-governmental organisations in these countries. It is 'hard to specify just how formal a grouping must be before it is considered an NGO' (A. Thomas, 1992: 123) but there are estimated to be between 10,000 and 20,000 NGOs in the South and 4,000 in OECD countries (Edwards and Hulme, 1992). In continuity with organs within the UN system, the World Bank has stated a commitment to work with NGOs as a means of fostering democracy and local empowerment:

> Many basic services are best managed at the local level. . . . The aims should be to empower ordinary people to take charge of their lives, to make communities more responsible for their development, and to make governments listen to their people. Fostering a more pluralistic institutional structure – including non-governmental organisations . . . – is a means to these ends. (World Bank, 1989: 54–5)

The number of World Bank projects within which NGOs are involved has risen from thirteen annually between 1973 and 1988 to fifty (one-quarter of all projects) in 1990 (World Resources Institute, 1992: 217). The pattern of collaboration with NGOs has also changed, away from international NGOs towards 'grassroots' organisations, such that although NGOs have worked primarily as implementors on World Bank projects, they are increasingly being involved during planning and evaluation stages as well.

Progress towards democracy, however, also depends on conditions of social equality. There is mounting evidence to suggest that the institutional changes demanded by SAPs may in fact be compromising democratic processes through compounding, rather than alleviating, socioeconomic and gender inequalities. A UNICEF report in the late 1980s concluded that 'the World Bank and IMF adjustment programmes bore a substantial responsibility for lowered health, nutritional and educational levels for tens of millions of Third World children' (Rich, 1994: 186). Figure 7.4 illustrates the processes through which institutional changes required under structural adjustment can lead to declines in child welfare, particularly among the poorest households. For example, households may

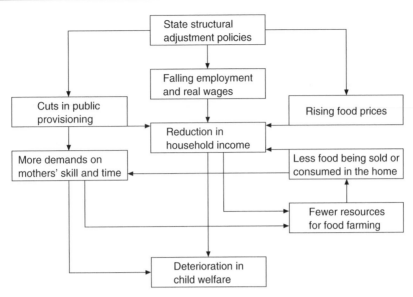

Figure 7.4 Child welfare: the impact of institutional changes under SAPs. Adapted from Messkoub (1992).

experience declines in income through employment changes (such as retrenchments from privatised industries or decreased public sector job opportunities) and declines in real wages as a result of devaluation. Furthermore, the cost of the most important purchases rise due to the removal of food subsidies and other shifts in government expenditures such as charging for primary health care.

It has been suggested that the market-only policies of SAPs have specific and 'probably negative' impacts on women 'given the complexity of gender relations, the invisibility of women's work both in the rural and urban sectors, and the multiple and complex roles women actually perform' (Pearson, 1992: 309). Reed (1996: 16) suggests that the fundamental problem with SAPs is that they are 'grounded in a gender ideology which is deeply, and fundamentally exploitative of women's time/work and sexuality'. Some of the profound and wide-ranging effects of the various policies associated with moving towards a more free market economy on the lives of women and girls are highlighted in Box 7.2.

It is evident that the relationship between structural adjustment policies, the environment, the alleviation of poverty and progress towards democracy are complex. The causal links may be indirect and are specific to local and national circumstances (Reed, 1996). There is no doubting, however, the importance of the World Bank and related international institutions in shaping

the nature and outcome of development policy worldwide. Hildyard (1994: 26) goes so far as to suggest that through the conditionalities attached to World Bank- and IMF-dictated structural adjustment programmes, these institutions 'virtually control the economies' of many developing countries.

The World Trade Organisation (formerly GATT)

International trade is the most important aspect of the global economy and a determining factor in the social and economic development of nations. The majority of world trade takes place according to a set of rules administered by a multilateral institution now known as the World Trade Organisation (WTO). The origins of this institution go back to 1948 when the General Agreement on Tariffs and Trade (GATT) was ratified by twenty-three countries. All contracting parties (as they were known) to the agreement committed themselves to upholding two 'liberal and unexceptional' principles (Hutton, 1993) designed explicitly to encourage free trade and to prevent the trade wars which had plagued the international economy in the 1930s. These principles refer to 'national treatment' under which all countries must treat participants in their economies the same as domestic firms, and to the 'most favoured nation', which aims to ensure that any concession granted to one trading partner is extended to all. Although the WTO refers strictly to international

BOX 7.2 The impact of structural adjustment policies on girls and women

1 Increased numbers of women are looking for income-generating work outside the home to compensate for decline in family purchasing power. In Bolivia, between 1976 and 1986, male employment increased by 0.1 per cent per annum, and female employment by 2.7 per cent. More women than men become unemployed under structural adjustment policies (SAPs) due to the nature of their employment; for example, women are often on temporary contracts and locked into apprenticeship grades.

2 Working conditions deteriorate with cuts in provision of day care for children or the duration of maternity leave (both affect women more than men). Women also enter forms of employment where security and benefits are few, such as domestic service, and where their vulnerability to abuse is often high.

3 There is much evidence to suggest that SAPs have led to a widening of wage differentials between men and women in all sectors of the economy. Women become poorer absolutely and in relation to men under these processes. Often poverty data is not disaggregated by gender, but several studies have shown that a sizeable proportion of poor households are headed by women. For example, 30 per cent of poor urban households in Brazil are headed by women, which suggests that the numbers of poor women (as household heads or 'sharing' poverty with partners) will be higher than the number of poor men. The UN found that in times of economic recession, progress in girls' education slows. Pearson (1992) cites work in Nigeria which has shown that with the introduction of school fees, there has been a marked fall-off in the enrolment of girls in schools to save costs and to care for younger children as women have increasingly had to take up income-generating jobs. There are also lots of research findings of women eating less in times of food insecurity. Subsequent declines in women's health may have effects on children if women are pregnant or lactating. Research in Dar es Salaam found that 58 per cent of households had reduced the number of cooked meals from three to two in years under structural adjustment (Pearson, 1992). In Tanzania, cutbacks in public health services have led to an increase in under-5 mortality from 193 per thousand in 1980 to 309 in 1987 (Pearson, 1992: 310).

4 The shift to export crops encouraged by SAPs often does not benefit women. For example, women may produce and trade more in restricted or local markets and potentially non-monetary markets; farm support services are often biased towards men and SAPs do not require any significant reform in terms of enhancing women's access to land, credit or other inputs, (i.e. SAPs do nothing to improve gender equity or reduce gender biases).

5 Women's unpaid work has increased. The greater unemployment, decreased purchasing power and cut-backs in social services result in women adopting strategies to make funds go further, such as time spent in shopping for cheaper alternatives, purchase of foods that have had less processing (leading to increased food preparation time) and supplementing income through the growing of vegetables for household and/or sale. In agricultural production, women's unpaid labour may increase as households try to cut costs and raise production.

Women may derive localised benefits from SAPs. Deregulation of market trade in Ghana led to decreased harassment of women by soldiers and police, and from Tanzania there is some evidence of women taking on employment and gaining enhanced status within the household. But the overwhelming balance is that SAPs have detrimental impacts on women.

Source: Sparr (1994).

trade policy, the agreements made within the institution have diverse implications for national development policy and impact on domestic economies throughout health, environment, agriculture, etc. The activities of the WTO have major implications for the working of markets worldwide, but also the prospects for sustainable development and the alleviation of debt and for progress towards democracy in the developing world.

GATT had its origins in the same statecraft of the 1940s which gave rise to the UN, the World Bank and the IMF. The intention at the Bretton Woods conference was to set up an international trade organisation intricately related to the other two institutions and all directed at the promotion of economic recovery and international stability. The International Trade Organisation, however, proved particularly controversial (largely due to US fears over free trade), such that it was not ratified. GATT was established subsequently as an interim measure, but persisted until 1995. GATT (like IMF and the World Bank) was part of the UN family, but its employers were not international civil servants and it was far removed from parliamentary control (von Moltke, 1994). The UN set up its own programme relating to trade, the Conference on Trade and Development (UNCTAD) in 1964. All UN members participate in UNCTAD which was intended to provide a parallel forum for GATT, in which developing countries could negotiate more powerfully with the 'North' through the formation of regional trading blocks. However, the WTO remains the most important international trading forum (Buckley, 1995); by 1996 it had 122 member states, with China and Russia as the largest non-members (Buckley, 1996).

The WTO, in continuity with GATT, hosts a series of multilateral trade negotiations, known as rounds, within which arrangements are negotiated and disputes resolved. Voting within the WTO is unweighted, each member getting one vote. In continuity with the tradition within GATT, decision making is generally based on consensus, although there are provisions for majority voting in certain instances. For example, a three-quarters majority is required for adoption of an interpretation of a multilateral trade agreement or for a waiver of obligations imposed on a particular member (Buckley, 1996). The length of each round gives insight into the challenges of achieving international consensus on trade matters; the Uruguay round took seven years to complete.

Following the creation of the WTO, the rules of the agreement have been expanded to include trade in the service sectors of banking, tourism, intellectual property rights and insurance (as well as the established merchandise products). Over the history of GATT, the value of merchandise trade has increased from US$57 billion to over US$3,500 billion (Buckley, 1994: 10) and trade in services has risen to approximately 20 per cent of world trade (p. 13). The WTO has twice as many committees and councils as GATT, largely to deal with the new areas of trade now covered but also to provide stronger procedures for dispute resolution and enforcement, often not forthcoming under GATT. The structure of the WTO and the procedures for settling disputes are shown in Figure 7.5 (overleaf).

There is no explicit attention within the WTO mandate to the relationship between trade policy and other major policy areas such as the environment. Although there are articles in the agreement granting exceptions to the general free trade requirements where the protection of human, animal or plant life or health and the conservation of exhaustible natural resources are concerned, environmentalists fear that much progress towards sustainable development is being compromised through WTO rulings. Although WTO rules do not prevent a country from regulating trade in certain products in order to protect the environment or health, any such measure must be applied to domestic and foreign firms and may not be used as a protectionist device: 'Thus tuna fish caught with nets which also take dolphin have to be treated in the same way as those caught with dolphin-friendly methods' (Buckley, 1994: 6). The WTO has a new Committee on Trade and the Environment, but its terms of reference focus entirely on the negative impact of environmental measures on trade, rather than the effects of trade liberalisation on the environment. The concern of the WTO is with environmental measures as distortions to free trade and the possibility of countries using such interventions to protect domestic firms. Table 7.2 (overleaf) highlights some of the key areas in which free trade and environmentalist views are divergent concerning the relationship between market economics and environmental protection.

A huge proportion of world trade is now controlled by a few transnational corporations (TNCs). For example, some 75–90 per cent of trade in the major 'tropical products' such as tea, coffee, cocoa, cotton and forest products are controlled by between three and six transnational corporations (Middleton, O'Keefe and Moyo, 1993: 99). Yet these companies remain largely untouched by any form of international regulation, including through the WTO, which 'makes no distinction between enterprises in terms of power,

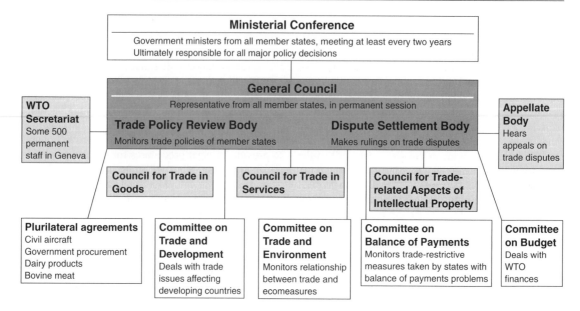

Figure 7.5 The structure of the World Trade Organisation. WTO agreements are normally multilateral and apply to all member countries; plurilateral agreements apply only to certain members. Adapted from Buckley (1996).

Table 7.2 Competing ideologies on trade and the environment.

Environmentalists	Free trade policy analysts
Trade liberalisation creates unchecked growth and pollution	Protectionism is inefficient and leads to even more wasteful use of resources
Environmental costs must be internalised into all WTO activities and decisions.	Trade distortions and protectionist intentions of environmental measures
Consideration of production processes must be part of WTO decisions	Trade can create finances for design and enforcement of environmental measures
Environmental solutions need to reduce consumption	Well-targeted environmental policy can control environmental impacts of free trade

impact or scale of operation' (UNRISD, 1995: 153). Stringent pollution controls and industrial regulations concerning production in the more developed economies were a primary factor in the expansion in the 1970s of TNCs into developing countries. Although many TNCs now have well-developed corporate strategies on the environment in response to fears that public sentiment against them will force tighter environmental controls in the developing nations and/ or prompt consumer boycotts in the developed world against their products, there remains much concern as to the prospects of sustainable development through such trade liberalisation. As Bryant and Bailey (1997: 127) conclude, 'transnational businesses have generally

had an adverse environmental effect, because these firms privilege profit maximisation over social justice and environmental conservation in their day to day operations'.

In continuity with other international institutions such as the World Bank and the IMF in the 1990s, the WTO prescriptions are centrally concerned with debt alleviation and the reinvigoration of the global economy. It is proposed that only through processes such as liberalisation, privatisation and deregulation will competition force the required changes in state and market institutions and overcome the generalised economic crisis. The Secretariat of the WTO estimates that real world income could increase by US$510 billion annually by the end of the implementation period of the Uruguay round (Harmsen, 1995: 26). Approximately US$116 billion is predicted to accrue to developing and transition economies, largely from the phasing out of the 'Multifibre Arrangement' which has served since 1974 to protect the clothing industries in the United States and Europe from competition with the developing nations.

However, the dominance of TNCs within international trade and the economic strength of these companies in comparison to whole nations within the developing world, have also raised fears concerning the transition to democracy in these countries. For example, Shell in 1990 had a gross income which exceeded the combined gross national products of Tanzania, Ethiopia, Nepal, Bangladesh, Zaire (as it was known), Uganda, Nigeria, Kenya and Pakistan in that year (Bryant and Bailey, 1997: 108). Transparency and the opportunities for public participation within transnational corporations are limited. Transnational corporations by definition are 'non-place-based actors' and therefore it could be suggested have no loyalty to any community, government or people. Indeed, trade-related investment measures (TRIMs) which had served to enable national governments to place conditions on foreign investors, including measures to ensure a degree of social responsibility such as in environmental protection or local participation, have now been curtailed under WTO rulings (UNRISD, 1995).

In continuity with structural adjustment programmes, under certain conditions the blanket prescriptions of the WTO may threaten democratic processes through the exacerbation of inequality and increased vulnerability within nations. B. Harriss and Crow (1992) found that deregulation of agricultural markets in three country studies prompted greater price variability. In all cases this had led to an increased vulnerability of the poorest groups such as nomadic pastoralists in Somalia, landless labourers in Bangladesh and smallholders in remote locations of Malawi. Fundamentally, markets are diverse and complex institutions which rarely operate to provide equal benefits to all participants. WTO rulings, however, do not distinguish between corporations, companies, individual producers, petty traders or consumers, nor do they recognise their differential power in accessing markets (which themselves are not uniform even within a region). Under certain conditions, it is increasingly evident that the trade liberalisation intentions of the WTO may be inappropriate and contrary to the pursuit of more equitable development patterns.

The role of the state

The very subject of 'development' was built on the idea of the state as the main lever for changing the economy and society. The ideology of 'developmentalism' and the concept of the interventionist state were inseparable in the optimistic post-war beginnings of development theory. (Mackintosh, 1992: 61)

In most developing countries throughout the 1960s, the state took a primary role in the development and implementation of development (Chapters 1 and 3). Many developing nations built up large state enterprises in public utilities, nationalised mining and agricultural enterprises to lead industrial development post independence. It was a stage of 'great optimism' (Mackintosh, 1992: 68) concerning both the benevolence and the competence of the state to work in the 'public interest'. 'By the 1960s, states had become involved in virtually every aspect of the economy, administering prices and increasingly regulating labor, foreign exchange, and financial markets' (World Bank, 1997: 23). However, by the late 1970s, disillusion had begun to set in concerning the ability of the state to match policy and resources and to promote development. Generally within the developing world during this period, government expenditures had grown faster than GDP, such that further state investment was only enabled by borrowing from commercial and multilateral sources. Despite the easy availability and increased uptake of international finance in the 1970s, the economic performance of many developing countries remained poor and the state was beginning to be seen as part of the problem rather than the solution: 'The oil price shocks were a last gasp for state expansion. . . . As long as resources were flowing in, the institutional weaknesses stayed hidden' (World Bank, 1997: 23).

By the 1980s, the capacity of national governments in the developing world to continue to make investments in development were severely limited by rising oil prices, mounting debt burdens and recession within the global economy. Substantial reforms of the state have since been held as central to global economic recovery, and as conditions for the receipt of further development assistance from multilateral and bilateral sources, and are acknowledged widely to be necessary for democratic and sustainable patterns of development in the future.

The state is a universal feature of the contemporary world and is seemingly necessary to the functioning

State, in its wider sense, refers to a set of institutions that possess the means of legitimate coercion, exercised over a defined territory and its population, referred to as society. The state monopolises rule making within its territory through the medium of an organised government.

Government has different meanings in different contexts:
- the process of governing, the exercise of power
- the existence of that process, a condition of ordered rule
- the people who fill the positions of authority within a state
- the manner, method or system of governing in a society, i.e. the structure and arrangement of offices and how they relate to the governed

Figure 7.6 Concepts of state and government. Adapted from World Bank (1997).

of both capitalist and socialist modes of production (Johnston, 1996). Figure 7.6 details and contrasts the concepts of state and government. Broadly, the state is a 'recognisably separate' set of institutions distinct from the rest of society (the private sphere) and is the supreme power within its territory and over the people therein. However, the state takes many different forms across the globe: 'Individual states have developed through conflict and accommodation, as various interest groups have contested for power within society' (Johnston, 1996: 146–7). States do not operate in any particular or predetermined way, rather (at least in democratic societies), it is people who interpret and define the roles and tasks of the state.

In operation, therefore, the role taken by the state in any arena of activity is influenced by a host of forces including inter- and intrastate conflicts. For example, in terms of environmental conservation, it has already been noted how international institutions are influencing national policy development in recipient countries. Bryant and Bailey (1997) suggest that SAPs have often simultaneously reduced the ability of states to respond to environmental problems and increased their seriousness, such as through pressures to increase primary commodity production but also to cut the budgets and staffing of environment departments. Box 7.3 shows how the British government intends in future to work in the areas of aid, trade and international development only with those governments who share their stated commitment to sustainable development. Further factors shaping the activities of the state include the mounting pressure from grassroots

movements within the developing countries (often in response to perceived shortcomings of the state) for environmental reforms and the devolution of powers to local communities, as discussed below.

The Brundtland Report identified a key role for the state in finding solutions to environmental degradation worldwide (WCED, 1987) and in ensuring that individuals within society behave in a responsible manner such that collective goods including resource conservation are achieved. Bryant and Bailey (1997: 55), however, assert that to date under both socialist and capitalist ideologies of development, 'rather than being an actor with possible solutions to environmental problems, the state has typically contributed to exacerbating those problems'. Often it is intrastate conflicts such as bureaucratic resistance and corruption which have limited the stewardship role of the state. For example, it may be that the most powerful groups have accessed that strength through their control over environmentally damaging activities, such as in mining and energy generation which they are reluctant to give up. The often close associations between political leaders and business interests in many developing countries provide further factors of inertia and affect the ability of states to undertake their stewardship role effectively.

It is in response to the economic crises of the 1980s that potentially the most far-reaching changes in state institutions have occurred. By the late 1980s a new economic orthodoxy, 'market triumphalism' (Peet and Watts, 1996), was gaining the sympathy of leaders in the industrialised nations, in which the market rather than the state was seen as the prime instrument of economic development. This monetarist thinking soon spread into the policies and practices of the multilateral institutions identified earlier. Prescriptions for institutional change towards the regeneration of the global economy and debt reduction (as encapsulated within SAPs), included reducing government spending, internal reforms of the state and the devolution of state activities and decision making.

Figure 7.7 (page 176) illustrates how government spending worldwide has expanded since 1960. In the last fifteen years, many governments of the developing world have implemented widespread changes in macroeconomic policy in response to external pressures (particularly within SAPs). Typically, the squeeze on state spending in such countries has been most severe where debts were most serious, such as in Latin America and the Caribbean. Furthermore, Mackintosh (1992) notes that the weaker the economy, the greater the 'developmental' part of public spending such as

BOX 7.3 The role of other states in development: the UK's Department for International Development

In May 1997 the new Labour government in Britain established a Department for International Development (DFID) with responsibility not only for aid (as within the former Overseas Development Administration), but also for trade with developing countries, global environment issues and debt. In November of that year, the first White Paper on development for twenty-two years was issued, entitled *Eliminating World Poverty: a challenge for the twenty-first century*. The Secretary for State, Clare Short, suggested that the time was right for a reorientation of international development efforts and for building new partnerships between Britain and other donors and with developing countries.

The targets set within the White Paper (see table) are those already established by the international community; they aim to create sustainable livelihoods for poor people, promote human development and conserve the environment. Cutting the proportion of people living in extreme poverty is considered to be the most fundamental factor in promoting sustainable development.

The White Paper identifies twelve strands to the future work of the department. A key role is envisaged for the British government to use its position to encourage both domestic and international understanding and the required political will, concerning these targets. The intention is to work in partnership with poorer countries who are themselves committed to those targets (strand 3).

The stated intention is to encourage economic growth which reaches the poor (strand 1) and to reduce the external debt of developing countries (strand 9). The British government is also committed to start to reverse the decline in UK spending on development assistance and to reach UN targets in these terms.

Strands 7 and 8 state that resources in future will be used to build transparent and accountable governments and to promote political stability and social cohesion in developing countries. 'Problems of international development can be resolved if there is the political will to address them in both poorer and richer countries. This government has that political will, and will seek to mobilise it elsewhere' (DFID, 1997: section 1.25).

International development targets.

Economic well-being
A reduction by one-half in the proportion of people living in extreme poverty by 2015

Human development
Universal primary education in all countries by 2015
Demonstrated progress towards gender equality and the empowerment of women by eliminating gender disparity in primary and secondary education by 2005
A reduction by two-thirds in the mortality rates for infants and children under age 5 and a reduction by three-quarters in maternal mortality, all by 2015
Access through the primary health care system to reproductive health services for all individuals of appropriate ages as soon as possible and not later than the year 2015

Environmental sustainability and regeneration
The implementation of national strategies for sustainable development in all countries by 2005, so as to ensure that current trends in the loss of environmental resources are effectively reversed at both global and national levels by 2015

Source: DFID (1997).

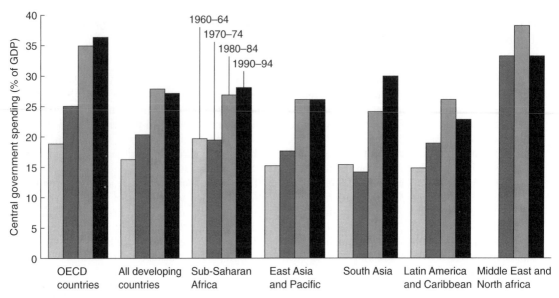

Figure 7.7 The expansion of governments worldwide. Adapted from World Bank (1997).

in physical infrastructure, agricultural support services, public utilities and in social services tended to be cut within structural reforms. In Ghana, for example, real health expenditure in 1982–83 had fallen to 20 per cent of 1974 levels (Mackintosh, 1992: 78). However, despite the widespread condition within SAPs for the reduction in state expenditure, the World Bank itself has failed to find any significant association between the size of the state and the rate of economic growth in the developing world (World Bank, 1988).

Conditions within SAPs for internal reforms of the state may include deregulation (e.g. giving up public monopolies in education), cost recovery (the use of fees for services such as health aimed at cutting demand and promoting efficiency) and the targeting of remaining tax-financed spending on the most needy (such as subsidies concentrating on primary education and higher-level services being charged at nearer cost). However, such state reforms have proven much more difficult to implement:

> State reforms dealing with regulation, social services, finance, infrastructure, and public works, cannot be accomplished so rapidly because they involve changing institutional structures established for different purposes, to fit different rules of the game. This kind of institutional reform involves wrenching changes in the way government agencies think and act, and often a complete overhaul of long-established systems of patronage and corruption. (World Bank, 1997: 13)

A further promoted means of saving on tax revenues, and thereby assisting with balance of payment difficulties in the developing world, is to devolve the activities of the state and its decision making to other institutions such as in the private or voluntary sectors. Under such circumstances, 'a patchwork of provision develops, each patch tended by an agency or NGO. The state function is reduced to a co-ordination, or more vaguely, an "enabling role"' (Mackintosh, 1992: 83).

Making state institutions more accountable for their actions and responsive to public need are functions which are being promoted currently on the grounds of democracy as well as economic efficiency. It has been suggested that the debate concerning 'good governance' in recipient countries initiated by the World Bank may be the most significant factor in promoting democracy in those countries (Sharp, 1992). However, the World Bank's emphasis on the rapid introduction of multiparty political systems as a condition for democracy and good governance, is not without controversy. For example, such prescriptions may not do justice to the often very contrasting sociopolitical contexts within the developing world and may deny political changes already under way. Sharp (1992) cautions that a move towards multipartyism may itself serve to 'manipulate and sharpen' ethnic and/or religious antagonisms. Progress towards democracy requires that transformations in the political process reach and strengthen all levels of civil society, i.e. go beyond the existence of free and fair elections: 'Political participation is not just a casting of votes. It is a way of life' (UNDP, 1993: 23).

In the 1990s, the rather sterile debate of state versus market is being replaced by a more nuanced assessment of the extent of state interventions, of their nature and of the role of the state in relation to other institutions of development. It is recognised from the ideological Left and Right that both state and market as institutions may 'fail to work well' (Mackintosh, 1992). The search is now increasingly for an appropriate mix of market and state activities as essential for 'sustainable, shared, poverty-reducing development' (World Bank, 1997: 4).

Chambers (1993) asserts that 'the state has often done those things which it ought not to have done, and has left undone those things which it ought to have done' (Chambers, 1993: 121). The suggested task is to overturn the 'normal bureaucracy' of the state (which centralises, standardises and simplifies) and which performs worst where needs are diverse and complex (now recognised to be characteristic of the livelihoods of the poor), to establish an 'enabling' state. For the World Bank, the challenge is to enhance the capability of the state to undertake and promote collective actions more efficiently and to provide an appropriate institutional foundation for markets (World Bank, 1997). Broadly, this depends on the state focusing what capability it has on those tasks it can and should undertake and on raising that capability through the strengthening of public institutions. For Chambers, a whole range of reversals 'of location, learning, explanation, values, control, authority and power to put first the poor and the periphery' (1993: 110) are required. In consequence, the pattern of state activities will necessarily be varied. However, understanding and supporting the diverse priorities of poor people, it is suggested, will be equitable in that people will be assisted in gaining what they want, will be efficient through the mobilisation of creative energies of people and also will be sustainable, through the provision of incentives for long-term self-reliant investments by the poor:

> The vision is then of a state which is not only protector and supporter, but also enabler and liberator; and of the 1990s as a decade for equity and efficiency through reversals and diversity.
> (Chambers, 1993: 121)

The final section of this chapter considers how local communities may be enabled to control their own development, including through new relationships with state (as well as voluntary and private) institutions.

Community participation and empowerment

Community, people's or grassroots organisations exist in many diverse forms and have a long history of working to serve members' interests and to pursue collective goals in local development. They are one form or level of non-governmental organisation (NGO), a wider term used to include also those voluntary organisations such as charities and campaigning organisations which exist to serve the needs of third parties (who themselves are not members of the organisation). Empowering such organisations is now promoted by various agencies as the route to an alternative development that is more democratic, efficient and sustainable. In the process, new relationships between community organisations and other institutions are being developed and changes are occurring within such institutions, including in terms of the nature and strategic orientation of activities undertaken.

Traditional forms of social organisation at the community level include burial societies, workteams, village cereal banks and squatter associations, existing to serve some mutual benefit of their members. They are typically highly localised and focused on specific issues (Craig and Mayo, 1995). They are dynamic, have varied degrees of formality and their formation may or may not have been stimulated from outside those communities. NGOs, more widely, have been defined as a 'residual' category, potentially including every formal association which is not government-led (Crow, 1992: 268), but excluding economic enterprises such as transnational companies (TNCs), political parties and informal networks that represent religious, ethnic or class interests. The activities of NGOs may be local, national or extend internationally. The uniting characteristic of these diverse institutions is suggested to be their 'almost universal slogan' of 'empowering the poor' and, as such, is the basis for the contention that NGOs constitute a potential source of an alternative development (Craig and Mayo, 1995).

The term *empowerment* is used by the development community to refer to 'something more than participation'. Since the early 1970s, fostering the 'participation' of local people in development through the activities of community institutions has been widely promoted by various agencies (Potter, 1985). In practice, however, participation has often meant little real influence in policy making or planning for local communities. Empowerment is a desired process by which individuals, typically including the 'poorest of the poor', become the agents of their own development (Craig and Mayo, 1995: 132). Empowerment entails creating power among local communities through consciousness raising, education and the promotion of an understanding within communities of the sources of local disenfranchisement and of the actions they may take. It may also (or involve the transfer of power from one group, such as the controlling authority, to another.

Box 7.4 outlines the example of the Campfire programme in Zimbabwe, which aims to provide the institutional setting in which communities can control the management of predominantly local wildlife resources and manage the benefits of these developments. One of the best-known programmes in Africa, it combines community empowerment, conservation, local resource management and development.

NGOs do not set development policy or pass legislation, but traditionally they have been able to influence these interventions and operate in development through lobbying, consultation and collaboration, giving advice

---

BOX 7.4 Some community organisations in development

Campfire, Zimbabwe

Campfire (Communal Areas Management Programme for Indigenous Resources) is a programme which combines largely wildlife resource management with community empowerment and development ends in those areas of Zimbabwe where land is held in trust by the community rather than under individual ownership. It provides the institutional structure in which community groups can receive the benefits, for instance through wildlife tourism and hunting or safari operations, as well as bearing the costs of living with wildlife. From the village through to the district level, coordination and management activities are carried out by elected committees. Many areas are recording financial successes, with revenues subsequently being allocated to individual households or for community purposes such as establishing a clinic or buying school furniture. Several government ministries, a number of Zimbabwean NGOs, World Wide Fund for Nature and the US Agency for International Development (USAID) have been important in providing technical support, training materials, assistance on issues of ecology and research on the programme. Further legislative and institutional changes are needed to strengthen local participation in decision making.

The Aga Khan Rural Support Programme

In Gujarat Province of India, the Aga Khan Rural Support Programme (AKRSP) is an NGO working to catalyse community participation in watershed management. At the outset, through techniques of participatory rural appraisal, the external team and local people work to assess the natural resources of the village, the indigenous and adapted practices and existing management systems and institutions. A village natural resources management plan is then prepared by the villages and local institutions reinvigorated or created anew to present to external agencies for funding (local communities also invest their own resources) and to implement the activities identified. Experience is showing that individual investments by farmers have risen since the initiation of the watershed programme. Village institutions have developed group credit schemes and taken on a number of operations such as ploughing and pooled marketing activities. Out-migration from participating villages has declined.

The Self-Employed Women's Association of India

The Self-Employed Women's Association of India (SEWA) is an organisation established to represent women working in the 'informal' sector in India, in hawking, petty vending and as home-based workers in the textile industry, for example. Before the formation of SEWA in 1972, such women were ignored by the organised unions. SEWA aims to improve women's working environments and to enhance their income-earning opportunities. Initially the focus was on work-related issues of women in urban areas. However, the organisation has since expanded into rural areas and training and welfare concerns. SEWA activities currently encompass savings and credit cooperatives, producer cooperatives, training courses (from radio repairs to midwifery), and banking and legal services.

Sources: Campfire from Olthof (1995); AKRSP from Shah (1994); SEWA from UNDP (1993).

Table 7.3 Four generations of strategic orientation for NGOs.

	Generation 1 relief and welfare	Generation 2 community development	Generation 3 sustainable systems development	Generation 4 people's movements
Problem definition	Shortage	Local inertia	Institutional and policy constraints	Inadequate mobilising vision
Time frame	Immediate	Project life	Ten to twenty years	Indefinite future
Scope	Individual or family	Neighbourhood or village	Region or nation	National or global
Chief actors	NGO	NGO plus community	All relevant public and private institutions	Loosely defined networks of people and organisations
NGO role	Doer	Mobiliser	Catalyst	Activist or educator
Management orientation	Logistics management	Project management	Strategic management	Coalescing and energising self-managing networks
Development education	Starving children	Community self-help	Constraining policies and institutions	Spaceship Earth

Source: D.C. Korten (1990).

(including to governments and business) and through undertaking projects themselves. The balance of activities in which NGOs are involved changes over time. Korten (1990) has identified three generations through which voluntary development organisations appear to evolve and which express varied strategic orientation. Although many NGOs intervene initially in the areas of relief and welfare, they may subsequently move towards actions focused on addressing the underlying causes of such deficiencies through catalysing broader policy and institutional changes (Table 7.3). Clearly, these generations 'overlap and co-exist' (Chambers, 1993: 90). 'Fourth generation' strategies of NGOs aim to go beyond the previous initiatives focused on changing specific policies or institutional subsystems. They encompass and reflect some of the most contemporary factors of institutional change centred on mobilising the energies of communities for voluntary action towards extending the successes already shown of people's movements in driving social changes:

> The job of the fourth generation voluntary organisation is to coalesce and energise self-managing networks over which it has no control whatever. This must be achieved primarily through the power of ideas, values and communication links. (Korten, 1990: 127)

The search for sustainable processes and patterns of development has been a primary force in the mounting attention through the 1990s given to promoting the empowerment of local communities using new and existing community institutions; 'an essential element of sound resource management is a local body that discusses, organises, plans, takes action and responds at a human scale' (Pye-Smith and Feyerarbend, 1995: 304). Poverty as a symptom of the disempowerment of local people (such as through a decline in access to resources on which they depend or modifications to their traditional tenure rights or the right to exclude outsiders) is a major constraint on sustainable development in the future (Vivian, 1992; J. Friedmann, 1992). As Pye-Smith and Feyerarbend suggest, much presumed 'lack of care' regarding the environment arises because 'people do not feel in charge of or, indeed, do not have the power to act' (1995: 303).

The pursuit of sustainable development has also been a very important factor in prompting changes within non-governmental institutions themselves (as well as promoting new relationships between multilateral or state institutions and NGOs as identified above). For example, although many NGOs were established around, or in response to, natural resource and environment issues, immediate relief and welfare needs often exceeded the capacity of these institutions to fulfil them. Furthermore, such first-generation strategies carried the danger of creating local dependence. Second-generation strategies therefore aim to create local self-reliance through developing the capacities of people to better meet their own needs. However, for such actions to continue beyond the duration of NGO presence or extend beyond a few favoured localities and to

Plate 7.2 A community-based organisation: sign calling people to form a cooperative for savings and credit, Kampala, Uganda (photo: Ron Giling, Panos Pictures)

be sustainable, it was recognised that NGO strategies needed to look beyond the local level to address the policy and institutional structures which often served to centralise control of resources and prevent essential services from reaching the poor. Creating an institutional setting which facilitates sustainable development now involves NGOs in 'working simultaneously to build the capacity of the people to make demands on the system and working to build alliances with enlightened power holders in support of action that makes the system more responsive to the people' (Korten, 1990: 121).

The readiness and evident ability to adapt is one characteristic of NGO activity which is suggested to make these institutions particularly suited to effecting sustainable development interventions. In addition, their relative smallness (Chambers, 1993), the tradition of working with the poorest groups, with women as well as men and from the grassroots (Craig and Mayo, 1995), and the calibre, commitment and continuity of staff (Conroy and Litvinoff, 1988) have proven to be essential characteristics in developing the relationships with local peoples required for sustainability. Further characteristics of projects which appear to be showing signs of sustainability (Conroy and Litvinoff, 1988) depend on and are part of the process of empowering communities. For example, successful projects have adopted a 'learning process' rather than a 'blueprint'

approach to planning and implementation, involving a continued dialogue between all interested parties with modifications being made as experience is gained during the course of the project (Chambers, 1993). People's, rather than outsiders', priorities are put first and communities have the required perceived self-interest in their involvement. In turn, this has demanded, for example, security of rights over productive resources.

As debt burdens within developing countries have forced cut-backs in state investments in development interventions, community-based organisations (CBOs) have increasingly been promoted as a route to alternative solutions which reduce state expenditure as discussed above (Plate 7.2). The promotion of community participation and empowerment by institutions such as the World Bank is not without its critics, however. Mackenzie (1992: 1) suggests that the call for greater community involvement in development may be cheaper financially and ideologically 'at a political moment when the more difficult options of challenging the protectionism of industrial economies and renegotiating terms of trade are ignored and when state intervention is made the scapegoat for policy failure'. Donations to NGOs may also prove cheaper than conventional bilateral aid for Northern governments in times of economic recession. For example, Middleton, O'Keefe and Moyo (1993) note that although NGOs are usually charities whose funds come from donations, the state is frequently the largest single donor. Since aid budgets are easy targets during slump periods, they suggest that many Northern governments may see donations to NGOs as a surrogate for taxation and the cheaper alternative to official development assistance.

The desire for democracy has also prompted new ways of working with community organisations. It is now recognised that community participation and empowerment for an alternative development requires more than the occasional incorporation of localised community actions into wider development processes. Democracy and empowerment depend on the reordering of social relations and the challenging of local and state power. As such, local communities must move beyond actions in their immediate practical or survival/ coping spheres (which are the traditional arenas for many people's organisations), to act strategically and to engage with the state and political processes at various levels (Taylor and Mackenzie, 1992).

Evidently, such processes towards democracy may be unpopular and may be resisted by those with local or state power. Korten (1990) identifies a number of critical areas in which NGOs can contribute towards local empowerment and community involvement in strategic development actions. They include the monitoring of and protesting against abuses of power, particularly on behalf of governments and businesses, and opening them for public scrutiny and action. For example, even where a free press and independent judiciary exist, their effective functioning depends on an informed and vigilant citizenry. NGOs may have a comparative advantage in promoting such developments through their own democratic principles and structure. For example, NGO membership is more often based on a commitment to a normative purpose than to a narrow self-interest, and their organisational frameworks tend to be more democratic than hierarchical: 'simple, human concern for other people as individuals and in very practical ways is one of the hallmarks of NGO work' (Edwards and Hulme, 1992: 14).

However, democracy also depends on the condition of social equality. Although community empowerment necessarily involves the challenging of social structures, it cannot be assumed that all interests

are served through 'community development efforts'. Vivian (1992: 62) notes that many traditional societies are repressive and even 'seemingly highly participatory traditional resource management systems' in reality often exclude large numbers of people. Evidently, societies are not undifferentiated in requiring answers from NGOs on difficult questions, such as who they represent. J. Friedmann summarises it like this:

> Many fault lines run through both rural and urban communities: religious, ethnic, social class, caste, linguistic. And the universally subordinate role of women requires us to identify yet another source of social tension and conflict that cuts across all of the others. (1992: 7)

Many varied expectations in development are being placed on NGOs and CBOs, and this will continue into the future. There is already evidence, however, to suggest that these pressures may well be competing under certain conditions. For example, as increasing aid to developing countries is channelled through NGOs rather than through the state, there is a danger that control of public policy will become concentrated in the hands of the most powerful NGOs based abroad, which may be 'more difficult to counter or even monitor than conditions on aid to governments' (Mackintosh, 1992: 83). The dependence of NGOs on money from official sources may also present problems of whether such NGOs are accountable to the donors or the grassroots communities. Furthermore, as service providers, NGOs are increasingly being required to operate in more market-driven ways. In order to be commercially competitive, the risk is that the smaller, perhaps more community-based organisations, will be driven out of that market. The need to focus on those services within such contracts may also compromise the broader watchdog or advocacy functions more traditionally associated with NGOs.

The relationship between NGOs and the state will be critical in determining the prospects for inclusive and sustainable development in the future. However, it is evident that the capacity for community organisations to influence development interventions and processes is intricately related to factors of change throughout the hierarchy of institutions considered within this chapter.

PART III

Spaces of development:
places and development

CHAPTER 8

Movements and flows

Introduction: unravelling complexities

In addition to examining the nature of specific development variables in particular locations, geographies of development should consider relationships between people, environment and places in different locations and at a variety of different scales, ranging from the micro-level, such as the individual or the household, through the local community level to the regional, national, international and ultimately the global level. These relationships are, however, by no means static. On the contrary, the nature and relative significance of these relationships are changing constantly, both through time and space, and are themselves determined to a large extent by complex movements and flows of people, commodities, finance, ideas and information. This chapter will attempt to elucidate and 'disentangle' some of these complex and interrelated movements and flows. Starting with a real-world example, in order to illustrate some of the interlinkages and flows which take place over time and space, we will then consider movements of people, before examining flows of commodities and finance through trade, aid and debt.

Movements and flows in the 'real world'

A real-world example should serve to demonstrate some of the many possible interconnections between people, commodities and finance across the world. Let us consider the case of a resource-poor farming household in Ivory Coast (Côte d'Ivoire), West Africa, whose main income is derived from producing coffee for export. Although Ivory Coast is one of Africa's major coffee producers, and in 1995 was the world's ninth largest producer, world coffee production in the 1990s is dominated by Latin American countries such as Brazil and Colombia, each producing four times more coffee than Ivory Coast, with 16.6 per cent and 14.5 per cent respectively of world production. The spread of coffee cultivation in Ivory Coast dates from the French colonial period when, after 1930 and again in the 1950s, France assured a guaranteed market at very high prices for large quantities of coffee.

The producer household in Ivory Coast is heavily dependent on receiving a good return from cash crop sales in order to buy food and clothes, pay school and medical bills and hopefully, over time, steadily improve its standard of living. But there is a wide range of factors which can affect household income in any given year. Some of these factors may be environmental, whereas others are social, political and economic. Inadequate rainfall, declining soil fertility and pest attacks are common, and usually unpredictable, environmental factors. Small farmers would generally receive little, if any, advice and practical help from agricultural extension officers, and they often lack the necessary finance to buy pesticides and fertilisers to raise productivity. The availability of labour is another key element, perhaps the most important factor in poor households with low levels of technology. Such farmers can rarely afford to hire wage labourers and are invariably totally dependent on the reserve of family labour. Family members between the ages of 15 and 40 years are likely to be the fittest and therefore particularly valuable for working on the farm. But if at crucial times in the cropping cycle, just one key member becomes unwell, or perhaps decides to leave the village to seek work in the city, then this withdrawal of labour can have a major impact on the productivity and therefore the general well-being of the household.

In addition, rural producers are also affected by a range of economic and political decisions that are well beyond their control. For example, farmers and others in developing countries have been affected to a greater or lesser extent since the mid 1980s by structural adjustment programmes (SAPs) imposed by major international donors such as the World Bank and the International Monetary Fund (IMF). In fact, it might be suggested that 'not since the days of colonialism have external forces been so powerfully focused in shaping Africa's economic structure and the nature of its participation in the world system' (Binns, 1994b: 163). Currency devaluation, raising interest rates and the removal of subsidies and price controls are just a few of the measures commonly introduced by SAPs (Mohan, 1996: 364). In the case of Ivory Coast, with primary products representing 83 per cent of exports, the country had a balance of payments deficit in 1993 of US$1, 229 million, a total external debt of US$19,146 million, and debt service representing 29.2 per cent of exports (World Bank, 1995). As far as the coffee farmer is concerned, currency devaluation could have a significant effect on returns from coffee sales. In fact, shorter- or longer-term fluctuations in the prices which producers receive for commodities such as coffee, can have a major impact on household economy and well-being. With the dominance of Brazil and Colombia in world coffee production, one or more of a variety of factors affecting production in these two countries could have a significant effect on the world coffee price. In simple terms, overproduction could lead to lower world prices, whereas a fall in production, perhaps due to perturbations in the Brazilian climate, might result in higher world coffee prices. This happened in early 1997, when prices reached a twenty-year high, due to extremely cold weather in Brazil, forcing producers to relocate entire coffee plantations to warmer areas. But in reality the situation is often much more complex.

The decision to invest in the production of 'cash crops' such as coffee is usually taken by smallholder producers when prices are high. It may involve a major reorientation of household activities and a significant switch of labour inputs from food production. There is evidence to show that in some cash crop producing areas, family nutrition has suffered due to a relative neglect of food production (Kennedy and Bouis, 1993). Furthermore, by the time the first cash crops are harvested, which may be up to ten years after planting, prices may well have fallen below the levels which existed at the time of planting. The situation is further compounded by the fact that Third World producers generally only receive a small fraction of the final selling price of commodities such as coffee. Within producing countries there are usually networks of buyers, agents and sub-agents, each taking their share of the price, to say nothing of

the substantial element taken by large transnational companies who process the final product in Europe or North America. An Oxfam study revealed that in Uganda, where coffee accounts for 90 per cent of all exports, coffee growers in 1993 received the equivalent of £0.08 (just 5 per cent) of the final value of a jar of coffee sold in UK supermarkets for £1.60. In sharp contrast, the shippers and roasters, who are generally part of one transnational corporation, received 65 per cent of the final selling price (Oxfam, 1994).

In an attempt to reduce the impact of fluctuating world coffee prices on small producers, Oxfam's Bridge programme attempts to provide a market for Third World producers, paying fair prices and purchasing through organisations which ensure that the bulk of the price reaches the producers. Bridge is involved with three other trade organisations in marketing a brand of coffee called Cafedirect, where the coffee is purchased directly from small farmers, who receive a price linked to the minimum floor price previously defended by the International Coffee Organisation. As Oxfam point out: 'When Cafedirect was launched during the trough in world coffee prices, producers were paid $1.20 per lb. Had they been selling in the international market, they would have been paid around 65 cents per lb' (Watkins, 1995: 148).

The purpose of presenting this case study, typical of many other situations, is to demonstrate that the smallholder coffee producer in Ivory Coast is just one element in a complex system of relationships involving local, national and international movements and flows of people, commodities, finance, ideas and information. Too frequently in the past, researchers have considered just one element in the system without appreciating the interconnectedness at different scales. As we have seen in earlier chapters, globalisation is one of the most significant features of the late twentieth century. With the increasing speed and frequency of international air travel since the 1960s, and most especially with the 'great leap forward' during the 1980s and 1990s in the transmission of information through satellite communication and the Internet, the world is indeed becoming an ever shrinking 'global village'. Unhappily, however, as in so many communities (and particularly those in the Third World), there is a wide and ever growing disparity between the wealth and living standards of the haves and the have-nots!

Although it would be impossible to identify, let alone discuss, the myriad of movements and flows across the globe, this chapter aims to shed light upon some of those which involve people, commodities and finance. We will firstly examine movements of people and then consider trade, aid and debt, emphasising wherever possible the changing nature of these movements and flows over time and space.

People on the move

Population movements, or migrations, have been taking place in various shapes and forms for centuries, and there are many detailed studies and publications on this topic. As was recognised in Chapter 5, the movement of people within and between countries has played an important role in determining population growth rates and in affecting other factors such as ethnic composition.

In broad terms, population movements may be divided into forced and voluntary, but then more detailed classifications commonly focus on such aspects as the distance covered and the frequency and time span over which the migration occurs. Some writers differentiate between migration and circulation (Drakakis-Smith, 1992; A. Gilbert and Gugler, 1982; W.T.S. Gould and Prothero, 1975). Although migrations are usually permanent or irregular and involve a lengthy change of residence, circulations are generally shorter, sometimes daily, periodic or seasonal. Much interest has also been shown in the decision making involved in the migration process, and in the consequences of migration for the well-being of the migrant and the migrant's family, as well as the problems and benefits which migration causes in both the source and reception areas.

Seasonal migration and circulation

Although some population movements have a long history, there seems little doubt that colonial policies played a key role in accelerating the process. In Africa the introduction by the colonial powers of taxation in the form of cash payments, together with the creation of many new towns, mines and cash-cropping areas, led to large-scale migration of wage labourers. In West Africa the movement of Mossi men from Burkina Faso, to help with the cocoa harvest in Ghana and Ivory Coast, has been going on for many years and the money they earn is a vital addition to their poor villages in Burkina.

In northern Nigeria, with a long dry season when little can be cultivated on non-irrigated farmland, there is a tradition of seasonal migrations called *cin rani*. Since the great majority of migrants in this strongly Muslim region are men, they are more correctly known as *masu cin rani* 'men who while [?eat] away the dry season' (Prothero, 1959). Commonly, men leave their homes in the dry season to visit relatives and/or to engage in craft industries, trade or irrigated farming. Before the 1930s, Prothero reports that there was little reference to seasonal migrations in the Sokoto region of north-western Nigeria, but a traffic census in 1928 recorded 3,500 migrants per month passing southwards through Yelwa on the River Niger and heading mainly for the large towns of Yorubaland in south-western Nigeria. Dry season migration seems to have accelerated during the 1930s, such that the *Annual Report for Sokoto Province* in 1936 notes that 'there has been a large seasonal migration to the Gold Coast [Ghana] and elsewhere by men in search of money to pay their tax and support their families' (Prothero, 1959: 22). A later survey, undertaken in Sokoto Province in the dry season of 1952–53, enumerated some 259,000, predominantly male, migrants, with south-western Nigeria and Gold Coast (Ghana) as the main destinations. Some 92 per cent of migrants were seeking to supplement their income in various ways, through such occupations as labouring, petty trading, fishing and craft work. It seems likely that, on their return home, migrants probably brought more money into Sokoto Province than they could have created without migrating (Prothero, 1959). But much time is spent in travelling and Prothero argued that the Province would gain a great deal if this labour could be diverted to local productive work, e.g. in the expansion of cotton and groundnut production for export. A particularly significant point, which has relevance for other poor migrant source areas is that 'the total number of migrants away from the Province for several months of the year must go a considerable way towards conserving supplies in the home areas' (Prothero, 1959: 34).

Seasonal migration is well established in other parts of the developing world. For example, much of the labour force for cutting sugar cane in north-west Argentina comes from neighbouring areas of Bolivia, where unemployment and poor wages provide an incentive to migrate. Whereas this generally involves migration on a seasonal basis, migrants have sometimes settled in Argentina, moving to cities such as Buenos Aires, where there are better employment opportunities (D. Preston, 1987).

Rural–urban migration

One particularly significant form of migration in many Third World countries is from rural areas to towns and cities, for reasons such as advancing education or to obtain specialist health care (Chapter 5). Young people are frequently attracted to towns by the 'bright lights' syndrome – the idea that the towns have modern facilities compared with what are often perceived as being backward and traditional rural areas. In many parts of the Third World, economic reasons play the most important role in drawing people to the cities. For many, and particularly for young males, a spell in the big city is seen as an opportunity to earn a good income and is also regarded as an initiation into adulthood and Western culture (Chapters 1, 3 and 10).

In Central America, Mexico City's population grew from 14 million to a massive 19 million during the 1980s, and much of this growth was due to migration. Meanwhile, on Mexico's once predominantly rural Yucatan Peninsula, there has been rapid development of tourism since the 1970s, which has led to the city of Merida tripling its population in twenty-five years, to reach 650,000 in 1992, which represents almost half of the state's total population. Ninety per cent of migrants to the city have been drawn from Yucatan's rural areas, whereas villages surrounding Merida have become dormitories for commuters working in the city (MoBbrucker, 1997).

A fascinating Indian study of some fifty years of migration from the rural village of Sugao in Maharashtra State to the city of Bombay over 150 miles away, reveals some interesting features about both the migrants and the process of migration. Significantly, over a long period of contact between village and city, a valuable city-based network of village relatives and friends has been established, and this plays a key role in locating jobs and providing shelter for newcomers. Most urban employment continues to be in the textile industry, though its relative importance has declined in recent years. Few rural families remain untouched by the migration process, which is overwhelmingly male-dominated. Remittances from city workers have had a major impact in improving conditions back in the village, such that almost half the households now have piped water and more than two-thirds have electricity. However, very few men actually leave Sugao permanently, and generally return home when their productive working life is over and it becomes too expensive to remain in Bombay (Dandekar, 1997).

Although Africa is still overwhelmingly a rural continent, with some 70 per cent of the population in sub-Saharan Africa living and working in rural areas, the average annual urban growth rate of 4.8 per cent between 1980 and 1993 was more rapid than any other part of the world, and much of this growth was due to rural–urban migration (World Bank, 1995). In the immediate post-independence period, rapid rural–urban migration was fuelled by significant increases in urban formal sector wages. It is estimated that real average urban wages increased by 40 per cent in Zambia between 1964 and 1968, compared with only a 3 per cent increase in farmers' incomes. In Tanzania the real urban minimum wage increased nearly fourfold between 1957 and 1972 (Jamal and Weeks, 1994).

Concern has been expressed about feeding these growing urban populations and the availability of employment, and terms like *overurbanisation* have been used (A. Gilbert and Gugler, 1982: 163; Gugler, 1997: 114–23). For example, in Ghana during the first decade after independence (1957–67), the urban population grew by 6.6 per cent, but employment in the modern sector increased by only 3.3 per cent, whereas registered unemployment rose by as much as 9.3 per cent each year (Knight, 1972). Another early study by Gutkind in Lagos, estimated that in 1967 approximately 21 per cent of all males over 14 were actively seeking work (Gutkind, 1969). While searching for work, many migrants relied on the support system provided by ethnic friends and relatives, who could assure a level of survival not radically different from that experienced in the rural village. Work is often gained through such ethnic and kinship links and the higher wages earned by a few manage to 'trickle down' to support those marginally employed and even unemployed. Others will in time find employment in the highly diverse 'informal sector', which might involve working in a family tailoring or carpentry workshop, or perhaps more likely begging or hawking on street corners, or even illegal activities such as prostitution and theft.

Governments and academics have given much thought to strategies for controlling rural–urban migration, notably the reduction of urban–rural differentials through the improvement of living standards in the rural areas and the redistribution and generation of more urban employment opportunities (Becker and Morrison, 1997). However, Riddell concluded in the late 1970s that 'there is no easy remedy for spatial differentials: urban incomes cannot be lowered because of the severe economic and political implications; jobs in the modern sector cannot be created in sufficient numbers because of the limitations upon the national economy; and rural agriculture cannot be transformed because of the scale of the problem and the very limited results likely to accrue' (Riddell, 1978: 260). Perhaps only in South Africa under the oppressive apartheid regime was rural–urban migration tightly controlled, with Black migrants carrying identification passes and their movements being closely regulated by police and security forces. However, during the 1980s and 1990s, rural–urban migration in South Africa accelerated considerably with the progressive relaxation of apartheid controls.

But in the 1980s and 1990s, the situation has changed dramatically in many countries, with structural adjustment programmes, falling urban incomes, deteriorating urban services and public sector retrenchments leading to a massive decline in urban living standards, which has resulted in urban growth rates and migration processes adjusting to urban economic conditions, in some cases leading to people actually moving back to rural areas. The rural–urban income gap collapsed in the 1970s and 1980s, and by the 1980s it seems that a 'new urban poor' had developed in many African cities (Jamal and Weeks, 1994). In Ghana, indexes of real minimum wages show a rise from 100 in 1970 to a peak of 149 in 1974 then a massive decline to 18 in 1984. In Tanzania there was a rise from 100 in 1957 to 206 in 1972, then a fall to 37 by 1989. The gap between rural and urban incomes in many countries either vanished or actually shifted in favour of the rural sector. In Sierra Leone the average non-agricultural wage in 1985–86 was estimated to have been 72 per cent less than average rural household incomes (Jamal and Weeks, 1994). Riley suggests that by 1986 the urban poor in Sierra Leone were 'a deprived group with fewer income or equivalent earning opportunities than the rural poor', which had adverse effects on levels of infant mortality and malnutrition (Riley, 1988: 7). Households have often responded by engaging more in informal sector activity, with wage-earners taking on additional cash-earning activities, and by growing food on any available pieces of land. Jamal and Weeks (1994) suggest that in Uganda another coping strategy is for urban residents, particularly from non-local ethnic groups, to migrate to rural areas. The 1980 Zambian census revealed a slowing down of urban growth rates and indicated a significant increase of urban–rural migration. Surveying the evidence, Potts suggests that a form of counterurbanisation has been taking place in a number of African countries, 'where

the number of urban residents opting to leave the city and move to rural areas has exceeded the number of ruralurban migrants' (Potts, 1995: 259). These return migrants are often the poor and unemployed, who are moving back to their rural homes, and this movement seems to be over and above the patterns of circulation between rural and urban areas which are such a long-established feature in many African countries.

International tourism and developing countries

Another type of population movement is represented by the phenomenal growth in international tourism since the Second World War, from some 25 million tourist arrivals in 1950 to 425 million in 1990. Although Europe remains the dominant source and destination of international tourists, the expansion of air travel and the quest for adventure and exotic places has led to a massive increase in long-haul travel, particularly to the Caribbean and Central America, to Asian countries such as China, Hong Kong, Malaysia, Thailand, and to African countries such as Morocco, Tunisia, Kenya, Zimbabwe and South Africa (D. Harrison, 1992); see Chapters 4 and 6.

Tourism makes a valuable contribution to the economies of some developing countries (Table 8.1, overleaf); for example, in 1988 tourism constituted 60.1 per cent of the value of exports in Barbados, 36.6 per cent in Egypt, and 33.5 per cent in Jamaica.

In countries such as Barbados, Egypt, Jamaica, Kenya, Morocco and Thailand, tourism makes an important contribution to both exports and GDP. In 1988 tourism actually earned more for Kenya than either of the two traditional exports of coffee and tea, and the 676,900 tourists who arrived in the country contributed no less than US$404.7 million to the economy. By 1995, despite exchange rate fluctuations, tourism accounted for 16 per cent of GDP and brought gross receipts of $486 million, compared with $350 million, from tea. Almost half of the international visitors to Kenya in 1995 came from Germany and the United Kingdom (Binns, 1994b: 146).

Tourism also provides a considerable amount of employment, both directly within hotels, and also indirectly in taxis and transport, craft industries, restaurants and entertainment. The spectacular tourism development on Mexico's Yucatan Peninsula since the 1970s has transformed a formerly sparsely settled agricultural region around Cancun and has resulted in considerable in-migration, which is linked to work

Table 8.1 Tourism balance in selected countries, 1988.

Country	Tourism monies (US$ million)			Tourism as a percentage of		Receipts per capita (US$)
	Receipts[a]	Expenditures	Balance	Exports[b]	GDP	
Bahamas	1 136	150	986	–	53.0[c]	4 544
Barbados	459	29	430	60.1	29.8	1 836
Singapore	2 399	930	1 469	–	9.7	905
United Kingdom	11 023	14 650	−3 627	5.7	1.3	193
Swaziland	18	14	4	4.1	2.9	24
Jamaica	525	45	480	33.5	16.5	214
Israel	1 343	1 130	213	8.9	3.2	305
Costa Rica	165	72	93	10.3	13.5	57
Morocco	1 102	132	970	26.3	5.0	46
Thailand	3 120	602	2 518	15.6	5.4	57
Mexico[d]	3 497	2 361	1 136	12.7	2.5	43
Argentina	634	975	−341	–	0.8[e]	20
Egypt	1 784	75	1 709	36.6	4.9	34
Kenya	410	25	385	21.9	4.8	17
Sri Lanka	79	60	19	4.3	1.1	5
India[f]	1 390	438	952	10.3	0.6	2

[a] Tourist receipts exclude payments for transport.
[b] Exports of all goods and services.
[c] Data from the Caribbean Tourism Organisation.
[d] 1987 figures.
[e] Data from the World Bank.
[f] 1986 figures.
All other data from the World Trade Organisation and the International Monetary Fund.
Source: D. Harrison (1992).

opportunities both in the construction industry and servicing the growing tourist population (MoBbrucker, 1997). In many developing countries there is considerable potential for strengthening the linkages between the tourist sector and local food and drink production and supply systems, thus reducing the need for expensive imports. However, in Fiji, the linkages have so far been limited, due to high imports of goods, hotel furnishings and services, and the repatriation of substantial profits by airline companies and foreign-owned hotels, which account for almost 60 per cent of bed capacity (Lockhart, 1993).

There are several concerns relating to the growth of tourism in poor countries. Hotel managers are often expatriates and a large proportion of tourism-related jobs are low paid; the wages of Kenyan hotel staff are lower than in many other sectors of the economy, except possibly agriculture and domestic service. Tourism may also be a strongly seasonal activity, such that during slack seasons staff may be laid off without wages. The impact of tourism on environment and society is a further controversial issue. The loss of valuable farm and grazing land and the 'dilution' of local cultures, with the transferance of Western values and patterns of behaviour, are among the major concerns. In small countries such as those of the Caribbean and the islands of the Indian Ocean, along with states such as The Gambia in West Africa, there is a danger of large numbers of tourists overwhelming the country and its people (Plate 8.1). The tiny Caribbean island of St Lucia has a population of only 140,000, yet in 1996 it attracted 356,000 visitors, of whom 195,000 were stayover tourists, the remainder being cruise ship passengers (154,000) and excursionists (6,700).

In Africa, The Gambia is the smallest mainland country and one of the poorest, with an average life expectancy in 1993 among its one million people of only 45 years. South of the capital, Banjul, the 30-kilometre Atlantic coastal strip is being steadily taken over for hotel building, to cater for the growing number of tourists wanting to escape the European

Plate 8.1 Globalisation and international tourism: a tourist hotel in The Gambia (photo: Tony Binns)

winter. In 1988–89 some 112,800 tourists visited the country and numbers are steadily growing, particularly with stability restored after the July 1994 coup. With an estimated average expenditure per tourist of US$400 in 1989, this amounted to about US$45 million from the tourist industry, more than 10 per cent of GDP. The great majority of The Gambia's tourist infrastructure is still confined to the coast, with most tourists seeking a beach holiday with the occasional day excursion. There are a few 'up-country' camps catering for more adventurous travellers, but so far they have had a limited impact on the rural hinterland. Indeed, there is some concern within the country about the possible negative effects of encouraging the penetration of larger numbers of tourists into the poor, remoter rural areas. Already in the coastal area, theft, begging and prostitution are commonplace, as relatively wealthy visitors, invariably with little local knowledge, are seen as easy prey to those who are desperate to make a living. Similar problems are evident on the Kenyan coast.

In Kenya, farmers and pastoralists have had their traditional lands incorporated into national parks, where wild animals have caused damage to crops and livestock. The traditional pastoral economy of the Masai has been severely constrained, whereas some critics have questioned the profitability of allocating large areas of Kenya for wildlife-based tourism. Like other countries which are experiencing increasing pressures from tourism development, Kenya is keen to promote ecotourism which, in theory at least, is designed to

have a more sensitive and sustainable approach to people and the environment. It remains to be seen whether this happens.

Perhaps Kenya and other countries could learn from the experience of the Central American state of Costa Rica, where ecotourism has been successfully developed. The Monteverde Cloud Forest Reserve (MCFR), in western Costa Rica was designated in the early 1970s and is an area of great biological diversity. Visitor numbers have increased from 471 in 1974 to 49,552 in 1992, representing an annual increase of 578 per cent. The MCFR is now one of the main ecotourism destinations in Costa Rica, and the tourism industry has led to the creation of over eighty different businesses, of which a large percentage are locally owned, including hotels, restaurants, craft shops and bookstores. Additionally, tourism has led to the improvement of local education and the conservation of natural resources, whereas successful community participation has resulted in Monteverde becoming one of the most prosperous and successful communities in Costa Rica (Baez, 1996).

Forced migration

In the case of forced migration, the decision to relocate is made by people other than the migrants themselves. The Atlantic slave trade was probably one of the most massive forced migrations in history when, from the late sixteenth to the early nineteenth centuries, more

Plate 8.2 Refugee camp for Rohingya Muslim refugees from Burma in southeast Bangladesh (photo: Howard J. Davies, Panos Pictures)

than 10 million Africans were transported to work on plantations in North and South America and the Caribbean. More recently, in West Africa, Nigeria shocked its neighbours by expelling 2 million foreign workers in 1983 and a further 700,000 in 1985. The purpose of the expulsion was supposedly to reduce unemployment among its own people during the slump following the 1970s oil boom, but other members of the Economic Community of West African States (ECOWAS), notably Ghana, whose nationals constituted the majority of those expelled, were appalled at Nigeria's action and argued that it contravened the 'spirit' of earlier ECOWAS agreements.

A further example of migration, where there is some controversy about the level of force involved, concerns the movement of over 6 million Indonesians from densely settled Java, where more than 60 per cent of the national population live on 7 per cent of the land area, to some of the other 13,000 less populated islands and national territories. Population resettlement began in 1905 under Dutch colonial rule, but accelerated after Indonesia gained independence in 1945, and became even more intensive in 1969 under the government's aggressive 'transmigration programme'. Hancock has described this migration as 'the world's largest ever exercise in human resettlement' (Hancock, 1997: 234). Furthermore, it has been supported by funding from a number of national and international development agencies. Between 1976

and 1986, the World Bank committed US$600 million to the programme, and further assistance has been given by the governments of the Netherlands, France and Germany, as well as USAID, UNDP, EEC, FAO, the World Food Programme and Catholic Relief Services.

The programme has been highly controversial in many ways, including migration to the island of Irian Jaya, where land has been taken by force for settlers from Java. This has fuelled a growing conflict between Indonesian armed forces and nationalist Irianese. Reports suggest that villages have been bombed, people tortured and shot dead, with over 20,000 Irianese fleeing their homes and seeking refuge in neighbouring Papua New Guinea. Additionally, the Indonesian government wants to settle and 'assimilate' all Indonesia's tribal peoples, including moving by 1998 (with force if necessary), Irian Jaya's entire indigenous population of 800,000 tribal people into resettlement sites on the island (Hancock, 1997: 235). Transmigration is also happening elsewhere in Indonesia, with East Timor being seized by the Indonesian army in 1975, to provide for further resettlement from Java. An estimated 150,000 indigenous inhabitants of East Timor have been either killed in fighting or have died of hunger.

Other examples of forced migration in Southeast Asia (Plate 8.2) are the massive outflow of Khmer and Lao fleeing genocide and invasions in the late

1970s and early 1980s, and the many refugees escaping from Vietnam during the 1980s, in some cases as 'boat people' journeying to places such as Hong Kong in search of asylum. Meanwhile, in 1990s Africa, streams of refugees have fled into neighbouring countries to escape civil wars in Liberia, Sierra Leone, Rwanda, Burundi and Mozambique.

The impact of migration on environment and health

The Indonesian resettlement programme has destroyed vast areas of rainforest in one of the most biologically diverse areas of the world. Sumatra alone has lost 2.3 million hectares, of rainforest and the cleared land has rapidly become severely degraded. More than 30 per cent of Sulawesi has been reduced to a similar state. Hancock suggests that some 300,000 people are now estimated to be living in 'economically marginal and deteriorating transmigration settlements,' and are recognised by the Indonesian government itself as 'a potential source of serious political and social unrest in the future' (Hancock, 1997: 237). Infrastructure in the shape of clinics, schools and roads is often inadequate and the settlers suffer from malaria and other diseases. The land has become so impoverished that many people are moving back to the towns

and cities. In recent years, although transmigration has been reduced from its aim of settling 20 million people, the scheme still continues with private companies running enterprises such as the Barito Pacific plywood factory on Mangole Island, and the creation of new settlements to supply cheap labour for agri-business, the timber industry and mining.

The potential environmental impact of the large-scale displacement of populations can be considerable, as was the case during the Ethiopian crisis in the 1980s (Box 8.1). However, two studies undertaken in the mid 1990s among Mauritanian refugees in the Sahel region of the middle Senegal River valley in Senegal (Black, 1997), and in areas of settlement by Liberian refugees in the remote forest zone of eastern Guinea (Black, 1996), revealed some decrease in woodland areas, but remarkably little other negative impact on the environment. In the Senegal study, Black questions whether changes in the fragile environment can indeed be attributed entirely to the influx of 50,000 refugees, and concludes that because the refugee population became dispersed over a large area and good relations were maintained with local populations, there has been little conflict over natural resources between the two populations. In Guinea a similar number of refugees were also spread over a wide area and, as in Senegal, they were largely of the same ethnic groups as the local populations. Black

---

BOX 8.1   The 1980s refugee crisis in Ethiopia

Some of Africa's poorest people are the millions who have been forced to flee from war, terrorism, persecution or natural disasters such as drought. It was estimated that in 1991 there were about 4 million refugees in tropical Africa (O'Connor, 1991), and in the same year Oxfam warned: 'In Africa as many as 27 million people now face starvation as a result of drought and war.... The famine threatening millions of Africans is on a greater scale than the famine of 1984/85.' Famine and refugees in 1991 were associated with Liberia, Angola, Mozambique, Ethiopia and Sudan, but since then people have been forced to leave their homes in Somalia, Sierra Leone and in the Great Lakes region of central Africa, most notably the small, but densely settled, states of Rwanda and Burundi.

Mohamed Amin's film and Michael Buerk's commentary from Korem, Ethiopia, in October 1984 produced what was probably one of the most powerful

pieces of television documentary ever; to millions of comfortable homes around the world it brought images of starving refugees. Here is part of the commentary:

Dawn, and as the sun breaks through the piercing chill of night on the plain outside Korem, it lights up a biblical famine, now, in the 20th century. This place, say workers here, is the closest thing to hell on earth. Thousands of wasted people are coming here for help. Many find only death. They flood in every day from villages hundreds of miles away, felled by hunger, driven beyond the point of desperation. Death is all around. A child or an adult dies every 20 minutes. Korem, an insignificant town, has become a place of grief. (Harrison and Palmer, 1986: 122)

This report had a crucial impact across the world and was instrumental in generating such popular

Box 8.1 continued

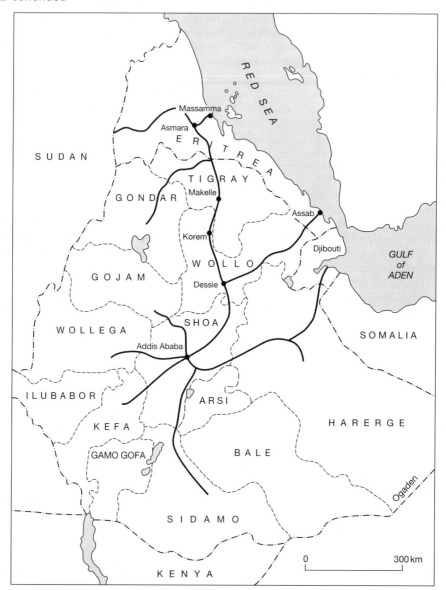

Map of Ethiopia. Adapted from Oxfam (1984).

fund-raising efforts as Band Aid and later Live Aid, which took place simultaneously in London and Philadelphia in July 1985 and broke new ground in worldwide satellite communications, raising over US$100 million by mid 1986.

The causes and effects of Africa's continuing refugee problem are highly complex and each case has its unique features. Nowhere is this more evident than in the case of Ethiopia in the early 1980s, where drought, rural impoverishment and armed conflict all played a role in generating a serious refugee problem (see figure). Ethiopia has a long history of famines. The great majority of rural Ethiopians in the 1960s were living in conditions which were similar to those of European peasants in the Middle Ages. Feudal landlords exacted heavy taxation and

Box 8.1 continued

other obligations from their subjects; quite apart from the sheer volume of produce leaving the peasants' hands, the amount of time spent working for the landlord cost dearly in lost production. It was the repercussions from droughts in Wollo during 1965–66 and 1972 that eventually brought to an end the 44-year rule of Emperor Haile Selassie in September 1974, with the take-over of the Provisional Military Government of Socialist Ethiopia, known as the Derg (see figure). A land reform programme was quickly introduced, but in the years immediately following the revolution, food and cash crop production was disrupted by the many uncertainties caused by the radical transformation in land tenure arrangements. A poor distribution of surplus grain supplies, combined with an inadequate transport infrastructure, severely hampered redistribution to food deficit areas.

But the origins of the refugee problem which developed in the 1980s must also be examined in the context of the Derg's heavy military expenditure and its involvement in a series of costly and protracted conflicts. The civil war in Eritrea was ongoing, having begun in 1962 when Haile Selassie dissolved the federation between the two countries and annexed Eritrea to Ethiopia. The war in Tigray started in 1975, and was concentrated in the densely settled central highlands and the important agricultural region in the west of the province. To compound the situation further, the Derg's forces were engaged in a war with Somalia in 1977–78 over the disputed Ogaden region. American military aid was withdrawn and Ethiopia requested Russian and Cuban aid, which had the effect of alienating many Western governments. In addition, the Derg increased massively its defence spending to US$378 million in 1981, representing a higher per capita military expenditure than any other Black African country and giving Ethiopia the second best equipped army on the continent, after South Africa (Harrison and Palmer, 1986: 94).

Against this background of conflict, poor rains in the early 1980s resulted in hardly any harvest in the northern regions of Wollo, Eastern Gondar and parts of Tigray in 1982. Refugees abandoned their homes and flooded across the Sudan border to await relief supplies. The main rains in July 1983 also failed, and it was estimated that more than 2 million people in Tigray and Eritrea were seriously affected by the drought and needed emergency assistance. The relief supply situation in both regions was further complicated by the lack of security, due to the separate armed conflicts with the government of Addis Ababa. The liberation fronts accused the government of witholding relief supplies and aid donors were also criticised for channelling all their relief through the government, when many of the people at risk were actually in areas not under government control.

The Relief and Rehabilitation Commission (RRC) of the Ethiopian government reported in May 1984 that the official population of 7,800 in the Wollo town of Korem had been swelled by some 35,000 displaced people gathering at the feeding centre there, with a further 110,000 people registered to receive emergency food supplies from the RRC store in Korem. The RRC argued that it was too stretched to distribute seeds to ensure people could plant a crop during the July rains, and other seed distribution schemes were ineffective due to the security situation. One long-term solution to the droughts in northern Ethiopia tried by the government was to move people out of the highlands and resettle them in new RRC settlements in the south. However, this policy proved unpopular, as highland farmers were reluctant to leave their lands and when they did move away, they would require food and other subsidies for some years before they could regain self-sufficiency. The situation in Ethiopia deteriorated during 1984 as the central highlands and areas near the Kenyan border in the south were also affected by drought. In the eastern province of Harerge, close to the Somali border, the RRC estimated that some 350,000 people, mostly pastoralists, were at risk.

The Ethiopian refugee crisis, which reached its peak in 1984, was caused by an inability to break the constant cycle of drought–famine–emergency feeding, largely due to the continuing conflicts in the northern provinces and the Derg's heavy military expenditure. Both relief aid and long-term development projects were severely hampered by the conflict situation, but there was also a need for the improvement of basic infrastructure, particularly roads, health services and water supplies, and for the introduction of better agricultural techniques and conservation measures such as terracing and reafforestation to reduce people's vulnerability during future droughts.

could find 'little or no evidence that refugees are using natural resources in a more "wasteful" manner than local people' (Black, 1996: 37). In both studies it seems that the degree of dispersal of refugees, the generally good relations between refugee and host populations and the existence of strong local institutions, seem to have been key factors in minimising environmental impact.

A further important consideration is the impact of refugee movements on disease transmission and health. Prothero (1994) has demonstrated the significance of interactions between a variety of diseases and population mobility in tropical Africa, just as the spread of falciparum malaria, which is resistant to chloroquine, has been facilitated by movements of people, particularly of refugees, in South Asia and Southeast Asia (Prothero, 1994). In African refugee camps, disease and high death rates have been associated with overcrowding, poor accommodation, inadequate water supply, sanitation and waste disposal, as well as the amount and quality of food available. More than half the deaths in the 'emergency phase' are due to measles, diarrhoeal diseases and acute respiratory infections. Malaria is a major hazard when refugees from areas of low endemicity are forced to move into areas of high endemicity. Health problems were caused by forced resettlement in Ethiopia and Somalia, where people were moved from the relatively malaria-free Ethiopian plateau above 2,000 metres, to lower areas in the west and south-west where malaria was endemic. Irrigation projects at altitudes below 2,000 metres also extended areas of endemic schistosomiasis (bilharzia). Increased contact between Ethiopian pastoralists and agricultural settlers also increased pastoralists' risk of infection with schistosomiasis. Migrants also experienced nutritional problems, since the traditional cereal crops of the Ethiopian plateau could not be grown at lower altitudes (Prothero, 1994).

The spread of AIDS and sexually transmitted diseases can also be linked to population mobility (Chapter 5). In Burkina Faso, from where there is much migration to Ivory Coast, AIDS is known as *la maladie (ou la diarrhée) de la Côte d'Ivoire*. In Abidjan, Ivory Coast's largest city, some 25,000 deaths occurred from AIDS-related illnesses between 1986 and 1992. The incidence of AIDS along important national and international routes in Ivory Coast, Mali and The Gambia, has been reported as being generally much higher than elsewhere in these countries, and lorry drivers, itinerant traders and prostitutes have higher than average levels of infection (Prothero, 1996). In Uganda,

Cliff and Smallman-Raynor (1992) examined a number of possible factors underlying the spread of AIDS, including proximity to major roads and migrant labour. They found that the recruitment and movement of Ugandan soldiers in the 1970s and 1980s, and their contact with prostitutes, were key influences in the diffusion of AIDS.

Communications and transport

As we approach the millenium, we find ourselves in a world where people living and working in rich Northern countries can increasingly perform their business and everyday activities without even leaving their homes, through an ever changing array of technology, including telephone, fax, electronic mail (email) and the World-Wide Web (WWW). Virtually instant contact by email is now possible across most of Europe and North America, and the WWW serves the needs of millions by providing vast quantities of information for business, education and entertainment.

For many in the privileged North, the world is shrinking day by day as new innovations come on stream. As we have seen in Chapter 4, the 'globalisation' of a wide range of processes and transactions is a key feature of the last decades of the twentieth century, which will presumably continue and even accelerate in the new millenium. Meanwhile, in West Africa, only six hours' flying time from Europe, millions of people are still without electricity and fresh water supplies; their health and education services are inadequate, and they must work long hours in the fields using low-level technology to produce enough food to satisfy family needs. West Africa is one of the world's poorest regions. For many poor people in the rural areas of the Third World, their main means of communication, in the absence of television and newspapers, is still by word of mouth. However, in some countries it is fair to say that the transistor radio has had a considerable impact in the transmission of knowledge and information. With the high cost and unavailability of motorised transport, poor people are forced to cover thousands of miles each year on foot, often carrying heavy loads. A 1980s study in Ghana estimated that village households typically spent some 4,830 hours per year in transport activities, particularly collecting water and fuelwood, with most of this work being done by women (Porter, 1996). In such communities there has generally been

Plate 8.3 Cyclists in Kunming, southern China (photo: Tony Binns)

Plate 8.4 Crowded water transport in Georgetown, Guyana (photo: Rob Potter)

little tangible improvement in living standards, and certainly no evidence of a shrinking world (Plates 8.3 and 8.4). It is astoundingly difficult at times to appreciate that people with such contrasting lifestyles actually inhabit the same planet!

It was during the colonial era that the first railways and surfaced roads were constructed in what are now the developing countries. In Africa, railways were built to ensure strategic and military control, but more especially to extract raw materials, whether cash crops

Plate 8.5 Traffic congestion in Bangkok, Thailand (photo: Mark Edwards, Still Pictures)

or mineral resources, and typically linked major source areas with coastal ports. English, French and Portugese colonial powers, far from collaborating in the development of transport infrastructures to 'open up' Africa, were actively competing with each other, such that rail links between neighbouring anglophone and francophone countries never materialised. Despite ambitious plans, single lines were the norm within countries, and only states such as Morocco, Nigeria and South Africa can be said to have anything resembling a rail network. Today, in many African countries, rail transport suffers from a lack of investment and maintenance, and in relative terms is much less important than in the colonial period. For example, the 1,146-kilometre line which links Abidjan in Ivory Coast with Ouagadougou in Burkina Faso carried 3 million passengers in 1988, but only 760,000 in 1993, due largely to the poor condition of the rolling stock. Freight tonnage also fell from 800,000 tons in 1980 to 260,000 tons in 1993 (Economist Intelligence Unit, 1996a). In 1995–96 the Nigerian rail system was effectively closed down, and awaiting rehabilitation by a team of Chinese rail engineers. In Latin America, by 1940, the railway systems of Argentina, Brazil and Mexico accounted for 75 per cent of the region's network, and in these three countries there was significantly more interlinkage of lines than in African countries. The railway system in São Paulo, Brazil,

was a key element in the state's industrialisation, facilitating the advance of the coffee frontier, increasing exports and the development of engineering enterprises linked to the railways.

Since the Second World War, road transport has increased in importance, and with growing populations and less traffic carried by rail, roads now have to shoulder a much greater burden (Plate 8.5). In many developing countries the road networks are similar to those constructed during the colonial period, though new capital cities such as Abuja (Nigeria), Brasilia (Brazil) and Islamabad (Pakistan) have necessitated further highway construction. In Nigeria, funds generated by the oil boom in the 1970s had a significant effect on the upgrading of the country's road network; but in the 1990s, like other African countries, Nigeria has had less to spend on transport, and road quality has steadily deteriorated as a result. The stringencies imposed by SAPs in countries such as Ghana and Nigeria have had a major impact on road transport, with less road maintenance, fewer vehicles and a severe shortage of spare parts (Porter, 1996). In tropical regions, road surfaces quickly become potholed in the rainy season, and unpaved roads can become completely impassable. Many Third World governments face the difficult dilemma of whether to invest limited funds in building a few all-weather highways to connect the main towns, or alternatively constructing and upgrading many more kilometres of unpaved rural feeder roads, which could actually improve the lives of a greater proportion of the population, in connecting them with clinics, schools and markets. In reality, it is probably most effective to try to achieve a balance between the two strategies. For many rural producers, their main concern is how to transport their produce to markets as easily and quickly as possible. Women vegetable farmers in The Gambia, while praising the assistance from the NGO Action Aid in sinking wells and supplying tools and seeds, were concerned about the lack of reliable and refrigerated transport to carry their produce to large urban markets (personal communication).

North and South:
an interdependent world

We live in an interdependent world where links and relationships have developed over time and space and where flows of commodities and finance reinforce these links. But it is also an unequal world, and many would argue that such issues as trade, aid and debt are crucial in perpetuating inequalities both between and within countries.

An important landmark in considering the ramifications of an 'interdependent' world was the Brandt Commission, established in 1978 under the chairmanship of Willy Brandt, former Chancellor of West Germany, and its influential report, entitled *North–South: A Programme for Survival* (Brandt, 1980). Possibly the most memorable thing about this report was the world map on its cover, across which a black line divided the rich North from the poor South. One of the main themes running throughout the report is the mutual interest of rich and poor countries in a better-regulated world economy.

The Brandt Commission covered many issues, but its most important conclusions concerned the international monetary system, the transfer of resources from North to South, better trading opportunities for countries of the South and the nature of aid. A well-argued critique of the International Monetary Fund (IMF) was presented, with the commission proposing a system that would place less severe restrictions on borrowing countries and also establish greater stability in exchange rates. Brandt argued for a massive increase in resource transfers between rich and poor countries, with the aim of Northern governments first reaching, and then substantially surpassing, the target

of allocating 0.7 per cent of their GNP to aid. A special initiative was proposed to cope with the world's thirty poorest countries, with a special programme of long-term and flexible financial and technical assistance. Within poor countries, Brandt recognised the need for social and economic reforms to reduce inequality.

Reviewing the question of food aid, Brandt suggested that the best long-term solution to food shortages is for food production in poor countries to be increased to meet most of their own needs. Meanwhile, external aid should be devoted mainly to improving the capacity for local food production, rather than shipping in food supplies which, whether free or subsidised, compete with local production and may actually discourage it by depressing prices. The report concluded that governments of poor countries must devote a large part of their development effort to increasing agricultural production. Brandt was also concerned that rich countries, while reducing trading restrictions among each other, should consider dismantling trade barriers which affect the import of goods from poorer countries. A new set of trade rules and the negotiation of new commodity agreements were advocated, and both rich and poor countries were urged to liberalise their trading policies. The powerful position of transnational corporations (TNCs) within the world economy was recognised, which in 1980 controlled somewhere between one-quarter and one-third of world production, with just a few corporations controlling production, marketing and processing of important food and mineral commodities. Brandt advocated the establishment of a new mutually agreed 'investment regime' to ensure that host countries, as well as TNCs, benefited adequately from investment, through contractually agreed arrangements covering such aspects as foreign investment, transfer of technology and the repatriation of profits, royalties and dividends. In addition, the commission favoured the introduction of legislation in each country to regulate the activities of TNCs in matters such as ethical behaviour, disclosure of information, restrictive practices and labour standards.

The Brandt Report received much attention at the time of its publication, and was regarded as visionary, though somewhat unrealistically idealistic, in the light of the strength and entrenched position of TNCs and other key actors on the world stage. In fact, its recommendations fell victim not only to apathy and intransigence, but also to international recession in the early 1980s. Just three years later, a sequel to the report commented:

Three years have passed since the publication of the Brandt Commission's Report: *North–South: A Programme for Survival* – years which have brought increasing economic hardship to the industrial countries, and little short of disaster to much of the developing world. . . . The Commission offered hope. It expressed the belief that national problems could be solved, but only with a degree of collaboration and wider vision which is still lacking in international affairs. . . . The Cancun Summit [October 1981], which brought world leaders together to consider North–South issues, was the first of its kind and was a direct result of the Report. The leaders present felt that their exchanges had been valuable, but while the Summit helped to keep alive the process of global negotiations within the United Nations, it did not make any immediate contribution to resolving the problems of developing countries; nor did it set up any continuing procedure to accelerate negotiations. Now, more than a year later, there is still little sign of action. The North–South dialogue remains much where it was when the Commission reported. . . . Meanwhile the world economy continues its dangerous downward slide, and the desperate situation of many developing countries finds no new hope of relief. Crisis induces an impulse to contract. (Brandt, 1983: 11–12)

Crisis and commodity dependency

Despite the good intentions of the Brandt Report, it certainly seems that little progress had been made in the world's poorest continent, Africa. An Oxfam report in 1993 presented an extremely depressing view, commenting:

Sub-Saharan Africa is on a knife edge. For more than a decade the region has been locked in a downward spiral of economic and social decline. That decline, unlike the tragedies of famine and drought, which dominate news coverage of the region, has been largely invisible to the outside world. Yet it has spread human suffering and misery on an unprecedented scale. Hard-won gains in health and education have been reversed; living standards, already among the lowest in the world, have fallen; hunger is on the increase. And the tragedy is set to deepen. On current trends, the ranks of the 218 million Africans already living in poverty will increase to 300 million – equal to half the region's population – by the end of the decade (2,000). (Oxfam, 1993: v)

So what has gone wrong? Earlier in the chapter the case of the Ivory Coast coffee producer was examined and the widespread implications of changing coffee prices for both national economies and poor rural households considered. In fact, it is the long depression in world commodity markets that has had such a profound impact on African economies, and therefore on the quality of life of Africa's people. The situation has been particularly serious where countries are heavily dependent for the generation of foreign exchange on a limited range of primary agricultural and mineral commodities such as coffee, cocoa, cotton and copper. Table 8.2 shows that Mauritania gains 99 per cent of its export earnings from primary commodities, notably iron ore and fish.

In certain countries a single product dominates export earnings; copper provides 98 per cent of Zambia's total export earnings, whereas coffee provides 95 per cent of Uganda's earnings. The prices of such primary commodities fell dramatically during the 1980s, but import costs continued to rise, leading to a fall of about 50 per cent in the purchasing power of sub-Saharan Africa's exports in the decade from the early 1980s. The situation in some countries was much worse than the average picture: 'In 1986, coffee provided Uganda with US$365 million in foreign exchange earnings and financed about 70 per cent of its imports. By 1991 it yielded only US$115 million, and financed less than a quarter of imports. . . . Overall, the slump in commodity prices cost Africa US$50 billion in lost earnings between 1986 and 1990 – more than twice the amount the region receives in aid' (Oxfam, 1993: 7).

The collapse in commodity prices and the deteriorating terms of trade, together with rising debt-service payments and a reduction in foreign investment, have made it even more difficult for many developing countries to purchase imports. Furthermore, these trends have seriously undermined structural adjustment programmes (SAPs) sponsored by the World Bank and the International Monetary Fund (IMF), which were so dependent on increasing exports. SAPs have probably exacerbated the situation by encouraging countries with a narrow range of exports to increase their production, for markets which are already saturated and have fixed levels of demand. This can be seen in the case of increased cocoa exports from Africa, the world's major producing region, which have led to the collapse in world prices (Oxfam, 1993). Oxfam suggests that Africa's trading prospects can only be improved by establishing an African Diversification Fund to promote the increased processing of raw commodities in African countries, plus reducing protectionist barriers against Africa's exports as well as ending the subsidised disposal of agricultural surpluses on world and regional markets (Oxfam, 1993).

Table 8.2 Commodity dependency of African countries

Primary commodities as a percentage of total export earnings	Country	Individual commodities as a percentage of total export earnings
99.9	Mauritania	(iron ore 45.0, fish 42.0)
99.7	Zambia	(copper 98.0)
97.9	Rwanda	(coffee 73.0)
97.9	Niger	(uranium 85.0)
95.1	Burundi	(coffee 87.0)
95.0	Uganda	(coffee 95.0)
95.0	Namibia	(diamonds 40.0, uranium 24.0)
94.7	Somalia	(live animals 76.0)
93.4	Malawi	(tobacco 55.0, tea 20.0)
90.0	Ethiopia	(coffee 66.0)
88.9	Burkina Faso	(cotton 48.0)
88.5	Sudan	(cotton 42.0)
84.3	Mali	(live animals 58.0, cotton 29.0)
83.3	Togo	(phosphates 47.0)
82.0	Guinea-Bissau	(cashew nuts 29.0, groundnuts 23.0)
79.3	Tanzania	(coffee 40.0)
76.3	Mozambique	(fish 27.0, prawns 16.0)
72.0	Chad	(live animals 58.0, cotton 29.0)
71.6	Senegal	(fish 32.0)
68.7	Zaire	(copper 58.0)
68.5	Ghana	(cocoa 59.0)
63.2	Sierra Leone	(diamonds 32.0)
61.5	Kenya	(coffee 30.0)
56.9	Zimbabwe	(tobacco 20.0)
48.0	Gambia	(groundnuts 45.0)
46.2	Lesotho	(mohair 24.0)
6.5	Angola	(95.8 inc. oil)

Note: Table shows primary commodities, excluding fuel, as a percentage of total exports for 1982–86 (showing the percentage of the individual commodity where it exceeds 20 per cent of total exports).
Source: Oxfam (1993: 8).

World trade: the changing scene

There have been major changes during the twentieth century in the geography of international trade. At the beginning of the century, Europe and the United States dominated the world scene. The European powers relied on their colonies in the developing world, but also places such as Australia, Canada and New Zealand, to produce raw materials to supply growing industries at home, industries that produced manufactured goods which could then be traded for more raw materials. Strong trading links still remain between many former colonies and their former European masters. Jamaica still exports most of its bananas to the United Kingdom; and in Africa there is still much trade between francophone countries and France and between the former British colonies and the United Kingdom. In 1994 some 33 per cent of Kenya's exports to non-African countries were destined for the United Kingdom; and the United Kingdom supplied just under 50 per cent of the country's imports (Courier, 1996: 24). The association between colonialism and export economies was emphasised by the concentration of the large-scale export trade in the hands of a few large, mostly European firms and by transnational corporations.

Although there have been colonial links between Western Europe and Pacific Asia – the British colony of Hong Kong was only returned to China in 1997 – international trade in the Pacific region is dominated by Japan and the United States. It has been suggested that 'the West Europeans withdrew from Pacific Asia in the post-war decades not just politically and militarily, but also economically' (Shibusawa, Ahmad and Bridges, 1992: 30). The dominance of Japan and the United States in the Pacific Asian region is reflected in the fact that the region's share of US imports was 37 per cent in 1988, whereas Japan's overall trade surplus with the eight major East Asian economies grew from US$1.2 billion in 1980 to US$20.4 billion in 1989 (Shibusawa, Ahmad and Bridges, 1992: 12).

The General Agreement on Tariffs and Trade (GATT), established in 1947 at a time of world reconstruction, was designed to bring some order to world trade and prevent the instability of the interwar years, at the same time advocating the pursuit of free trade policies. The reduction of tariffs, prohibition of quantitative restrictions and other non-tariff barriers to trade, and the elimination of trade discrimination, were the main objectives of GATT. The Uruguay round of GATT negotiations began in September 1986 and were only concluded in April 1994, after which GATT was replaced by the World Trade Organisation (WTO). The talks focused more on debates between Europe and the United States on agricultural subsidies, whereas issues of greater relevance to poor countries, such as gaining better access to developed world markets, were sadly rather neglected. The expansion of commerce through the deregulation of markets is the main aim of WTO, and trade liberalisation in the shape of measures such as removing tariff barriers, quotas, price supports and subsidies is also increasingly central to economic policy in developing countries. As Watkins observes, 'By contrast, issues of sustainable resource management, the regulation of commodity markets, and poverty reduction strategies, are conspicuous by their absence from the international trade agenda' (Watkins, 1995).

Two important trends in world trade have developed during the 1980s and 1990s. Firstly, with the end of the Cold War and the collapse of Soviet communism in 1991, trade between Eastern Europe and the Western capitalist countries has accelerated. Western investment is playing an important role in the restructuring of the former Communist bloc countries, and some would argue that these countries are receiving both the attention and the funding which the world's poorest countries urgently require. In the early 1990s, over 400 agreements were signed between Western businesses and bodies in the newly democratised Czechoslovakia, Poland and Hungary, much of this investment going to the major cities and heavily industrialised regions.

A second and very significant development in world trade has been the increasing power and participation of the export-oriented newly industrialising countries (NICs), such as Hong Kong, Malaysia, South Korea, Singapore and Taiwan, in addition to the already powerful Japan (Chapter 4). Industrial employment in South Korea increased by 77 per cent between 1974 and 1983, and by 75 per cent in Malaysia during the same period. 'Trade, trade and more trade was what propelled the so-called Pacific Rim states out of agrarian destitution or post-World War II destruction and decline into world economic prominence' (Aikman, 1986: 10). The growth of international trade in the countries of the western Pacific Rim has been remarkable (Figure 8.1). From 1982 to 1988, the growth of export volume from East Asia (even excluding Japan) was over 12 per cent per annum, a rate almost double that of South Asia, three times that of the Middle East, North Africa and Latin America, and about six times higher than in sub-Saharan Africa (Hodder, 1992: 67). The western Pacific Rim's share of world trade increased from 14.3 per cent in 1971 to 22.8 per cent in 1984. This reflects the success of export-led growth strategies and the readiness of the peoples of these countries to undertake programmes of rapid structural adjustment, the development of new products and the exploitation of new markets (Dicken, 1993). However, in 1997 and 1998, many of the Asian NICs suffered a serious economic downturn which had widespread implications both domestically and internationally.

The western Pacific Rim countries conduct most trade with the United States and Canada, representing some 15 per cent of world trade. International trade as a whole is dominated by Japan and the United States, with Japan supplying 30 per cent of the US car market. In most of Southeast Asia, except Singapore, there is still evidence of the colonial pattern of trade in which countries export raw materials and primary products and import most of their manufactured goods. The reverse is true in Japan, where 55 per cent of its imports are crude materials and primary products and 99 per cent of its exports are manufactured goods. In the case of Hong Kong and South Korea, a clear majority of both imports and exports are manufactured goods (Hodder, 1992). There is considerable potential for tapping into the vast Chinese market, and Southeast Asia's 'overseas Chinese' have already established a close network of business links with the Chinese in Taiwan, Singapore and Hong Kong, and they are well placed to take full advantage of potential business opportunities. Hong Kong has provided about 80 per cent of investment in southern China's Guangdong Province, which has experienced

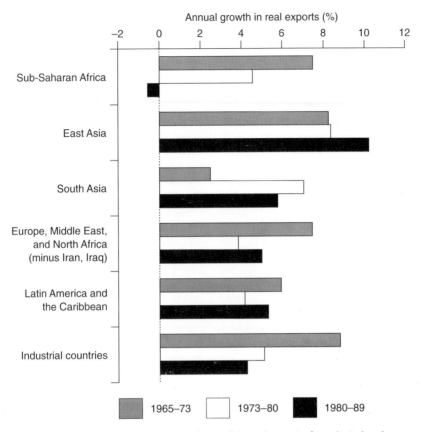

Annual growth in real exports (%)

Figure 8.1 Estimated percentage annual growth in real exports for selected regions, 1965–89. Adapted from Hodder (1992).

BOX 8.2   Female migration to southern China

The reforms introduced since 1979 under Deng Xiaoping's rule have had a major impact on population movements within China. Although migration did occur before 1979, it was generally involuntary and much more centrally controlled than in recent years. Resettlement programmes took place, most notably from the densely populated provinces of the east to the sparsely populated western regions. In the case of Xinjiang Province in the far north west of China, there was a deliberate government policy to change its ethnic balance. Xinjiang has many minority groups, but in-migration of Han Chinese from the east increased the proportion of Han within the population from under 10 per cent in 1949 to 40 per cent in 1982. Furthermore, many Han were appointed to important administrative and political

posts in Xinjiang, so Beijing is able to maintain strong central control over the region.

Another wave of migrations occurred during the period of the first five-year plan (1953–57), when millions of peasants moved into towns looking for jobs during a phase of intensive reconstruction and industrialisation after the Second World War and the subsequent civil war. As a result, between 1949 and 1957, China's urban population increased by 60 per cent, whereas the rural population grew by only 13 per cent (Jowett, 1990). During the Great Leap Forward (1958–60), despite a strong emphasis on rural industrialisation, this initiative was thwarted by widespread famine in 1959–61, leading to 'surplus' population being moved back into the countryside. Later, during the Cultural Revolution of the 1960s

Box 8.2 continued

and 1970s, the Chinese government imposed strict controls on rural–urban migration, and urban youths were sent to work in the countryside in the so-called rustication programme. During this period the country's largest cities scarcely grew through migration; for example, there was a relatively small net gain of 350,000 migrants in Tianjin during the thirty years between 1950 and 1980, and China's largest city, Shanghai, actually experienced a net loss of 1 million people through out-migration during the same period.

The reforms of 1979, however, had a major effect on population movements within China. Probably the most important reform was the replacement of people's communes with individual farming units under the 'household responsibility system'. This new approach to agricultural production led to the collapse of collective farming and gave rise to abundant rural surplus labour, due to a substantial increase in the efficiency and productivity of the agricultural sector. Controls on internal migration were relaxed, such that surplus rural workforce could move freely without the need for permanent registration. The late 1970s and early 1980s were characterised by a considerable increase in migration, and it was estimated that in Shanghai alone there was a net in-migration in 1979 of 264,800 (Jowett, 1990).

A second significant element in the reforms of 1979 was the establishment of the first four special economic zones (SEZs). Located on China's southeastern coast, close to Hong Kong and Taiwan, and with a series of tax inducements, these areas were designed to attract foreign investment and 'joint ventures' between Chinese and overseas companies, to manufacture export goods which would generate foreign exchange. The SEZs were also seen as 'social and economic laboratories, in which foreign technological and managerial skills might be observed and adopted' (Phillips and Yeh, 1990: 236). Two of the first SEZs, Zhuhai and Shenzhen, are located in the Pearl River Delta zone of Guangdong Province. Shenzhen, the largest SEZ (327.5 square kilometres) is situated adjacent to the Hong Kong border and has grown spectacularly during the 1980s and 1990s, from being just a small rural town to an ultramodern city with over a million people, which is less than an hour's travel from the 'throbbing heart of capitalism' in Kowloon and Hong Kong Island. China's southern coast has become an innovative capitalist periphery for the new international division of labour and capital flowing into the region from all over the world (see Figures 1 and 2, pages 205 and 206).

These two factors – the changes in the rural production system and the creation of the SEZs – have provided an important stimulus for the massive increase in migration which China has experienced in recent years. The 1990 population census revealed there were 525 million population movements during 1982–85, 740 million during 1985–87 and 660 million during 1987–90. Whereas in many developing countries migration is usually dominated by males, China's 1990 census indicates that 56 per cent of migrants were male and 44 per cent female. However, if intraprovincial migration alone is considered, females were in the majority at 66 per cent, due largely to migration for marriage, since brides commonly move to their husband's home. But aggregate data mask striking differences at the local level and, in fact, females dominate migration from some counties of the vast, and predominantly rural, Sichuan Province (China's most populous province), to the growing industries of Guangdong Province, particularly in the Pearl River Delta region (Davin, 1996). Regulations relating to labour migration were further relaxed from 1984, and from July 1985 peasants were allowed to be temporary residents in urban areas, applying for six-monthly permits through their work units. As a result, in Guangdong Province the 'temporary resident' population rose from 280,000 in 1982 to 3.3 million in 1990, representing nearly a twelvefold increase (So, 1997).

A study undertaken in the mid 1990s of female migrants from the rural areas of Sichuan to the industries of Guangdong, found that only 12 per cent of household heads were strongly against the migration of their female members (So, 1997). The decision for a household member to migrate was taken by the entire family, but migration was seen as important in generating cash income to pay for such items as education, marriage, consumer goods, farming inputs, house building and maintenance. Migration was also seen as a 'risk aversion' strategy to diversify sources of income (So, 1997: 161). The considerable attraction of migration is summed up in a popular Chinese phrase, *dongnanxibeizhong, facaidaoGuangdong* 'East, south, west, north or central; to get rich, go to Guangdong Province.' However, rural households were concerned about the social problems of cities, such as prostitution, robbery and rape, and the potential vulnerability of

Box 8.2 continued

Figure 1 Province-level administrative divisions of China (Xinjiang, Tiber and Inner Mongolia are autonomous regions). Adapted from So (1997).

their women in a strange environment. In relation to the income of rural households, the economic cost of migration can be considerable, with very long journeys by bus and train. A typical journey from Sichuan to Guangdong would take up to three days and cost more than 100 yuan; 1 yuan = US$0.12 (12 cents) in 1996.

Food and accommodation in Guangdong are provided by the factory, although the employer may charge for certain services. It appears that female migration had little effect on the rural household's productivity. On the contrary, the income from migrants' remittances was regarded as very valuable and far outweighed the impact of the loss of labour. Furthermore, the status of migrant women improved after they engaged in wage labour, and young migrants felt they had 'grown up' by learning new skills, experiencing a different environment and earning their own wage before getting married. The great majority of female migrants (79 per cent) were between the ages of 16 and 24, and a similar

proportion were unmarried (So, 1997: 174, 189). The Chinese government has started to regulate the flow of labour migration by coordinating efforts between the migrant-sending and migrant-receiving provinces, in order to relieve pressure on the transport system and reduce social problems. Coordinating offices have been established since 1991 to regulate the flow of labour from Sichuan and other provinces to Guangdong, and factories recruit migrant workers through the local labour bureaux. Potential migrants find out about job prospects through labour offices or industrial enterprises, which may even contact their villages. Well over half (57.8 per cent) of the female migrants interviewed in 1994 found out about job opportunities through families and friends, and many already had contacts in Guangdong, contacts who played a key role in helping new migrants adjust to the new ways of living and working (So, 1997: 180).

Employment was mainly in producing electrical goods, toys, clothing and shoes, many of the prod-

Box 8.2 continued

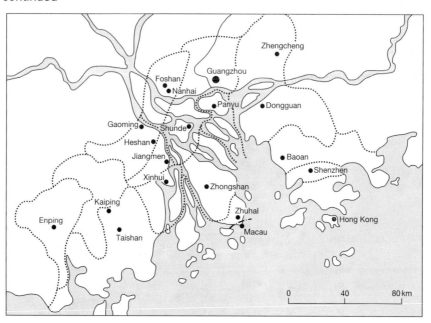

Figure 2 The major cities in Guangdong Province. Adapted from So (1997).

ucts being destined for the export market. Working conditions were difficult and the work was manual and highly repetitive; there were many industrial accidents. Women migrants typically worked for 70 hours or more in a seven-day week and additional overtime at a rate of 1 or 2 yuan per hour was common. Wages averaged 10–15 yuan per day, with a typical monthly income of between 250 and 450 yuan, including bonuses. From this, factories then often deducted as much as 100 yuan for provision of meals and accommodation (usually in dormitories), leaving a monthly net income of 150–350 yuan. The migrants were seen by the indigenous population as poor peasant workers and in a much inferior position compared with local residents, who invariably earned higher wages and held more responsible positions in the factories. Upward mobility for migrants was therefore very difficult, and workers coped with the poor conditions through a mutual support network known as the *tongxiang* system (people from the same village or county).

Migrants retain strong links with their villages by exchanging letters. However, the Chinese New Year holiday provides a valuable opportunity for migrants to return to their villages, taking money, consumer goods and much information about life and work in Guangdong. Between 70 per cent and 90 per cent of Sichuan migrants return home at least once a year, usually for the spring holiday, which lasts about sixty days around the New Year period (Davin, 1996). However, transport costs increase at this time and there is much competition for places. Typical remittances from migrants varied between 100 and 200 yuan per month (1200–2400 yuan per year), which is quite considerable in relation to the average annual net income in 1993 of only 698 yuan for a Sichuan peasant. An estimated 5 billion yuan is remitted each year to Sichuan province from migrant workers, a substantial proportion of whom were working in the coastal cities, predominantly in Guangdong (So, 1997: 237).

The advantages and problems of migration on such a massive scale within China are highly complex, and the extent to which these factors are reflected in the fortunes of individuals and their rural households needs further careful investigation. However, one is inclined to agree with Davin's conclusion that 'migration in the form it takes at present in China has the potential to return human and financial resources to the villages, and thus helps prevent the gap [between the poor countryside and the prosperous urban areas] becoming even wider' (Davin, 1996: 665).

massive industrial development (Box 8.2), whereas Taiwan has invested heavily in Fujian Province, opposite Taiwan on China's eastern coast.

Whereas in the 1960s and 1970s, the United States was the dominant world trading power, in the 1990s a 'multipolar' system has developed, with power concentrated in three blocs: North America, Europe and increasingly the Pacific Rim. Trade relations are being transformed by such features as increased flows of foreign investment, the globalisation of production under the auspices of TNCs and trade liberalisation in developing countries. Institutional structures have also changed, such as the new customs union between Brazil, Argentina and Uruguay created in 1995, and the Asia-Pacific Economic Cooperation (APEC), established in 1993, which could lead to links between the Pacific Rim states (Japan, China, South Korea, Malaysia, Philippines and Thailand) and the North American Free Trade Agreement (NAFTA), comprising the United States, Mexico and Canada (Watkins, 1995: 113).

Transnational corporations

It is often assumed that trade is an activity which is conducted between countries, each of which controls its own economic destiny. However, as noted in Chapter 3, world trade flows are in reality dominated by incredibly powerful TNCs. Dicken has examined the nature of TNCs and concludes that 'because these big companies are often based in a single country, though they operate in at least two countries, including the firm's home country, they are now usually termed "transnational" rather than "multinational". All multinational corporations are transnational corporations, but not all transnational corporations are multinational corporations' (Dicken, 1992: 47). The role of transnational corporations in world trade should not be underestimated, indeed they are vital actors in the global economic and trading system. Until the 1960s, most TNCs were either of US or UK origin, but in recent years, Japanese, German and other companies have become important on the global scene. Although TNCs are by no means homogeneous, they typically have their headquarters or strategic base in one country, but with a variety of production sites and subsidiary operations in other countries (Plate 8.6). TNCs also generally have a good amount of geographical flexibility, enabling them to shift resources and operations from one global location to another as production factors change, in order to seek new competitive advantages. The relocation of operations

Plate 8.6 TNC headquarters: Lonrho building, Nairobi, Kenya (photo: Tony Binns)

in developing countries may be because labour costs are cheaper and there is less militancy among labour unions. Abundant, low-cost and largely illiterate workforces, with little industrial tradition, are very attractive. Schneider and Frey (1985) found that other factors, such as the size of the home market, price and exchange rate stability, and political and institutional stability, were also important considerations in TNC overseas investment decisions (Clayton and Potter, 1996). The power and influence of TNCs continue to grow, facilitated both by governments withdrawing controls on foreign investment, and thus encouraging the greater mobility of capital, and also through government support of WTO trade rules, which limit the rights of governments to control TNC activities.

TNCs and the globalisation of fresh food

During the 1990s, with a growing trend towards trade liberalisation and the associated change in the global

Plate 8.7 Rose-growing for export, south of Nairobi, Kenya (photo: Tony Binns)

regulatory network, improvements in transportation technology and changing consumer demand have resulted in an increasingly integrated global food production system that is dominated by TNCs. Over-production of staple crops in the European Union under Fordist production systems and favourable government subsidies through the Common Agricultural Policy (CAP) have meant that, during the 1980s, trade in cereals and sugar, as well as tropical beverages from developing countries, has declined (Dixon, 1990). One element of the global food trade that has shown a spectacular increase in recent years is the export of high-value crops, such as fresh fruit, vegetables and cut flowers. In 1989 the trade in these items comprised 5 per cent of global commodity trade and was equivalent in volume to trade in crude petroleum (Watts, 1996b; Jaffee, 1994). Developing countries contributed one-third by value to this lucrative trade, twice the value of their traditional agricultural exports of cocoa, coffee, cotton, sugar, tea and tobacco.

In 1990, twenty-four low- and middle-income countries, mainly in Asia and Latin America, exported annually in excess of $500 million of high-value, fresh horticultural products. The main producers are Chile, Argentina, Brazil and Uruguay in South America; Malaysia and Thailand in Asia. Elsewhere, in order to maintain their foreign exchange earnings, as well as to diversify their economies, some African countries have been giving more attention to the production and export of high-value horticultural produce, most notably in Egypt, Kenya, post-apartheid South Africa and Zimbabwe. Africa has been part of the

global food market for centuries, with efficient and well-integrated marketing chains developing from the late nineteenth century to move cash crops such as tropical beverages, sugar, cotton and tobacco from African producers to consumers in Europe. However, these traditional marketing chains are not suitable for the export of highly perishable items such as fruit, vegetables and cut flowers (Plate 8.7). New chains have therefore evolved which, perhaps more than anything, reflect the considerable power of the large European retailers in responding to changing consumer demands (Box 8.3).

Until the beginning of the twentieth century, urban populations in Europe and temperate regions of North America could only eat fresh produce seasonally and had to rely on canned and, later, frozen foods. The major change came with bananas, a tropical fruit that could withstand a long transportation link between producer and consumer, provided the temperature could be controlled. Early experiments in the banana trade began in the 1870s, when nationally based British, French and US specialist firms produced bananas in their tropical colonies, or 'semicolonies' in the case of the United States, for consumption in Europe and the United States. As the banana industry grew, the US firms became extensively involved in the internal politics of states such as Cuba and the Central American 'banana republics' (Friedland, 1994). Three firms involved in the early production and trade of bananas are Dole, Chiquita and Del Monte Tropical Products. All three firms also have major stakes in food labelling and transportation, with refrigerated cargo ships.

BOX 8.3   The globalisation of food: horticultural exports from Kenya

Trade in fresh fruit, vegetables and flowers from
sub-Saharan Africa to the European Union (EU) has
increased dramatically during the 1980s and 1990s.
The European consumer now demands high-quality
fresh commodities throughout the year, and with
a flight time of about nine hours from Europe to
Kenya, major supermarket chains have established
links and organised production to ensure these items
can be on their shelves less than twenty-four hours
after harvest.

Exports of fresh horticultural produce from Kenya
have grown steadily since independence, such that in
1994 they accounted for 10 per cent of total export
earnings and were the third most important agricultural exports after tea and coffee (Economist
Intelligence Unit, 1996b). The value of horticultural
exports rose from Ksh 3780 million (about US$66
million) in 1991 to Ksh 10,620 million (about
US$186 million) in 1995. Over the last five years,
the horticultural industry has undergone dramatic
change, coinciding with economic liberalisation.
Huge private investments have been made, particularly by the country's ten largest producers in
the cut flower and prepackaged vegetable sectors.
This has been in response to increased demand from
Europe, and especially to attract and keep lucrative
supply arrangements with large UK supermarkets.
Kenya is a major supplier of green beans, mange-
touts, avocados, mangos and cut flowers, as well as
a significant range of Asian vegetables. Green beans
are the most important vegetable crop exported,
although quantities declined in 1995–96. However,
this decline has been compensated by adding value
to green bean exports through sorting and packaging in Kenya. The country faces increasingly stiff
competition from other producers, such as Egypt for
green beans, South Africa for avocados and mangos
and Israel for avocados and cut flowers. The growth
of the cut flower sector has been spectacular, and
in 1996, Kenya overtook Israel to become the leading supplier of cut flowers to the Dutch auctions.
Although vegetable exports have stabilised at about
26,000 tonnes per annum, the export of cut flowers
has increased by about 20 per cent per annum since
the early 1990s. In 1995 the tonnage of cut flower
exports exceeded that of vegetables for the first time
(Barrett et al., 1997). Kenya exports most of its fruit
and vegetables to the United Kingdom and France,
whereas cut flowers are destined for the Netherlands
and Germany, as well as the United Kingdom.

The growth of the Kenyan horticulture industry
has been due in no small measure to government
support since the late 1960s. The Horticultural Crops
Development Authority (HCDA) was formed in
1967, through which state policy and support for
the sector has been channelled. The government has
generally restricted itself to the role of facilitator and
has not interfered with market mechanisms or pricing
policy, a role which has been endorsed by a series
of structural adjustment programmes since 1979.
Sessional Paper 1, 1986 specifically emphasised that
agricultural growth was to be achieved by higher
productivity, the expansion of high-value crops (such
as fruit and vegetables) and improved export com-
petitiveness. The promotion of the horticultural industry is seen as a partnership between government
departments, quasi-parastatal organisations, such as
the Export Promotion Council, and the main growers,
such as Sulmac – Brooke Bond's flower-growing
subsidiary and the country's largest producer of cut
flowers – and Homegrown, the largest producer of
vegetables. Meanwhile under structural adjustment,
the influence of agricultural marketing boards has
been greatly reduced. Two duty exemptions have
been particularly helpful: from 1991 the exemption
of imported packaging materials used in the industry,
and from 1994 the exemption of fertilisers, tools and
greenhouse sheeting.

Two distinct marketing chains can be identified in
the Kenyan export horticulture industry. One chain,
which has developed since the 1960s, involves mainly
small and medium-sized growers and accounts for
31 per cent of Kenya's horticultural exports. Small
farmers may either supply medium growers, who
then sell on to exporters, or alternatively sell to intermediaries and agents who then sell to exporters.
There is some criticism at various points in this chain
about the quality and reliability of produce supply.
This chain supplies large quantities of vegetables for
the Asian market and strong links have developed
between Asian exporters in Kenya and Asian importers and retailers in the United Kingdom. Importers of Asian vegetables are keen to have a wide
variety of produce, but in smaller quantities than the
supermarkets. Consumers are more concerned with
flavour and value for money, rather than presentation

Box 8.3 continued

or packaging. Produce for this market is imported and sold loose, not in prepacks, and it does not have to meet strict supermarket specifications.

In sharp contrast, a fully integrated chain, which has developed since 1990, links Kenya's largest producers to major companies in the EU. Virtually all their vegetable produce is sold under contract to supermarket chains, mostly in prepacks, which are processed in packing stations where standards exceed EU requirements. The packs use approved materials, and are bar-coded and priced in Kenya as directed by the supermarkets, which supply the pricing and other stickers. Many flowers are also sold to supermarket chains, and bouquets are made up in packhouses on the farm. The rest of the flowers are sent in bulk to Dutch flower auctions, where they constitute 23 per cent of all flowers sold. The major producers have all invested heavily in EU-standard pack-stations, refrigerated trucks and cold stores.

The large UK supermarket chains are at the top of the power hierarchy in this business, but shoulder few of the risks until produce actually reaches their shelves. Supermarkets depend on UK importers, through their associated exporting companies in Kenya, for getting produce out of Kenya and into the United Kingdom. The requirements of the UK Food Safety Act (1990) have had a major effect on production and marketing, since the Act calls for 'due diligence', requiring importers to know exactly where and how the crops were produced (including fertilisers and pesticides used) and there must be documentation to prove it. Traceability is now as crucial in the horticultural trade as quality, reliability and price, and logistically this favours dealing with a few large commercial farmers who can maintain strict standards and detailed records, rather than with many smallholder producers among whom there could be much variability. UK consumers are highly sensitive to issues concerning toxic chemicals or the perceived exploitation of local labour.

Two factors are absolutely crucial in exporting horticultural produce, the freight space and the cold chain. The larger exporters have more control than smaller exporters over freight space, because they can negotiate guaranteed space on aircraft and either fill it with their own produce or sell it on. Kenya's largest horticultural exporter has a prebooked arrangement for 25 tonnes on British Airways' nightly airfreight service to London. Maintenance of the cold chain is also vital in dealing with such highly perishable goods in a tropical climate. Exporters with their own dedicated cold storage facilities, at Jomo Kenyatta International Airport in Nairobi, run much less risk of breaking the cold chain than if they have to rely on using the general cold-store facilities.

If Kenya's export-oriented horticulture industry is to expand further, these constraints need to be addressed, particularly in relation to the possible incorporation of more smaller producers into the trade. In addition, the improvement of road infrastructure and the expansion of airport facilities are necessary. In 1997 a new airport at Eldoret was nearing completion, which should both reduce the pressures on Nairobi airport and also create potential for exporting more produce from the country's north-western region. Other possible measures might include the establishment of an agency to monitor controls and standards for all export crops, including banning sales of chemicals not permitted under EU regulations. Codes of conduct between growers, exporters and freight agents might also be enforced by trade associations, which also work to promote the industry. But in the context of a poor developing country, surely a key question is the extent to which poverty can be reduced among small producers engaged in export-oriented production. If horticultural production is to have a meaningful impact on poverty alleviation in Kenya, then more attention needs to be given to small-scale producers – perhaps coordinated within producer groups that help them gain access to export markets – as well as controlling quality, post-harvest handling and marketing techniques.

Dole is a US-based transnational, known until 1991 as Castle & Cooke. It originally began as a merchant firm in the Hawaiian Islands and then became involved in food processing, real estate and fresh fruit and vegetable activities. Chiquita is also originally US-based and was formerly known as United Brands, and before that as the United Fruit Company. The Del Monte Fresh Produce Company, originally a US-based company, was known as Del Monte Tropical Products until late 1992.

Although bananas were important in the early history of these companies, they have, like other TNCs, diversified considerably since the Second World War. Chiquita bought seven lettuce-producing firms in California in 1969, and integrated them into a single subsidiary, Interharvest, which dominated US lettuce production and distribution. The Dole Food Company emerged in its modern form in 1961, when Castle & Cooke acquired the Dole Company, a pineapple producer. It expanded into bananas from 1964, as it bought an increasing share of the Standard Fruit Company, a banana producer for the North Amerian market. In 1967, when Dole owned 87 per cent of Standard, Standard supplied 31 per cent of North American banana requirements. Ten years later, in 1977, Dole followed Chiquita into lettuce production, by purchasing Bud Antle, the second largest lettuce producer in the United States. Subsequently, it was from the Bud Antle base that Dole expanded into a wide variety of other commodities. During the 1980s, Chiquita, Del Monte and Dole all expanded substantially into global sourcing and distribution based on their banana operations.The recipe for success of these TNCs has involved (1) attracting capital from investors to make the initial purchases; (2) continuously generating new capital and demonstrating good profit levels; and (3) making good acquisitions, which have to fit into an overall strategy. Acquisitions should ideally be clustered geographically, rather than be spread all over the world, and fully consolidated before venturing into new areas (Friedland, 1994).

Developing countries and the debt crisis

The debt of developing countries soared from US$658 billion in 1980, to US$1,375 billion in 1988, and to US$1,945 billion in 1994. The origins of the debt crisis are complex, but undoubtedly major factors were the long-term effects of rising oil prices in the 1970s, compounded by developed countries adopting monetarist policies in the late 1970s following the 'second oil shock' of 1979, which forced up interest rates on debt repayments. Added to this was the collapse of commodity prices in the early 1980s, such that in 1993 prices were 32 per cent lower than in 1980; and in relation to the price of manufactured goods, they were 55 per cent lower than in 1960. As a result, there was a sharp deterioration in the terms of trade affecting developing countries (ICPQL, 1996). Facing massive debt repayments, the IMF and World Bank, which were created in part to transfer the savings of

surplus countries to deficit countries, then imposed structural adjustment programmes on these countries, which required deep cuts in public spending, often with little concern for local circumstances and human welfare. As Watkins observes, 'In Latin America, the epicentre of the debt crisis, average incomes fell by 10 per cent in the 1980s and investment declined from 23 per cent to 16 per cent of national income, causing widespread unemployment and poverty' (Watkins, 1995: 174).

The debt crisis occurred suddenly in the early 1980s, with the financial collapse of Mexico in August 1982, and affected other middle-income developing countries which were heavily dependent on commercial lending, particularly Brazil and Argentina. The poorest countries were also badly hit, but since commercial banks had been reluctant to lend to them, most of their borrowing has been through public sector aid programmes. During the 1980s, the question of rescheduling the massive debts of certain developing countries became a major issue, since some were unable to repay the interest on the sums borrowed, let alone reduce the basic sum.

The newly industrialising countries (NICs) of East Asia and Southeast Asia have not experienced such problems, due to the relative buoyancy of their economies, although in 1997 and 1998, many Asian NICs were experiencing a serious downturn in their economies. Countries such as South Korea, while borrowing heavily, have been able to service their debt due to a high level of exports. During the 1990s, private and often highly speculative capital flows, mainly in the form of direct foreign investment, have benefited China and middle-income countries such as Argentina, Malaysia, Mexico and Thailand. Furthermore, in 1989 the Brady Plan assisted middle-income countries by recognising that commercial debt could be reduced with IMF and World Bank support and by extending repayment periods.

But the world's poorest countries, particularly those in sub-Saharan Africa, continue to suffer from a huge debt crisis, and are heavily dependent on official aid flows for their financial survival. As the flow of aid declines in real terms, trade and debt reform assume a greater importance. Between 1980 and 1993, sub-Saharan Africa's debt more than tripled to around US$183 billion and, although considerably less than that of Latin America (US$446 billion), the region's debt increased from the equivalent of 28 per cent of its GNP to 109 per cent, compared with 37 per cent for Latin America. Incredibly, sub-Saharan Africa therefore owes more than it earns. Between 1985 and

1992, Africa disbursed US$81.6 billion in debt payments, diverting government spending from vital expenditure on education, health and other urgent priorities (Oxfam, 1993: 13). A number of countries have had to reschedule their debts, which has contributed to the steady build-up of arrears. Between 1989 and 1991 the official creditors cancelled some US$10 billion worth of debt to sub-Saharan countries, but the debt problem remains severe.

So what can and should be done about Africa's plight? As Oxfam observes, 'It is difficult to avoid being struck by the contrast between the urgency with which Western governments have responded to the financial problems of Eastern Europe and Russia, and their neglect, for more than a decade, of Africa's far deeper problems' (Oxfam, 1993: 17). Oxfam advocates a fresh approach on the part of Northern governments, arguing that Africa's problem is not a temporary problem, but rather a serious matter of bankruptcy which must be recognised 'in placing debtors' *ability to pay* above the claims of creditors.' Furthermore, Oxfam is calling on the industrialised countries to 'agree to the cancellation of between 90 per cent and 100 per cent of *all* non-concessional debt' (Oxfam, 1993: 17). Oxfam is also highly critical of the IMF, stating that it is 'not an instrument for providing long-term concessional development finance; and it is governed by apparently immutable orthodoxies entirely inappropriate to African conditions. The time has come therefore either fundamentally to reform the IMF, or to extricate it from Africa. In either case, measures to write off obligations due to it from low-income African countries are long overdue' (Oxfam, 1993: 17). It seems that the recovery of Africa will depend on substantial investment of foreign capital, since in 1993 sub-Saharan Africa only received about 3 per cent of worldwide private investment; amazingly, this is rather less than the amount received by Portugal. With the lack of private investment, Africa is becoming increasingly dependent on government and multilateral agency development assistance, such that aid accounts for 80 per cent of all financial resource flows into the region, and for 11 per cent of the region's entire GDP. It is the question of aid to developing countries which we will now consider.

Aid to developing countries

In 1970, through Resolution 2626 on the International Development Strategy for the Second United Nations Development Decade, the UN General Assembly set out for the first time agreed targets for finance resource transfers and flows of overseas development assistance – aid. The UN urged developed countries to achieve an allocation level of 0.7 per cent of their gross national product in overseas aid by 1975. As we have already seen, this figure was also emphasised in the Brandt Report in 1980, and more recently it has been reaffirmed at numerous world gatherings such as the Earth Summit in Rio de Janeiro in 1992, the Conference on Population and Development in Cairo in 1994, and the World Summit on Social Development in Copenhagen in 1995. However, as we approach the millenium, there has unfortunately been little progress on this, and in some cases governments of developed countries, which are members of the Development Assistance Committee (DAC) of the Organisation for Economic Cooperation and Development (OECD), have actually reduced their aid budgets substantially. For example, the United Kingdom's allocation of overseas development assistance reached an all-time high in 1979 at 0.51 per cent of GNP, but then declined steadily to an all-time low of 0.27 per cent in 1990. There was a slight increase to 0.32 per cent in 1991, but at the time of the Labour Party's general election victory in May 1997, the UK aid budget had fallen again to 0.27 per cent. Since the incoming Labour government pledged not to raise key taxes for two years, there seemed little prospect of an increase in the UK aid budget before the year 1999. The US record has also been disappointing. Despite giving the largest total amount of overseas aid, some US$11.2 billion in 1993, the proportion of GNP given as aid by the US has stagnated at about 0.2 per cent, and in 1991 actually fell to 0.17 per cent, the lowest rate for all the DAC countries (Figure 8.2). In 1991 only five countries – Denmark, Finland, the Netherlands, Norway and Sweden – had reached, and indeed exceeded the target. Norway, with 1.14 per cent of GNP, was well ahead of other countries, and has shown a steadily increasing commitment to development assistance since 1975.

In addition to the quantity, it is important to consider the nature of the aid and the reasons for giving it. Whether overseas development assistance takes the form of short-term disaster relief, longer-term development aid, food aid or military aid, each has many complex ramifications and implications, both in relation to the successful alleviation of poverty in developing countries and also to the priorities, and frequently ulterior motives, of donor countries. During the 1980s and early 1990s, Britain tied a higher

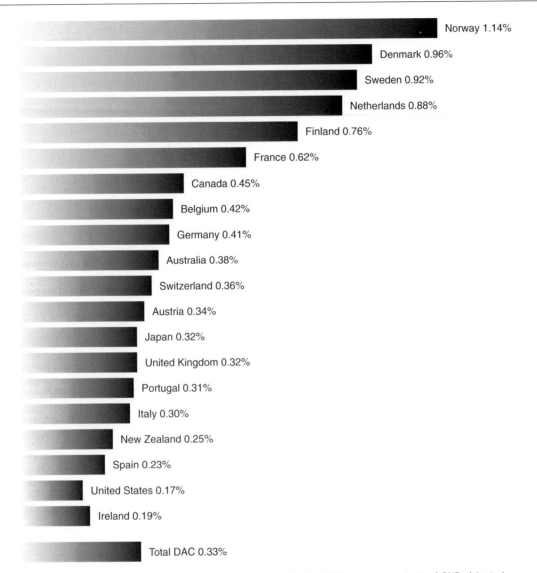

Figure 8.2 Net development assistance from OECD countries in 1991 as a percentage of GNP. Adapted from Oxfam (1993).

proportion of its aid than most donors; 74 per cent of bilateral aid in 1991 was tied to the purchase of British goods and services (German and Randel, 1993). The construction of the controversial Pergau Dam in Malaysia in the early 1990s, costing US$350 million, was the largest single project ever financed under the UK aid programme. However, critics argued that it was an expensive and inefficient source of power, and a waste of money for both Malaysian consumers and British taxpayers. Nevertheless, the project went ahead, because it was linked to large quantities of British exports to Malaysia, including over US$1 billion of arms exports.

A report written in 1993 (German and Randel, 1993) and reflecting on the policies of the Reagan and Bush administrations in the United States, noted that US aid was generally directed to two broad categories of countries: countries of strategic importance and poor but 'politically correct' countries. Egypt and Israel, both of strategic importance, respectively received 32.1 per cent and 8.3 per cent of aid in 1990–91. During the civil war in El Salvador,

Central American countries received about US$1 billion of bilateral aid from the United States each year. One-third of US bilateral aid in 1991 was spent on food aid. Interestingly, the primary stated purpose of US aid is to 'advance US interests by helping co-operating countries to expand their economies and the opportunities they offer their people' (German and Randel, 1993: 59). Poverty alleviation is not a specifically stated goal of US aid, though is an element of certain programmes.

Since the 1990 *World Bank Report on Poverty*, some governments have made a stronger commitment to poverty alleviation, and DAC is keen to monitor donor performance in this area. NGOs are concerned that too much aid is used to promote exports and subsidise domestic industry, rather than to benefit the poorest people in developing countries. Even Norway, the country with the greatest proportion of GNP allocated to aid, and with a strong poverty alleviation focus, was keen in the early 1990s to expand Norwegian commercial interests. With Prime Minister Gro Harlem Brundtland's high-profile involvement on the world development stage, environmental issues have become a particularly important element in Norwegian development aid.

Food aid continues to be significant in commodity flows from developed to developing countries, and although most food aid is provided bilaterally on a government-to-government basis, the creation of the World Food Programme (WFP) in 1961 added a significant multilateral dimension, such that it is now the main provider of international food aid for development and disaster relief. Food aid is a controversial form of development assistance, since political and economic motives may be significant in sustaining flows and determining their direction (Shaw and Clay, 1993). Food aid may be divided into three types:

Programme food aid is usually given as a grant or a soft loan on a government-to-government basis to fill the gap between the demand and supply of food from domestic production and any commercial imports. Such aid may reduce the amount of foreign exchange a country needs to spend on buying imports and if the food is sold it provides additional local currency for the government.

Project food aid is primarily aimed at satisfying the nutritional needs of the poor, mainly in rural areas, and is given as a grant, with the food aid closely targeted. Although the WFP is the main provider, other government and NGO bodies may also be involved. WFP is also involved in a variety of rural projects concerned with health centres for mother-and-child care, primary education and training, and food-for-work programmes.

Emergency food aid is a response to sudden disasters such as drought, floods and pest attack, as well as civil war. Emergency food aid is provided both bilaterally and multilaterally, mainly by the WFP. The 1980s crisis in Ethiopia, the severe floods in Bangladesh and the refugee crisis in Africa's Great Lakes region in the 1990s, all have received large quantities of emergency food aid. In the period 1987–91, Bangladesh dominated the recipient countries, followed by Pakistan, India and Tunisia (Shaw and Clay, 1993).

The world food aid system is highly complex and diverse, with the United States being the largest single provider; some of the other countries involved are Canada, Australia, Japan, Norway and Sweden. The European Union operates a union-wide programme, in addition to 11 separate national programmes.

Much has been written about the merits and problems of food aid. Although most commentators would agree with the importance of food aid as part of a disaster-relief package, there is more concern about longer-term food aid, which might affect local production and disrupt food-marketing systems. It is suggested that food aid can lower local food prices, encourage governments to neglect the drive to food self-sufficiency, create a dependency mentality and change eating habits. However, although potential problems in moving to food self-sufficiency are recognised, 'the widespread professional view of practitioners and economists [is] that disincentive effects are avoidable' (Shaw and Clay, 1993: 15).

With the end of the Cold War, it has been suggested that some of the so-called peace dividend might be allocated to overseas development assistance aimed at poverty alleviation, rather than on military expenditure. It is both ironical and deeply disturbing, however, that although military expenditure within developed countries has fallen, the sale of arms and military hardware to the poorer countries of the world is still big business. In 1987 the United States gave US$5.4 billion in worldwide military assistance, and the former Soviet Union gave US$13.5 billion. These figures declined to US$3.4 billion (United States) and zero (former Soviet Union) by 1993, but there was still a total military assistance of US$4.6 billion, some

74 per cent of this coming from the United States. The United Nations Development Programme (UNDP) argued that 'military assistance to the Third World formed one cornerstone of the cold war, . . . and also had commercial motives, helping sustain the output of the arms industry by subsidizing exports and unloading outdated weaponry' (UNDP, 1994: 53). Military assistance has many damaging effects for poor countries. Even after conflicts are resolved, large quantities of weaponry within developing countries pose a continuing threat to internal stability, and considerably strengthen the army and its ability to seize power. UNDP advocates the phasing out of military assistance and tighter controls imposed on the arms trade. In 1994, 86 per cent of conventional weapons exported to developing countries came from (in descending order); the former Soviet Union, the United States, France, China and the United Kingdom, all permanent members of the UN Security Council. Two-thirds of these arms were sold to ten developing countries, including Afghanistan, India and Pakistan. A comprehensive policy for arms production and sales is urgently needed, with special emphasis placed on cutbacks in the production of chemical weapons and land-mines (UNDP, 1994). In Angola and Cambodia it is estimated that millions of land-mines have been planted, causing continual suffering among local populations. The land-mine issue came to the fore in 1997, and many governments, including the UK government, have now agreed to ban their sale overseas.

In conclusion, there needs to be an improvement in the quality as well as the quantity of overseas development assistance if poverty alleviation, leading ultimately to poverty eradication, is to be achieved throughout the world. There have been proposals to make development assistance obligatory, perhaps by introducing a form of international tax on the rich countries (Watkins, 1995). Another proposal is a tax on international currency transactions, which would deter vast flows of speculative capital, estimated at US$1 trillion per day, which are destabilising economies in both the developed world and the developing world. National and international action would be needed to bring financial markets under more effective control. Some reform of the IMF is also necessary, to help developing countries with serious foreign exchange shortages to increase their reserves, without resorting to deflationary measures or constraining growth by cutting essential imports (Watkins, 1995).

Oxfam urged the UK government in 1993 to take the lead in providing more assistance to the poorest countries in Africa, and suggested that the United Kingdom should persuade the European Union to take urgent action too. Aid should focus on the poorest people in the poorest countries, and the nature of future development assistance must place greater emphasis on key aspects such as health and education, rather than emphasising the potential for exports from donor countries. There is a need to discuss aid and development priorities with local communities, and also to reduce the costs of delivering aid through an army of relatively well-paid expatriate consultants and developers (Oxfam, 1993). Perhaps most important of all, development assistance should concentrate more on achieving a sustainable improvement in the quality of life.

If nothing else, it is hoped this chapter has emphasised the significance of many different types of movements and flows which take place at different points in time and space. It is impossible to fully appreciate the character and underlying causes of different geographies of development without an understanding of the ways in which people and places are connected through movements of people themselves, as well as flows of commodities, finance and knowledge.

CHAPTER 9

Urban spaces

Urbanisation and development: an overview

Over recent history it has generally been assumed that urbanisation – the increase in the proportion of a given population that is to be found living in urban spaces – goes hand in hand with the process of 'development', however the latter is defined. Thus, through time, since the emergence of the first cities some 6,000–9,000 years ago (Potter and Lloyd-Evans, 1998), it has been assumed that the processes of urbanisation, structural change, development, and latterly industrialisation, are fundamental correlates.

It is for this reason that it can be argued that urbanisation is one of the most significant processes affecting societies in the late twentieth century and beyond (Devas and Rakodi, 1993; Drakakis-Smith, 1987; A.G. Gilbert and Gugler, 1992; Potter 1992b, Potter and Lloyd-Evans, 1998). As stressed in Chapter 3, dualistic conceptualisations regard the development process as one of endeavouring to change what are regarded as traditional, rural, agrarian-based societies into so-called modern, urban-industrial societies, following the model provided by European nations. Hence urbanisation and industrialisation are conflated as synonymous processes.

However, today it is the countries of the developing world which are experiencing the fastest rates of urbanisation, and as this chapter will serve to demonstrate, their urban proportions are increasing much more rapidly than those of European countries at their fastest. This trend is very different from the past, largely because it has involved a change whereby the growth of large cities is now firmly associated with poor countries. In rich countries, many central cities are declining in size due to inner city decay and counter-urbanisation, whereby people are moving from the cities to the suburbs and indeed the rural areas.

Thus, Dwyer (1975: 13) observed that 'in all probability we have reached the end of an era of association of urbanisation with Western style industrialisation and socio-economic characteristics.' (Plate 9.1) This is best illustrated by the disparity which now

Plate 9.1 Part of the commercial centre of Tijuana, Mexico with peripheral squatter settlements in the distance (Photo: Rob Potter).

characterises the relationship between levels of urbanisation and industrialisation in Third World societies. In 1970 the non-communist, less developed countries, taken as a whole, showed a level of urbanisation of 21 per cent, whereas 10 per cent of the active population was employed in manufacturing, yielding an excess of urbanisation over industrialisation of 110 per cent (Bairoch, 1975). By contrast, Europe in the 1930s showed a 32 per cent level of urbanisation, but at this point some 22 per cent of the active population were engaged in manufacturing. Thus, the excess of urbanisation over industrialisation was appreciably lower at around 45 per cent.

Urbanisation in the contemporary Third World

The very rapid growth of towns and cities that is occurring in the so-called Third World can be illustrated in a number of different ways. One is the increase that is occurring in cities which have a population of a million or more persons in the tropical world. Statistics show that in the 1920s, twenty-four of the world's cities had more than one million inhabitants (Table 9.1). By the early 1980s, the number of such million cities had increased to 198. But more significant, during each decade between these two dates, the average latitude of million cities had moved steadily toward the equator. This had changed from 44°30' in the 1920s, to 34°7' in the 1980s (Table 9.1). Thus, million cities are increasingly associated with the tropical regions of the world (Mountjoy, 1976). In 1950 there were 31 cities of a million or more inhabitants in developing countries, but by 1985 there were 146. It is estimated that by 2025 there will be a staggering 486 cities of a million or more in developing countries (Harris, 1989).

The trend toward rapid urbanisation in the Third World is also shown by the league table of the largest urban places in the world. This is given for the years 1950 and 2000 in Table 9.2. In 1950 the three largest cities in the world, New York, London and the Rhine–Ruhr conurbation, were all to be found located in

Table 9.1 The world distribution of 'million cities': cities with more than one million inhabitants.

Date	Number of million cities	Mean latitude north or south of equator	Mean population (millions)	Percentage of world population living in million cities
Early 1920s	24	44°30'	2.14	2.86
Early 1940s	41	39°20'	2.25	4.00
Early 1960s	113	35°44'	2.39	8.71
Early 1980s	198	34°07'	2.58	11.36

Source: Potter (1992).

Table 9.2 The largest cities in the world, 1950 and 2000.

1950			2000		
Rank	City	Population (millions)	Rank	City	Population (millions)
1	New York	12.3	1	Mexico City	31.0
2	London	10.4	2	São Paulo	25.8
3	Rhine–Ruhr	6.9	3	Shanghai	23.7
4	Tokyo	6.7	4	Tokyo	23.7
5	Shanghai	5.8	5	New York	22.4
6	Paris	5.5	6	Beijing	20.9
7	Buenos Aires	5.3	7	Rio de Janeiro	19.0
8	Chicago	4.9	8	Bombay	16.8
9	Moscow	4.8	9	Calcutta	16.4
10	Calcutta	4.6	10	Jakarta	15.7
11	Los Angeles	4.0	11	Los Angeles	13.9
12	Osaka	3.8	12	Seoul	13.7
13	Milan	3.6	13	Cairo	12.9
14	Bombay	3.0	14	Madras	12.7
15	Mexico City	3.0	15	Buenos Aires	12.1

Source: United Nations (1985).

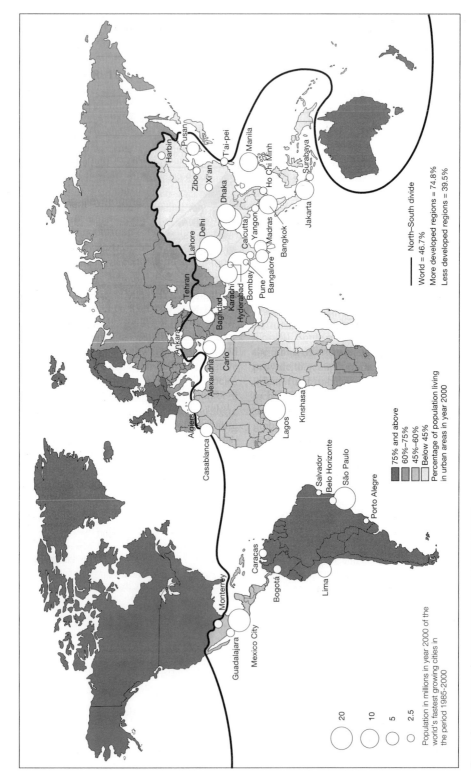

Figure 9.1 Aspects of contemporary world urbanisation.

developed countries. By 2000, Mexico City, São Paulo and Shanghai, all with more than 23 million residents, will be the world's largest cities. And by 2000, out of the fifteen largest cities listed in Table 9.2, twelve can be described as being in the Third World. This compares with only five in 1950.

The world's fastest-growing cities during the period 1985–2000 are shown in Figure 9.1, along with broad levels of urbanisation by continental division. The map also indicates the commonly accepted geographical definition of the South or Third World. It is noticeable that the fastest-growing cities are all located to the south of the line which divides the rich 'North' from the poor 'South'. Thus, cities such as Karachi in Pakistan, as well as Lagos, Kinshasa and Addis Ababa in Africa, grew by more than 50 per cent between 1975 and 1985, and will continue to grow rapidly. This diagram gives a very clear indication of the contemporary link between urbanisation and urban growth in the less developed realm.

In fact, cities are growing so fast in the Third World that in certain areas they are merging together to form large compound urban regions. These regions are called megalopolitan systems and at least eight of them can be recognised in the Third World today (Gottmann, 1978). They are based on Third World urban regions such as Mexico City, São Paulo, Lagos and Cairo. The terms *super-city*, *giant city, conurbation* are also used to indicate these large sprawling urban complexes. The concept of the World City, as reviewed in Chapter 4, relates primarily to the international functional importance of certain contemporary cities, irrespective of their size, whereas *mega-cities* are generally defined as those with populations in excess of 8 million (A.G. Gilbert, 1996; Oberai, 1993). Gugler (1996) has mapped 43 urban agglomerations with more than 4 million inhabitants, of which around 29 are located in less developed countries (Figure 9.2, overleaf).

Figure 9.3 (page 221) summarises the overall increase that has occurred in the level of urbanisation between the years 1950 and 2025 for the more developed and less developed regions of the world. The clearest feature depicted in the graph is that, although it has flattened out for the developed world since 1965, the urbanisation curve has increased very sharply in the less developed world. By the year 2000, 74.8 per cent of the population of more developed areas will be urban, but this will be true of only 39.5 per cent of those living in less developed areas. The situation is summarised by the data in Table 9.3 (page 221).

However, the percentage figures tell us only one part of the story, and it is the total numbers involved that are truly awe-inspiring. This is illustrated in Figure 9.4 (page 222). Up to 1970 the absolute number of city dwellers was larger in the more developed regions than in the less developed areas. But from that date onwards, the number of city dwellers has risen dramatically in the Third World. During the period 1950–2025, the number of urban dwellers in developing countries will have increased 14 times, from 300 million to a staggering 4 billion. By 2000 there will be two city dwellers in the Third World for every one in the more developed world. Table 9.4 (page 222) cogently summarises this situation. This fact stresses once more the degree to which urban living in the modern era has come to be associated with the poorer countries of the globe.

These data can be disaggregated by major world region, as shown in Table 9.5. By 2025 it is believed that just under 60 per cent of all Africans will be living in urban settlements, as will just over half of all those living in Asia. In the same year, nearly 85 per cent of all Latin Americans will be living in towns and cities. This is all a very far cry from the situation in 1920, when United Nations data indicate that less than 10 per cent of Africans and Asians, and only 22 per cent of Latin Americans were urban dwellers (Table 9.5, page 222).

Politicians, planners and development experts from all over the world must grapple with these facts during the coming decades. As illustrated above, this very rapid rise in urban living is occurring in the regions of the world where socioeconomic conditions are generally at their poorest and where industrial production is relatively low. In these areas, resources are very limited, so enormous pressure is being exerted on existing socioeconomic systems, and especially on the children, women and men who live in such poor areas and regions. Indeed, dealing with the challenges which are presented by these fundamental changes represents one of the major tasks faced by planners and politicians in the twenty-first century.

Causes of rapid urbanisation in the Third World

Urbanisation can be defined as the process which leads a higher proportion of the total population of an area to live in towns and cities. It is thus a relative measure, recording the percentage of the total population of a nation or a region that is to be found in towns and cities. This should not be confused with the absolute

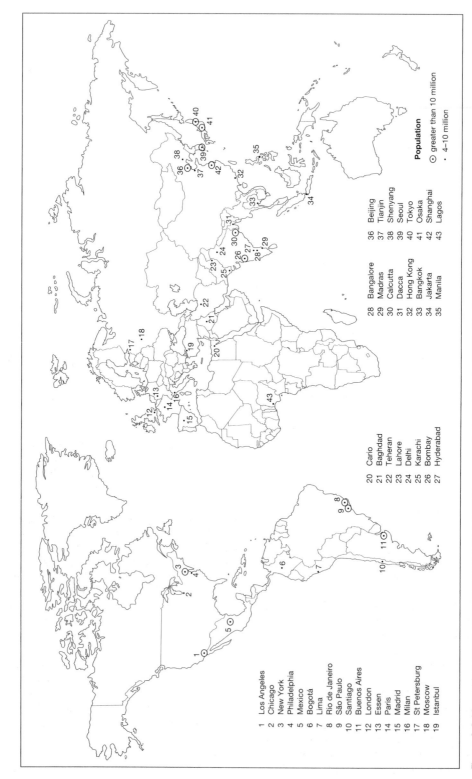

Figure 9.2 Megacities: urban agglomerations with populations exceeding 4 million in 1990. Adapted from Gugler (1996).

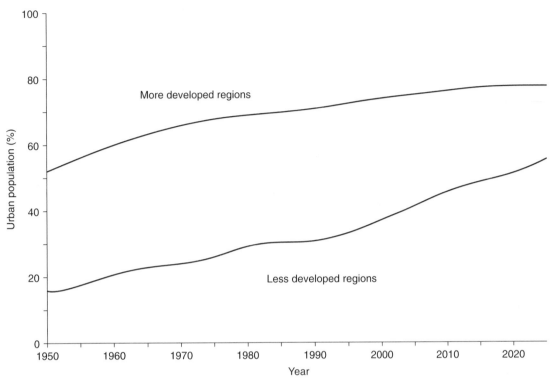

Figure 9.3 The proportion of population residing in more and less developed regions, 1950–2025. Adapted from United Nations (1989).

Table 9.3 Percentage of total population living in urban areas, 1970–2025.

Date	World	Less developed regions	More developed regions
1970	37.2	25.5	66.6
2000	46.7	39.5	74.8
2025	60.5	56.9	79.0

Source: United Nations (1989, 1996).

growth of urban areas and urban populations. These are best described by the term *urban growth.*

Throughout the Third World, people are migrating from the rural areas to towns and cities. Normally about half the growth of cities reflects rural-to-urban migration. For example, during the 1960s, the World Bank estimated that migrants as a percentage of total population increase amounted to 50 per cent in Caracas, 52 per cent in Bombay, 54 per cent in Djakarta, 50 per cent in Nairobi and 68 per cent in São Paulo. In the Philippines in the 1970s, the inmigration rate to cities was 1.9 per cent per year, out of an urban population

growth per annum of 3.9 per cent. For Brazil during the same decade, inmigration accounted for 2.2 per cent per annum out of a total rate of 4.4 per cent per annum (United Nations, 1989; Devas and Rakodi, 1993).

But why are migrants currently travelling to cities in the Third World in such large numbers? This mainly stems from the widespread existence of poverty, un-employment and deprivation in the rural areas of many Third World countries. Data show that where jobs do exist, rates of pay are higher in urban areas, and also that average incomes increase with city size in a progressive manner (Hoch, 1972). In addition, social and health facilities are better in the principal towns and cities (Phillips, 1990), although access to them is a major problem for the poor in society (Potter and Lloyd-Evans, 1998: Ch. 5). But just as important is the fact that, with so many people and activities around, it is literally possible to make a living for yourself, doing what are known as informal sector jobs.

The informal sector is also called the tertiary refuge sector; it is made up of jobs such as street hawking, shoe shining, ice-cream or 'snow cone' vending, car

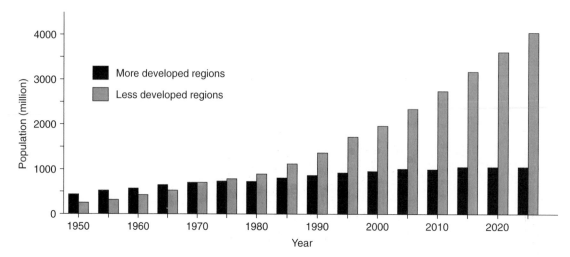

Figure 9.4 Total urban populations of more and less developed regions, 1950–2025.

Table 9.4 Number of people living in urban areas, 1970–2025 (millions).

Date	World	Less developed regions	More developed regions
1970	1 374	675	698
2000	2 916	1 971	944
2025	5 118	4 050	1 068

Source: United Nations (1989, 1996).

Table 9.5 Percentage of population living in urban area by major continental region, 1920–2025.

Region	Percentage of total population living in urban places			
	1920	1970	2000	2025
World	19	37.2	46.7	60.5
More developed regions	40	66.6	74.8	79.0
Less developed regions	10	25.5	39.5	56.9
Africa	7	22.9	41.3	57.8
East Africa	–	10.3	30.1	48.0
Middle Africa	–	24.8	47.6	64.7
Northern Africa	–	36.0	49.9	65.3
Southern Africa	–	44.1	61.7	74.2
Western Africa	–	19.6	40.7	58.9
Latin America	22	57.3	77.2	84.8
Caribbean	–	45.7	65.5	75.5
Central America	–	54.0	71.1	80.5
South America	–	60.0	81.0	87.5
Asia	9	23.9	35.0	53.0
Eastern Asia	–	26.9	32.6	49.0
Southeast Asia	–	20.2	35.5	54.3
Southern Asia	–	19.5	33.8	52.6
Western Asia	–	43.2	63.9	76.3

Source: United Nations (1969, 1988, 1996).

washing, taxi driving and many others (Plates 9.2, 9.3 and 9.4, pages 223–4) (Santos, 1979; T. McGee, 1979; Portes, Castells and Benton, 1991). Such activities subscribe to the notion that 'even to eat crumbs you have to be sitting at the table' (Jones and Eyles, 1977). The perceived advantages of living in large urban places are of great importance. This is because in the past, where they did exist, factories, roads, infrastructure and other facilities focused on the major urban areas in most developing countries. Later on we examine in detail the structure and role of the informal sector.

Another very important factor has been enhanced medical facilities in the postwar period, while birth rates have remained at traditionally high levels. Thus, Third World cities are growing by natural increase, as well as by migration, although the proportions have varied from region to region. This was not true of cities in Britain during the Industrial Revolution. Such areas tended to be far less healthy than the surrounding countryside, and were thereby regarded as death-traps. Quite simply, health conditions are much better in the urban areas where the large teaching hospitals, clinics and most doctors are to be found. Third World populations are growing at an average of 2 per cent per annum, so urbanisation is working on much larger base populations.

Plate 9.2 Motorised rickshaw driver, Delhi (photo: Rob Potter)

Plate 9.3 Informal sector furniture production, Castries, St Lucia (photo: Rob Potter)

This is shown if we juxtapose the demographic transition and the cycle of urbanisation (Figure 9.5, page 225). The countries of the developed world experienced a gradual process of demographic change, as shown in Chapter 5. The rapid urbanisation that occurred in Western Europe and the United States during the late nineteenth and early twentieth centuries was also associated with the rise of the factory system and industrialisation. Figure 9.5(a) and (c) show just how gradual this has been. Birth and death rates both fell relatively gradually from 1800 onwards. In less developed countries, the birth rate has generally continued at traditional levels of 40–45 per 1,000 of the population. But since around 1950 the crude death rate has fallen very dramatically, leading to very rapid rates of total population increase (Figure 9.5(b)). This is frequently known as the telescoping of the demographic transition by developing countries.

Plate 9.4 Female hawkers in Georgetown, Guyana
(photo: Rob Potter)

As noted in Chapter 3, some people also point to the fact that various organisations and governments in Third World countries have tended to follow the example set by rich industrial countries. Initially this was due to colonial rule and influence, but even since independence, such domination has frequently occurred in the form of imperialism. This is as true today as in the past, in terms of emulation, where ways of doing things and lifestyles from overseas are valued over and above those from the home country. This is seen as being intensified by the import of foreign goods and the activities of multinational companies, as discussed in Chapter 4.

The arrival of large numbers of tourists from wealthier countries can also give rise to what are described as demonstration effects. Together, these two influences are seen as giving rise to the continuing dependency of many Third World countries on those of the First World. It can be argued that too much reliance has in the past been placed on overseas goods, and on ways of doing things that have been derived from abroad. This has also come to be associated with an overconcentration in urban areas, and industry not agriculture has been stressed as the means for development (Chapters 1 and 3).

City systems and development: questions of urban primacy, regional inequalities and unequal development

The next section begins to examine conditions within Third World cities. But before that, we focus attention on some of the most important discussions about the sets of towns and cities which make up the so-called urban system of nations and regions.

It has frequently been argued that urban primacy – the eminence of one or more cities – is characteristic of Third World urban systems. An urban system can be defined as a set of interrelated towns and cities which together comprise the urban settlement fabric of an area. Table 9.6 (page 226) shows that for the world as a whole, 15 per cent of the urban population is to be found living in capital cities; the fraction is considerably higher for several of the world's developing regions. Thus, the proportion of the total urban population living in the capital is as high as 33 per cent for sub-Saharan Africa, and approximately 25 per cent for Latin America, the Middle East and North Africa.

But despite this broad association, when we turn to the level of urban primacy recorded in individual

Figure 9.5 also shows that, in a similar fashion, urbanisation is occurring at a much accelerated rate in Third World countries. The gradual increase in urbanisation which occurred in the more developed world is described as the cycle of urbanisation. This takes the form of an attenuated or squashed S-shape curve (Figure 9.5(c)). However, urbanisation is occurring much more rapidly in Third World countries. The very rapid rise in the urban proportion (Figure 9.5(d)) occurs at the same time as the massive spurt in population.

Real growth statistics for several nations are graphed in Figure 9.6 (page 226). The urban proportion for England and Wales increased gradually from around 25 per cent in 1800 to approximately 80 per cent in 1975. The swiftest rise came in the period 1811–1851, and the rate of increase dropped somewhat after that. In comparison, countries such as Brazil, Egypt, South Korea and India have shown very rapid rates of urbanisation in the relatively short period since 1945.

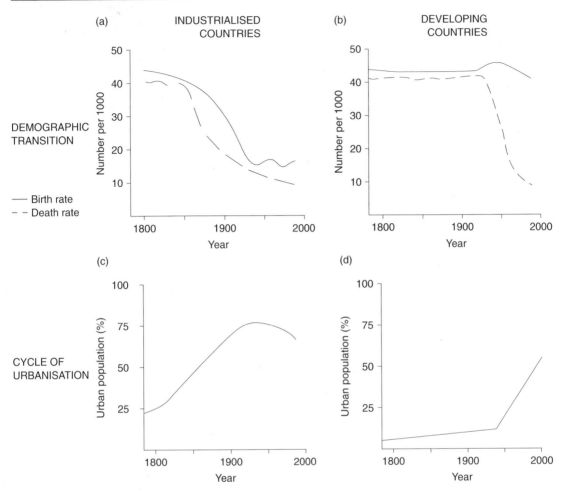

Figure 9.5 The cycle of urbanisation and the demographic transition for developed and developing countries. Adapted from Potter (1985, 1992b).

countries, the issue becomes more complex. In a pioneering paper, Berry (1961) showed that, at the level of nation states, there is no clear statistical relationship between a nation's city size distribution and either its level of urbanisation or economic development, as measured by GNP per capita. In the face of these negative findings, Berry speculated that a whole complex of forces influence relative city size distributions. In particular, it was posited that if a few strong forces operate, a primate distribution will be the outcome (Figure 9.7(a), page 227). It was argued that fewer forces are likely to influence the urban situation for a smaller country, a shorter history of urbanisation, a simpler economic and political life, and a poorer overall degree of development. The opposites of these cases suggest that a country will develop a

range of specialised cities performing a variety of functions. The smooth distribution of cities that results can be called a rank–size distribution (Figure 9.7(b), page 227) or a log-normal city size distribution (Figure 9.7(c), page 227).

Berry's line of argument was taken up afterwards in several research papers, notably those by Mehta (1964) and Linsky (1965), both of whom took essentially the same approach, correlating a number of variables against degree of urban primacy for a sample of nations. The results of this approach are summarised in Table 9.7 (page 227).

Linsky prespecified the predicted relationships between urban primacy and six variables. He suggested that primacy was positively associated with the degree of export-orientation of the nation, the proportion of

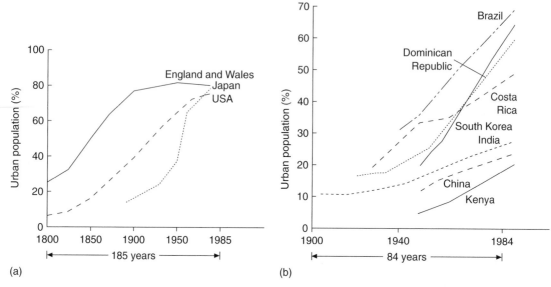

Figure 9.6 Examples of urbanisation curves for developed and developing countries. Adapted from Potter (1992b).

Table 9.6 An overview of urban development in the Third World.

Region	Percentage of total population classified as urban	Capital city as a percentage of urban population	Percentage of urban population living in cities over 1 million population
Sub-Saharan Africa	29	33	34
East Asia/Pacific	29	12	37
South Asia	25	8	38
Middle East/North Africa	55	26	41
Latin America	73	24	46
World	42	15	38

Source: World Bank (1994), Dickenson *et al.* (1996).

the workforce employed in agriculture, and the overall rate of population growth. The areal extent of dense population and per capita income levels were envisaged as being negatively correlated with primacy. Finally, somewhat curiously perhaps, an open verdict was initially pronounced on the association between primacy and former colonial status.

The analysis showed that all the hypothesised associations between the variables and urban primacy were as expected. In addition, former colonial status was positively related to levels of urban primacy. This might be expected, for colonial status involves a strong coastal–mercantile orientation, with the attendant urban polarisation that goes with this, as mapped into the mercantile model (Chapter 3). Significantly, the strongest relationship was the negative correlation between urban primacy and the size of countries ($q =$

-0.37 in Table 9.7). The other statistically significant relationship was the positive one existing between urban primacy and the overall rate of population growth ($q = +0.33$). Thus, Linsky's work was significant in confirming that although urban primacy is characteristic of small nations which have low per capita incomes, a high dependence on exports, a former colonial status, an agricultural economy and a fast rate of population growth, it is certainly not precluded elsewhere. For example, contemporary Thailand exhibits few of these features, yet it presents extreme urban primacy.

At virtually the same juncture, essentially similar conclusions were being arrived at by Mehta (1964), a demographer (Table 9.7). He found that the strongest association was a negative one between urban primacy and size of population (-0.29), followed by a negative

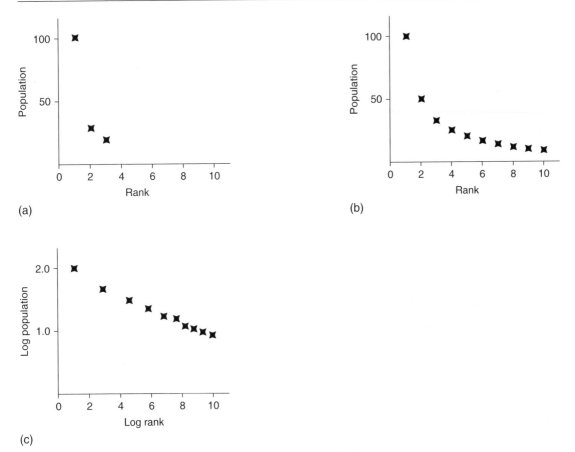

Figure 9.7 Urban settlement distributions: (a) primate, (b) rank–size (c) log-normal.

Table 9.7 The relationships between urban primacy and a selection of other variables.

A. Linsky (1965)

Variable	Expected relation with primacy	Correlation
Areal extent of dense population	negative	−0.37[a]
Per capita income	negative	−0.22
Export orientation	positive	+0.22
Colonial history	?	+0.21
Number in agriculture	positive	+0.14
Rate of population growth	positive	+0.33[a]

B. Mehta (1964)

Variable	Expected relation with primacy	Correlation
Gross national product	negative	−0.08
Level of urbanisation	negative	−0.12
Overall population density	negative	+0.02
Export dependency on raw materials	positive	+0.19
Area of country	negative	−0.28
Size of population of country	negative	−0.29

[a] Correlation statistically significant.
Source: Potter and Lloyd-Evans (1998).

Table 9.8 Vapnarsky's cross-classification of city size distributions.

High closure Low interdependence	No clear pattern
Low closure Low interdependence	Primate distribution
Low closure High interdependence	Primate city superimposed on a rank–size pattern
High closure High interdependence	Rank–size distribution

Source: Based on Vapnarsky (1969).

Table 9.9 Regional inequalities in a sample of Third World and developed countries.

Country	Ratio of gross regional product between the richest and poorest regions
Developed countries	
United Kingdom	1.43
France	2.09
Italy	2.20
Netherlands	1.56
Japan	2.92
Developing countries	
Brazil	10.14
Thailand	6.34
Iran	10.07
India	2.24
Venezuela	5.72

Source: Renard (1981).

association with the areal extent of countries (−0.28). Urban primacy was also positively related to the degree of dependency on raw materials shown by countries (Table 9.7). Mehta's findings again showed no correlation between urban primacy and gross national product, level of urbanisation or overall population density.

Just a few years later, Vapnarsky (1969) in an historical–empirical study of Argentina argued that the primate and rank–size distribution patterns are not to be seen as the extremes of a continuum, i.e. separate and mutually exclusive. Rather, he argued, the two distributional types are produced by different sets of circumstances. On the evidence offered by Argentina, Vapnarsky (1969) regarded urban primacy as being positively associated with the degree of closure of the economy, i.e. the degree of dependence on overseas trade. With increasing closure, urban primacy was believed to reduce, other things being equal. In contrast, Vapnarsky saw the distribution pattern as being affected by the level of interdependence existing in a country, i.e. the extent to which its various regions are interlinked by virtue of flows of people, goods, capital, etc. It was believed that as internal interdependence increases, the log-normal pattern will progressively be approached.

This gives rise to the argument that various admixtures of the primate distribution and the log-normal distribution can be recognised. The basic divisions are shown in Table 9.8. The classic primate distribution is seen as the outcome of low closure together with low interdependence; this may prevail for longer in small countries than elsewhere. The classic log-normal distribution and the classic rank–size distribution result from high closure together with high interdependence (Table 9.8).

But undoubtedly, the most important point is that primacy really occurs and should therefore be examined, at the regional scale. Thus, large developing countries such as India, China and Brazil display low levels of primacy at the national scale precisely because they are comprised of several primate regional urban areas, such as Calcutta, Bombay, Delhi and Madras in the case of India. Thus, primacy should be seen as one expression of the wider existence of regional inequalities and spatial polarisation in development. As reviewed in detail in Chapter 3, the mercantile model shows how global trade and capitalism since the 1400s have led to development being articulated through coastal gateway cities. Such places are the concentration points of social surplus product. It is likely in a large territory that a whole series of coastal gateways will have developed, thereby giving rise to a series of strong regional primate distributions; and this leads to a national log-normal distribution (Figures 3.9–3.11).

Another way of looking at this important issue is by means of stressing that it is really regional imbalances – often between an array of urban areas and their associated regions on the one hand, and the rural areas on the other – that are most pronounced. Here the evidence shows quite clearly that the regional inequalities that characterise developing countries are generally far more marked than those which exist in the case of more developed nations. For example, if the gross regional products of the richest and poorest regions of nations are compared, then developed countries typically show ratios between 1.5 and 3.0 (Table 9.9). Examples are France, where the richest region is 2.09 times better off than the poorest, the United Kingdom (1.43) and

Table 9.10 Urban–rural incidence of poverty for a range of developing countries.

	Population in poverty, 1980–1990		
	% total population	% rural population	% urban population
Botswana	43	55	30
Burundi	84	85	55
Ghana	42	54	20
Kenya	52	55	10
Morocco	37	45	28
Mozambique	59	65	40
Nigeria	40	51	21
Rwanda	85	90	30
Tanzania	58	60	10
India	40	42	33
Malaysia	16	22	8
Nepal	60	61	51
Philippines	54	64	40
Thailand	30	34	17
Bolivia	60	86	30
Brazil	47	73	38
Dominican Republic	55	70	45
Guatemala	71	74	66
Haiti	76	80	65
Honduras	37	55	14
Mexico	30	51	23
Panama	42	65	21
Peru	32	75	13
Venezuela	31	58	28
St Kitts and Nevis	46	50	40
Papua New Guinea	73	75	10

Source: United Nations Centre for Human Settlements (1996).

the Netherlands (1.56). With respect to developing nations, the ratio between richest and poorest region may be as high as 9 or 10 times, as Table 9.9 shows for Brazil (10.14) and Iran (10.07).

Perhaps a clearer way of looking at the nature of the differences which currently characterise rural and urban areas is to consider the proportions of the rural, urban and total populations that are designated as living in poverty. These data are shown for a range of developing nations in Table 9.10. For Brazil, 73 per cent of the rural populace are classified as living in poverty, but the proportion is only 38 per cent for urban areas. Even for Rwanda – the nation in Table 9.10 that displays the highest national level of poverty – the incidence of poverty is given as 30 per cent of the urban population and 90 per cent of the rural population (United Nations Centre for Human Settlements, 1996). However, as migration occurs, it has to be recognised that urban poverty is growing faster than rural poverty in some nations.

Evidence suggests to many commentators that differences between regions of this sort are showing relatively little sign of decreasing over time, and with development (see A.G. Gilbert and Goodman, 1976; Stöhr and Taylor, 1981; J. Friedmann and Weaver, 1989, A.G. Gilbert and Gugler, 1992; Potter and Lloyd-Evans, 1998). This is despite the fact that the classic view is that regional inequalities increase at first during the early stages of economic growth, but decrease with time thereafter. This U-shaped patterning of inequality over time was first recognised for European countries by Williamson (1965), and others have taken the same essentially laissez-faire stance (Alonso, 1968; 1971; Mera, 1973; 1975; 1978).

However, A.G. Gilbert and Goodman (1976) argue that for every Third World country that can be cited as showing a tendency towards regional equality, another can be found that is showing increasing disequilibrium. Indeed, A.G. Gilbert and Gugler (1992) and Potter and Lloyd-Evans (1998) argue that for a number of reasons it seems that regional income and welfare convergence is likely to be weak and slow in developing nations. This is in no small measure due to the magnitude of the differentials which now exist in Third World countries as well as the vast base populations that are involved.

A more contemporary view of the issues of primacy and inequality can be derived if we go back to the approach presented in Chapter 4. There it is argued that cities are the key points in the dual processes of global convergence and divergence. Following this approach, urban areas in Third World countries are seen as the points of introduction and diffusion of global norms and patterns of consumption (convergence), as discussed in Chapters 1 and 4 when dealing with globalisation. At the same time, cities in the Third World are regarded as the spatial localities at which production, capital and decision making are increasingly being concentrated (divergence). The divergence processes, it is argued, are increasingly coming to be controlled by a relatively small band of large transnational companies and concerns. It was in this light that Armstrong and McGee (1985: 41) described Third World cities as simultaneously acting as both 'theatres of accumulation' and 'centres of diffusion'.

It certainly has to be acknowledged that simple formulations – where urban systems are seen as developing from a primate distribution to a log-normal distribution and a central-place hierarchical structure – are rather unrealistic in the contemporary world. Such formulations can also be regarded as Eurocentric and linear in conception. Urban systems are better understood as being affected by a larger number of complex forces in the modern and post-modern times in which we live. And the global processes of convergence–divergence offer an interesting framework for exploring the multiplicity of meanings involved in the process of Third World development and change.

Urban and regional planning in Third World countries

The foregoing discussion demonstrates that the crucial point is whether there is faith that the free market will lead to the spread of growth and the equalisation of production, incomes and welfare throughout the national space. Referring back to Chapter 3, it was Myrdal, in contrast to Hirschman, who believed that such equalisation or 'polarisation reversal' would not occur spontaneously. J. Friedmann (1966) had

explicitly taken up this theme in his four-stage core–periphery model (Figure 3.7). Specifically, Friedmann believed that the transition from the second stage of the model, associated with a single strong national core, to the third stage, witnessing the emergence of peripheral subcores, would only come about in developing societies as the result of direct state intervention.

This debate is reflected just as clearly in the literature of the 1970s in what amounted to a debate about the growth of large cities. In the 1970s, several economists and regional economists followed the neoclassical approach, basically arguing that any attempt to retard the spontaneous growth of large cities would, by definition, be counter-productive, because it would serve to retard national rates of economic growth (Richardson, 1973, 1976; Alonso, 1971; Hoch, 1972). However, in a series of exchanges with Richardson, A.G. Gilbert (1976, 1977) took issue with the laissez-faire doctrine of unrestrained growth. From a primarily economic viewpoint, A.G. Gilbert (1977) questioned the assumption that higher productivity in big cities is brought about via economies of scale. Gilbert argued that such growth is essentially achieved at the expense of productivity elsewhere. It was maintained that if 'infrastructure of the same quality were provided in medium-sized centres, the productivity in these centres would rise' (A.G. Gilbert, 1976: 29). This is tantamount to saying that capitalist development has promoted an essentially circular argument. It has enhanced the productivity of large cities by concentrating investment in them, and has subsequently argued that only large cities are productive (Potter and Lloyd-Evans, 1998). Hence development theory ends up with 'modernisation surfaces'.

Thus, the degree to which the state should become involved in regulating urban growth, and redirecting it at the national scale, is one of the crucial development planning issues (A.G. Gilbert and Gugler, 1992; Potter and Lloyd-Evans, 1998: Ch. 4). Many social commentators have inferred, or stated overtly, that only socialist states have seriously endeavoured to reduce urban and rural imbalances in national change and development. For example, avowedly anti-urban policies were implemented in South Vietnam between 1975 and 1980, and policies of zero urban growth were followed in China periodically from the late 1950s. But it is Cuba since the socialist revolution in 1959 that is cited as the best example (Box 9.1).

But it is vital to recognise that urban and regional planning policies must be based on economic, political, even moral and ethical considerations, not just on economic foundations. Indeed, although Richardson

BOX 9.1   Cuba: urban and regional planning in a revolutionary state

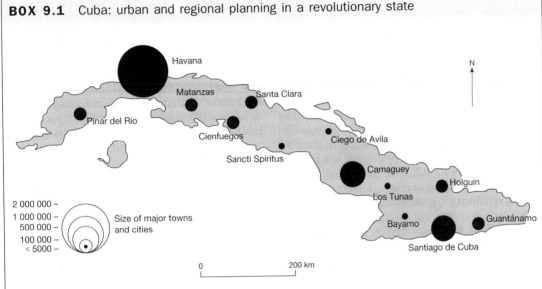

The principal towns and cities of Cuba.

Cuba, the largest of the Caribbean islands, was discovered by Columbus in 1492 at the dawn of the mercantile period. With the exception of a brief spell of British rule in 1762, Spain retained its colonial power over Cuba until defeated in the Spanish-American War of 1898. This represented the start of a period during which the island was dominated by the United States of America, first militarily and then economically, after independence in 1902.

During the first half of the twentieth century, the country was governed by a series of dictators, the last one being Fulgencio Batista whose corrupt regime had ruled the country from 1933. After a two-year guerilla campaign, law student Fidel Castro and his followers ousted Batista from power in 1959. It is generally accepted that the leaders of the revolution were not initially communists, but fervent nationalists who were opposed to the corruption and inequalities that had existed before. But the antagonistic stance taken by the United States after the revolution resulted in the Cubans increasingly turning to the Soviet Union. Before the revolution, Havana, the capital, was a classic primate city (see figure). Most of the wealth and activities of the country were concentrated there. However, it was also characterised by shanty towns, poverty, gambling and vice. By 1953 the Greater Havana area had grown to 1.2 million people, containing 21 per cent of the country's total population. At this time 75 per cent of all industry

was found in Havana and 80 per cent of the nation's exports passed through the port, serving to stress the dependent relation of the country to the United States. Most of the country's health care facilities, schools, colleges and cultural organisations were also situated in and around Havana.

Castro and his followers regarded the city as representing capitalist (American) interests and overprivilege. From around 1963, Havana was increasingly discriminated against. Its physical fabric was left to decay so as to make it less attractive to potential rural migrants. Two key policies were adopted: the decentralisation of people and activities from Havana; and the reduction of the striking differences which had come to exist between the urban and rural areas of the nation.

Thus, since 1959, promoting a more even geographical pattern of development has been the express aim of the state. The growth of provincial towns having populations between 20,000 and 200,000 has been encouraged. At the next level down, the regrouping of villages into rural new towns (comunidades) has occurred. Each rural new town has been developed with its own food and clothing stores, nurseries, primary schools, small clinic, social centre, bookshop and cafes. By 1982 some 360 comunidades had been created. Control has also been exercised over migration, with ministerial permission being required in order to move to a job in Havana.

Box 9.1 continued

Most important, massive efforts have been made to develop primary, secondary and tertiary health care facilities throughout the country. Treatment at the centres is free. Primary health care is available throughout Cuba, whereas secondary and tertiary facilities are located in towns and cities. There have also been great improvements in education. In 1971 only seven out of 478 secondary schools were to be found in rural areas; by 1979 this had changed to 633 rural schools out of a total of 1,318. All students are expected to work in agriculture at some stage in an effort to reduce elitist attitudes and values.

Today Havana has nearly reached the 2 million mark and is certainly much larger than Santiago (351,000) and Camaguey (252,000), the second and third largest cities (see figure). Some 69 per cent of the total population of Cuba is to be found living in urban settlements.

Cuba has done much to reduce the differences between town and country, although critics of the Marxist approach which has been followed suggest that the same could have been achieved without the state apparatus that controls all sectors of the economy. They also argue that much unemployment is disguised, that rural–urban differences still exist and that elite privileges have re-emerged.

Whatever the ultimate judgement, there can be no doubting that Cuba's 10 million inhabitants are now highly dependent on external controlling factors. In 1983 the total Soviet assistance to Cuba was US$4,100 million, consisting of US$1,000 million in development aid and US$3,100 million in trade subsidies. Thus, Cuba has remained a dependent state, even though this is now a socialist rather than a capitalist form of dependency.

Source: Potter (1992b).

Table 9.11 Richardson's categorisation of national urban development strategies.

Concentrated urbanisation
1 Free market or do nothing
2 Polycentric development of the primate city
3 'Leap-frog' decentralisation within the primate city

Deconcentration and decentralisation
4 Development corridors and axes
5 Growth poles and growth centres
6 'Countermagnets'
7 Secondary cities
8 Provincial capitals
9 Regional centres and hierarchy
10 Small service centres and rural development

Source: Richardson (1981).

had argued from a strongly pro-large-city standpoint, in the early 1980s, in reviewing national urban development strategies, he noted for the first time that the key goals were the same as societal goals in general, and that such strategies need to be highly country-specific (Richardson, 1981). In other words, there is no panacea or general solution to urban and regional problems.

Richardson's paper was particularly useful in itemising the wide range of policies which can be employed in attempts to decentralise people, jobs and social infrastructure away from primate cities and congested core regions. The range of policy reactions is shown in Table 9.11. These range from three

policies of continued concentrated urbanisation to seven representing genuine interregional deconcentration and decentralisation. The first policy of concentrated urbanisation is the laissez-faire policy of letting the market take its course. If, however, problems of congestion and imbalance are recognised in the primate city, efforts may be made to decentralise, but merely within the core region. Thus, a polycentric pattern of growth on the edge of the primate city, or a form of leap-frog decentralisation to the edge of the existing core may be envisaged (Table 9.11).

Strategies of genuine deconcentration can be categorised into seven generic types, as shown in Table 9.11. Development corridors or axes can be designated, leading from the core region, and growth can be focused upon them. Alternatively, growth may be channelled into what are regarded as dynamic growth poles or growth centres. A variation on essentially the same theme sees the strengthening of a few distant major nodes as countermagnets. Other forms of decentralisation can be created by the promotion of a limited number of secondary or intermediate cities, or the establishment of provincial state and departmental capitals. Yet another variant involves the promotion of regional metropolises and an associated hierarchy of urban places. At the far end of the spectrum, a dispersed policy of small service centres and associated rural development throughout the periphery may be pursued. Of course, these strategies are not mutually exclusive and several of them are very similar; various elements of these strategies can be

---

BOX 9.2   Nigeria: urban and regional planning in a top-down context

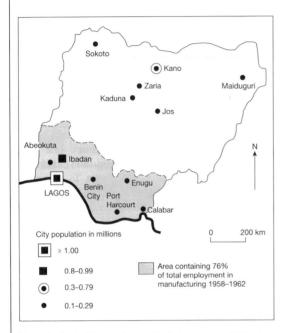

City population in millions

- ■ ≥ 1.00
- ■ 0.8–0.99
- ◉ 0.3–0.79
- • 0.1–0.29

Area containing 76% of total employment in manufacturing 1958–1962

0   200 km

The principal cities and manufacturing zone of Nigeria.

Nigeria is the largest nation in Africa and currently has a population of 94 million. In 1471 the Portuguese were the first Europeans to visit what is today the Nigerian coast, and they were followed by visitors from other European countries. British colonial rule dated from 1900. From the colonial era to the present, policies have tended to be top-down or from above, and development has been concentrated into a limited number of areas. Planning strategies, since their introduction in 1946, have been essentially market-oriented, concentrating on the production of agricultural crops for export and import substitution industrialisation. Investment and industrial plants have focused on the cities.

Today the twelve major cities of Nigeria account for nearly 77 per cent of all industrial establishments in the country and 87 per cent of the total industrial employment. However, in 1985 only 23 per cent of the population lived in towns and cities. Of the total employment in manufacturing, 76 per cent is to be found along the coastal belt (see figure). The capital, Lagos, which currently has a population nearing 2 million, accounts for well over 50 per cent of the nation's industrial wages, nearly 60 per cent of its gross output, 49 per cent of all industrial employment and 38 per cent of total industrial plans. Within each of the states making up the country, services and jobs are also strongly concentrated into the state capital. For example, in the north of the country, in Kano State, the Kano metropolitan area contained 71 of the 73 industries that were operating in 1971, and 11 of the 12 banks.

In such circumstances it is perhaps not surprising that rural-to-urban migration has been very strong and the main cities have grown extremely quickly. For example, during the period 1952–1963, Port Harcourt grew at the exceptional rate of 10.5 per cent per annum, whereas Lagos and Kano increased their populations at 8.6 and 7.6 per cent per annum respectively. A long search for oil proved to be successful in the mid 1950s, and by 1963 oil accounted for 3 per cent of government revenues. In 1982 oil represented 90 per cent of the value of the country's exports. However, many people maintain that the oil monies have been used inefficiently, leading to massive imports of expensive foreign goods.

Although many agree that the Nigerian economy has grown, others maintain that it has not developed. They suggest that the majority of the population are not better off and that deep regional inequalities still characterise the country. These critics claim there have been relatively few trickle-down effects of growth from the urban areas to the rural areas. Too much emphasis has been placed on sectoral growth – the promotion of different areas of the economy, such as industry – but little regard has been paid to the geographical consequences. Agriculture has been neglected, the drift from the land to the cities has not been reduced, and the country remains strongly dependent on the nations of the West.

Source: Potter (1992b).

---

combined into any number of hybrid forms. Boxed examples of the ways in which Cuba and Nigeria have applied national urban development strategies like these are provided in Boxes 9.1 and 9.2.

In conclusion, it is re-emphasised that arguments about urban and regional systems planning cannot sensibly be based on economic reasoning alone. As demonstrated by the boxes, strategic social, political

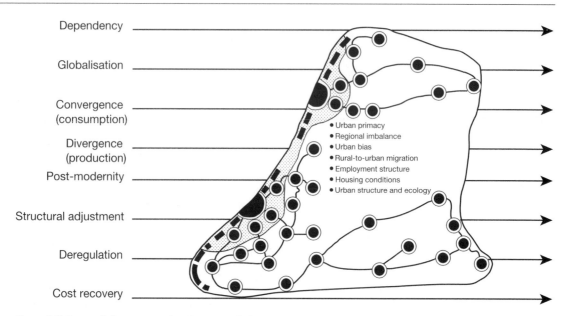

Figure 9.8 Current influences on the character of city systems. Adapted from Potter (1995a).

and ideological issues are just as important in the equation. This is also apparent when it is stressed that decentralisation and deconcentration are relative concepts and policy instruments. The choices are socio-political and moral, so once again we encounter a classic position where it must be accepted that there are many urban and regional geographies of the future which may be promoted by the state or other respons-ible agencies.

These circumstances are reflected in the current policy arena. Thus, although much of the literature over the past ten to fifteen years has stressed the importance of bottom-up and grassroots approaches to national planning, the World Bank and the United Nations Development Programme have been return-ing to the argument that urban growth and large cities are the keys to development and change (World Bank, 1991; UNDP, 1991). This argument is strongly based on the success of the Asian newly industrialising coun-tries (NICs), and on what is regarded as the overall failure of rural-based development programmes. The policies of the New Right, involving deregulation, privatisation, the rolling back of the state, and export-based programmes of industrialisation, along with the austerity measures associated with structural adjust-ment programmes (SAPs) are all signifiers of what some have called the new urban management pro-gramme of the World Bank and other bodies (Harris, 1992; W.T.S. Gould, 1992; T. McGee, 1994; Rojas, 1995; Yeung, 1995; Potter and Lloyd-Evans, 1998:

Ch. 10). Some of the forces affecting urban systems in the Third World are summarised in Figure 9.8. They provide the contemporary context in which aspects of urban–rural imbalances and urban primacy need now to be studied.

Whatever the balance of policies followed in the global urban and regional arena, massive urban growth and development are inevitable. Some evidence does now exist to suggest that the rate of growth of major cities may have lessened in Latin America and the Middle East in the 1980s, partly due to the effects of austerity packages; but on the other hand, secondary and intermediate cities appear to be growing more rapidly than ever before (A. Gilbert, 1993; Portes, Dore-Cabral and Landolt, 1997), so that continued urban growth and urbanisation are the order of the day.

Inside Third World cities

Understanding the processes

Although many cities in developing countries exhibit the signs of urban growth discussed earlier in this chapter, their individual characteristics as cities vary enormously according to a wide range of factors such as cultural context, the legacy of colonialism and the role the city and nation play in broader regional and global economies.

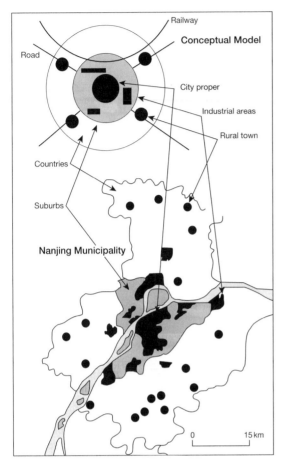

Railway

Conceptual Model

Road

City proper

Industrial areas

Rural town

Countries

Suburbs

Nanjing Municipality

0   15 km

Figure 9.9 Spatial structure of Chinese cities: conceptual model and a case study of Nanjing Municipality. Adapted from Yeh and Wu (1995).

Scholars have sought to identify common denominators in the complex process by constructing models of urban development. Anthony King (1976, 1994) is well known for his theories on the nature of the colonial city. But because of their complexity, few have attempted to construct a model for all Third World cities, so the most apposite studies have investigated city formation in particular regions such as Southeast Asia (T.G. McGee, 1967), post-apartheid South Africa (D. Simon, 1992a) and contemporary China (Yeh and Wu, 1995) (Figure 9.9); see also Potter and Lloyd-Evans (1998: Ch. 5). Although each of these models is used to attempt to explain changing processes and their impact on the city, their use often seems to restrict the scope of the discussion, giving pre-eminence to function and form when cities are above all agglomerations of people, many of whom have migrated there

in search of work. Few find the well-paid jobs they seek and face enormous problems in meeting the basic needs of themselves and their families.

In order to understand fully the range of problems faced by urban managers and the processes which give rise to them, we need a more comprehensive, more flexible and less ideographic conceptual approach. Such an approach has emerged in the 1990s and is concerned with sustainable urbanisation (Pugh, 1996; Burgess, Carmona and Kolstree, 1997). This must not be conflated with sustained growth, in which the city is seen to have a pivotal role in initiating and maintaining national economic growth. Economic growth is a vital component of sustainable urbanisation, but it is only one component in the array of interlinked processes which make up the contemporary city.

Sustainable urbanisation can and should constitute an important goal for any urban management team, irrespective of the level and nature of economic development. Figure 9.10 indicates some of the major components of sustainable urban development – Drakakis-Smith (1995, 1996) has given them fuller discussion – and, more importantly, the ways in which they interlink, illustrating the complexity of many urban problems. Thus the economic dimension of urban development is not just related to the role of the city in the national economy but also to the repercussions of the economy on the residents of the city in terms of employment, incomes and poverty at the household level, as well as its impact on the urban environment and social issues, such as workers' rights.

The management of urban development in order to achieve sustainable rather than just sustained growth is clearly a complex task. Nevertheless, many of the issues raised by such an approach have been well researched over the past two decades, but separately rather than interlinked. Moreover, urban management for sustainable urban development requires a new set of attitudes towards the objectives of intervention. So, in addition to seeking to create and sustain economic growth, there must be other, equally important priorities. These might be

- the pursuit of equity and social justice
- the satisfaction of basic needs
- recognition of social and ethnic self-determination and human rights
- environmental awareness and integrity
- appreciation of the interlinkages across space and time

Whether these principles do, or could, form a basis for urban management will be discussed at the end of

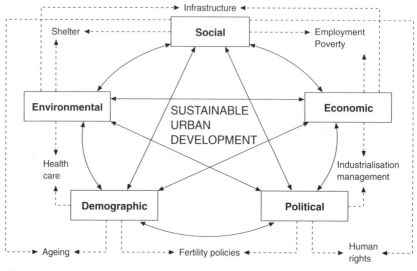

KEY
Main component
Issue

Figure 9.10 The main components of sustainable urbanisation.

this chapter. The following sections pick up on the main components of sustainable urbanisation as outlined in Figure 9.10 and discuss some of the main issues that might relate to appropriate policies.

Demographic factors

Some of the main demographic issues have already been alluded to. Indeed, urban growth by migration has probably been one of the most researched topics in urban studies. Many of the studies have attempted to construct predictive models, most of which have concurred with Todaro (1994) that much migration is motivated by perceived differences in economic opportunities. However, there is immense variation in the relationship between urban migration and economic circumstances. For example, large-scale rural poverty in Bangladesh has induced nowhere near the scale of migration to its cities as it has in southern Africa. History, culture and even the nature of communications and transport all have a role to play in the decision to migrate. Certainly the last of these factors has induced considerable change in the nature of both internal and international migration.

With regard to internal migration, in most countries it is almost as cheap and easy to access the major cities

as it is to move to smaller local centres; consequently, the smaller local centres are often bypassed in the rural–urban migration process. This has refocused attention onto the role that small and intermediate towns can and should play in the development process (Baker and Pedersen, 1992; Aeroe, 1992). David Simon (1992b) states that no such role has yet emerged; and unfortunately, the attention of most researchers has been diverted away from small towns *per se* and towards the phenomenon of mega-urbanisation.

Mega-urbanisation is a process whereby large cities expand rapidly along major lines of communication to envelop villages and villagers *in situ*, creating multinodal settlements, where rural and urban become very blurred, obfuscating the nature of movements between the two (T.G. McGee and Robinson, 1995; Potter and Unwin, 1995). Increasingly, the term *extended metropolitan region* is being used to describe such vast compound urban–rural zones. T.G. McGee (1989) used the word *Kotadesasi*, juxtaposing the Indonesian words for town and country, to describe such emerging urban forms. In other places McGee has referred to *desakota* regions. Some see this as an inevitable trend which needs to be managed rather than prevented, in the process implying that 'hard decisions [will have to be taken] against fostering small town development and rural industrialisation' (T.G. McGee and Greenberg, 1992: 7).

One of the reasons why mega-urbanisation is occurring in those countries with rapid economic growth is the expansion of international urban migration. In Pacific Asia the waves of economic growth which have rippled from Japan, through the four tigers to the industrialising ASEAN states have been followed by streams of job-seekers moving in search of work in manufacturing and tertiary activities. It is estimated that such cross-border urban migration amounts to some 2 million people in the region (Dixon and Drakakis-Smith, 1997). The result is an extremely complex situation in which multiple nationalities move between the various countries, particularly in Southeast Asia. In Africa there is much less cross-border migration specifically to cities. Indeed, such is the deterioration of economic circumstances over much of Africa, resulting from structural adjustment, that in some countries not only has rural–urban migration slowed considerably, it is occasionally being accompanied by reverse migration to rural areas (Mijere and Chilivumbo, 1987).

However, there is more to the demographic aspects of sustainable urbanisation than the nature and management of migration, not the least of which is the two-way relationship between urbanisation and fertility. For example, living in the city not only raises the cost of rearing children but also increases access to family planning programmes, and yet cities in developing countries have overwhelmingly young populations (Boyden and Holden, 1991), with cities as geographically distinct as Bogotá, Delhi and Jakarta all having half their population aged 15 years or less. On the other hand, in some of the cities of Pacific Asia ageing populations, and their impact on the labour force and social welfare (Graham, 1995), provide different but equally pressing issues. Other demographically linked issues related to sustainable urbanisation could encompass household composition and the roles of women in generating income and meeting basic needs, or ethnicity and the ways in which migration has created more complex and tense situations in the competition for limited urban resources (Box 9.3).

BOX 9.3   Urban migration and ethnicity

The fact that migration to cities is being drawn from increasingly extensive geographical areas has often resulted in a broader diversification of ethnic groups. In most countries this ethnic complexity is spontaneous as people from economically, geographically and ethnically marginal regions are drawn to capital cities. Almost two-thirds of the migrants to Bangkok are from the Lao-dominated north-east of Thailand. But in Malaysia, increased urban ethnic diversity has been the consequence of deliberate government policies designed to increase Malay participation in urban economic activities (Eyre and Dwyer, 1996). Although some might argue this process has occurred without increasing ethnic tension, it was in fact prompted by ethnic tensions in the first place. In other cities where ethnic mixing has accompanied migratory growth, tensions have increased markedly; for example, in many African cities where national politics reflect tribal antagonisms.

This growing urban ethnic diversification and its consequences have been exacerbated by the increased internationalisation of labour movements. In Southeast Asia this has produced a particularly complex pattern of movement. Singapore was the initial magnet for migrants from Malaysia, Indonesia and the Philippines, for factory work and domestic work. More recently, construction labour has come from Thailand and India. And as it has developed its own economy, Malaysia has recruited both legal and illegal workers from Thailand and, more particularly, Indonesia. There are now an estimated one million Indonesians working in Malaysia, most of whom are illegal, and local resentment at narrowing access to jobs has increased substantially. Meanwhile, illegal workers from the transitional socialist economies of Southeast Asia are also beginning to flow across weakly policed borders, into the regional capitals of Thailand, which have lost migrants to Bangkok.

This increasingly complex pattern of ethnodevelopment is threatening urban sustainability in a variety of ways. Although it may provide a larger labour pool for economic growth, it is also leading to growing ethnic and class antagonism and exploitation. In Singapore, for example, there are increasing reports in local newspapers of crude racism, especially against domestic workers (Teo and Ooi, 1996).

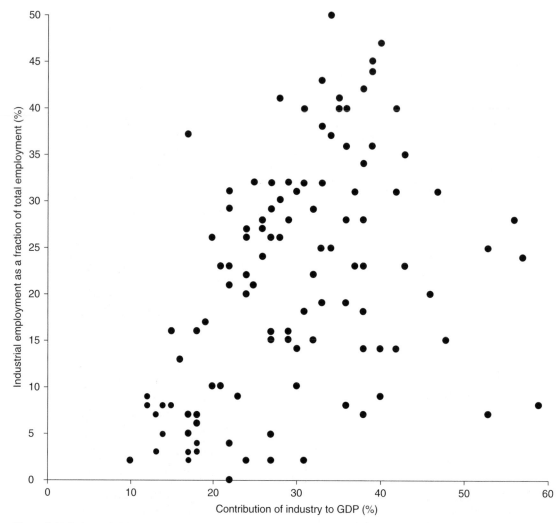

Figure 9.11 Industrial contributions of Third World nations: contribution to employment versus contribution to GDP.

Economic matters

Urban-based industrial growth fuelled by TNC investment has been the focus for many development policies in Third World countries, as noted in Chapters 1, 3 and 4. However, it is well known that TNC capital has been very selective in where it chooses to invest, and the countries which have experienced urban-based industrialisation have been relatively few (Dicken, 1991): Mexico, Brazil, the four tigers and some members of the Association of South East Asian Nations (ASEAN). But even in these countries the factories are not large, and throughout the Third World,

the contribution of industry to GDP is far greater than its contribution to the labour force in terms of jobs (Figure 9.11). Indeed, even in the four Asian tigers, most firms tend to be small to medium-sized, a prime factor in promoting industrial flexibility.

In many countries, therefore, those migrants who come in search of work in the city are not usually incorporated into the formal sector, particularly given the downsizing that has occurred in government employment as a result of structural adjustment. Most find work in what has come to be known as the informal sector (Plates 9.5 and 9.6). This is difficult to define adequately and an informal activity may possess any one or more of a number of features allegedly

Plate 9.5 Rubbish collection in Hanoi, Vietnam (photo: David Smith)

Plate 9.6 Hairdressing in the street, Hanoi, Vietnam (photo: David Smith)

typical of it: semi-legal, small-scale, family-oriented, traditional technology (Table 9.12, page 240). Certainly, there is considerable overlap between the two sectors (Santos, 1979). For example, many trishaw owners work in the formal sector and rent their machines out to the riders, whereas domestic outworkers are often essential to small businesses in helping them absorb fluctuations in demand.

For many years the activities of the informal sector, although clearly useful, have been anathema to urban

Table 9.12 Characteristics of the two circuits of the urban economy in Third World countries.

	Upper circuit (formal)	Lower circuit (informal)
Technology	Capital-intensive	Labour-intensive
Organisation	Bureaucratic	Primitive
Capital	Abundant	Limited
Labour	Limited	Abundant
Regular wages	Prevalent	Exceptional
Inventories	Large quantity and/or high quality	Small quantity and poor quality
Prices	Generally fixed	Negotiable (haggling)
Credit	Banks and institutions	Personal and non-insitutional
Profit margin	Small per unit, but large turnover and considerable in aggregate	Large per unit, but small turnover
Relations with customers	Impersonal and/or on paper	Direct, personalised
Fixed costs	Substantial	Negligible
Advertising	Necessary	None
Reuse of goods	None (waste)	Frequent
Overhead capital	Essential	Not essential
Government aid	Extensive	None or almost none
Direct dependence on foreign countries	Considerable	Small or none

Source: Santos (1979).

management as they mar the modernising image it is trying to create in order to attract investment. Many regulations have helped to keep the informal sector in check, despite the 'unconventional wisdom' of the 1970s, which saw support for the informal sector as a way of improving basic needs provision (Richards and Thomson, 1984). However, few urban governments were as enthusiastic as the experts and relatively little was achieved, particularly in the field of improved employment opportunities. However, the 1990s has witnessed a revival of interest in the informal sector, particularly in the wake of the reduced employment opportunities and reduced incomes which have followed structural adjustment programmes, particularly in Africa (Gibbon, 1995). The reaction was to introduce a social component to adjustment. As poverty was due to limited job access, and as the informal sector seemed capable of creating employment, it was reasoned that removing some of the constraints on the informal sector (deregulation) would help to expand work opportunities and absorb those in poverty (Urban Foundation, 1993).

This approach has not yet been the success it was hoped. There is a limit to the capacity of the informal sector to involute and create employment and income. Moreover, very few small informal firms have the capacity to upgrade and 'formalise' on their own without assistance from the state. Furthermore, the alleged deregulation created undesirable knock-on effects as employers were given freedom to exploit their workers

further (Wilson, 1994), and to pollute without undue fear of prosecution. In short, as Parnwell and Turner (1998) note, the urban informal sector does not equate with the flexible specialisation that has emerged in the West, despite structural similarities; it is much more a survival mechanism than an engine of growth.

The consequence of these urban labour market problems has been increasing poverty. By the 1990s, conflicting views were coming from the global institutions concerned with development. The World Bank (1993) was alleging that poverty was being rapidly eliminated by the spread of the market economy. On the other hand, many other institutions, such as UNDP (1991) and even the World Bank (1991) itself, were noting a growth in urban poverty. Part of the explanation for this contradiction lies in the difficulties of defining poverty and consequently the quality of the data, particularly for making comparisons (Drakakis-Smith, 1996; Rakodi, 1995; Wratten, 1995). But despite doubts about the data, in general terms it is possible to discern a steady shift in the distribution of poverty associated with the growth of urban populations. Table 9.13 therefore reveals that in countries with higher levels of urbanisation, there are now more urban poor than rural poor, in terms of absolute totals.

Bringing the discussion around to poverty further humanises the debate on urban economies and employment. We must not think of labour simply as an input into the development process; it comprises many different groups of people, and some of their specific

Table 9.13 Absolute poverty in urban and rural areas (percentage below poverty line).

	Urban areas	Rural areas	Percentage urban	Ratio of rural poor to urban poor[a]
Africa				
Botswana	30.0	64.0	29	4.8
Côte d'Ivoire	30.0	26.0	41	1.3
Egypt	34.0	33.7	47	1.1
Morocco	28.0	32.0	49	1.2
Mozambique	40.0	70.0	28	4.5
Tunisia	7.3	5.7	55	0.7
Uganda	25.0	33.0	11	10.3
Asia				
Bangladesh	58.2	41.3	17	3.4
China	0.4	11.5	60	18.9
India	37.1	38.7	27	2.8
Indonesia	20.1	16.4	31	1.8
South Korea	4.6	4.4	73	0.3
Malaysia	8.3	22.4	44	3.2
Nepal	19.2	43.1	10	21.0
Pakistan	25.0	31.0	33	2.5
Philippines	40.0	54.1	43	1.8
Sri Lanka	27.6	45.7	22	6.1
Latin America				
Argentina	14.6	19.7	87	0.2
Brazil	37.7	65.9	76	0.6
Colombia	44.5	40.2	71	0.4
Costa Rica	11.6	32.7	48	3.2
Guatemala	61.4	85.4	40	2.1
Haiti	65.0	80.0	29	3.1
Honduras	73.0	80.2	45	1.3
Mexico	30.2	50.5	73	0.6
Panama	29.7	51.9	54	1.3
Peru	44.5	63.8	71	0.6
Uraguay	19.3	28.7	86	0.2
Venzeula	24.8	42.2	85	0.3

[a] In absolute totals, less than 1.0 indicates more urban poor than rural poor.
Source: United Nations Centre for Human Settlements (1996), World Bank (1994), Drakakis-Smith (1996).

problems and needs often overlap with those of other groups. Women have been incorporated into the urban labour market in many different ways that depend on local economic and social conditions. However, as McIlwaine (1997) has noted, this enhanced value in the workplace has not always reduced gender inequalities in society in general or within the household. The changing links between gender and urban economic growth need therefore to be followed through other dimensions of urban sustainability, such as those related to basic needs provision and human rights.

Children too form an identifiable group whose specific needs must be taken into account in any review of sustainable urban development. The value of child labour is well recognised and exploited by employers. For example, it has been estimated that in Thailand the child labour force is approximately the same size as the female labour force. One-third of this 1.5 million work in urban factories where they receive about half the adult minimum wage. Although it is true that in many developing countries children are often important income earners in the family (A. Gilbert, 1994; Clifford, 1994) and are often proud of their household role, there are ways in which their conditions of work and their life as a whole can be improved without threatening household survival

Table 9.14 Urban household strategies for coping with worsening poverty.

Changing household composition
Migration
Increasing household size in order to maximise earning
   opportunities
Not increasing household size through fertility controls

Consumption controls
Reducing consumption
Buying cheaper items
Withdrawing children from school
Delaying medical treatment
Postponing maintenance or repairs to property or
   equipment
Limiting social contacts, including visits to rural areas

Increasing assets
More household members into workforce
Starting enterprises where possible
Increased subsistence activity such as growing food or
   gathering fuel
Increased scavenging
Increased subletting of rooms and/or shacks

Source: Rakodi (1995).

strategies (Lefevre, 1995). But as long as the use of child labour is seen as a 'comparative advantage' which 'humanitarian measures' should not threaten (Silvers, 1995: 38), the sustained and sustainable improvement in the quality of life of this segment of the labour force will not occur.

It must not be thought, however, that those who find themselves disadvantaged in the labour market are passive acceptors of their fate. Low-income households display a wide range of coping mechanisms (Rakodi, 1995), some of which are indicated in Table 9.14. Not all strategies are available to all households, depending on individual and local circumstances, but the ways in which poor families sustain themselves in the city ought to be the basis on which policy responses are formulated. These strategies will be discussed in the final section of this chapter.

Meeting basic needs and human rights

Basic needs is a rather fluid idea, with many of the early approaches in the 1970s encompassing what would now be regarded as environmental issues, together with employment and poverty. At present, the term *basic needs* tends to refer to those core areas of personal and household needs, such as education,

health care and housing, in which it is possible for both state and community to make a contribution towards improving the situation (Chapter 3). Indeed, such partnerships became almost the norm in the 1970s and 1980s as international agencies attempted to encourage an improved response to a deteriorating urban situation. Before the emergence of what became known as the 'basic needs approach', most governments had neither the funds nor the interest to invest in the social overheads of welfare programmes, preferring instead to use their resources to encourage and sustain economic growth, as stressed in Part I. The late 1980s and 1990s have seen a return to this position as debt crises and imposed structural adjustment have forced governments to withdraw from social welfare programmes of all kinds, shifting responsibilities on to the 'market' or the poor themselves. To illustrate the changing circumstances affecting access to, and provision of, basic needs, the remainder of this section will focus on housing, but the issues raised are often common to other basic needs.

Housing poverty (Pugh, 1996) has been well researched since the 1960s when John Turner and William Mangin first drew attention to the positive qualities of squatter settlements. Internal movements within the city and their links to shelter have been particularly well explored (Figure 9.12). In spite of three decades of admittedly varied responses to shelter needs, the problems seem to be as widespread as ever. Even in Pacific Asia, where economic successes have occurred, it is estimated that by the year 2000 some 60 per cent of the region's urban population will still be living in slum or squatter settlements (Pinches, 1994). Although much of the statistical information on housing poverty is unreliable and incompatible, it is useful in illustrating trends over time and between regions. In some cities, renting is far more usual and acceptable than ownership, yet the policy responses of much of the last thirty years have been based on the assumption that tenure security through ownership is the fundamental desire of most low-income populations.

Certainly, the discussions about shelter itself have evolved into a sort of dualism. On the one hand are the debates about the role of shelter provision in the development process as a whole, a debate which in recent years has increasingly been conducted at global rather than national level, with the main international development agencies dominating the discussion, the funding and hence the policy. Increasingly distinct from these events are the national and urban debates about programmes and projects for the real world. In recent years this has tended to focus upon the

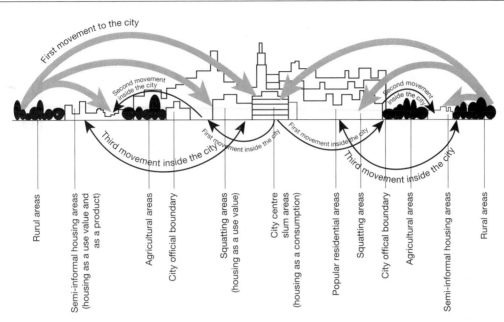

First movement to the city

Second movement inside the city

Third movement inside the city

First movement inside the city

First movement inside the city

Second movement inside the city

Third movement inside the city

Rurul areas

Semi-informal housing areas (housing as a use value and as a product)

Agricultural areas

City official boundary

Squatting areas (housing as a use value)

City centre slum areas (housing as a consumption)

Popular residential areas

Squatting areas

City offical boundary

Agricultural areas

Semi-informal housing areas

Rural areas

Figure 9.12 Mobility of the urban poor to and within the city of Alexandria. Adapted from Soliman (1996).

practical ways the various stakeholders involved can help improve access by the poor to better housing. As Ward and Macoloo (1992) observe, these two sides of the housing debate have been moving further apart; less and less is practical planning informed by the conceptual debates on the role of housing in sustainable urbanisation, and vice versa.

For many years the debate on practical responses to housing poverty has revolved around aided self-help (ASH) programmes, in which the energies and ambitions of the poor themselves are combined with tenure, material and land inputs from the state to produce developments which are largely self-built, but which have the support and approval of the state. Although many low-income households benefited substantially from such schemes, they were nevertheless subject to considerable criticisms (Plates 9.7 and 9.8, pages 244–5). Pinches (1994: 118), in particular, claimed that aided self-help schemes 'served the narrow economic interest of states, elites and international agencies' by offering cheap solutions to demands for housing, containing restive populations and formalising part of the informal sector. However, in the 1990s, enthusiasm for these approaches has diminished substantially, partly because there was an enforced retreat of the state from welfare programmes under structural adjustment, partly because the scale of the housing problem was not reducing substantially and partly because funds from the international agencies have dried up.

As urban populations have continued to grow, so housing poverty has remained an important issue related to sustainable urbanisation. In some cities, particularly in Africa, this has meant a resurgence of squatter settlements; in others, market forces have produced a rapid expansion of renting and sharing (A. Gilbert, 1992). The growing research interest in these phenomena has indicated a wide range of types and circumstances. In Latin America, for example, A. Gilbert (1992) argues that most landlords operate on the small scale and are not exploitative; but in many African cities there is widespread exploitation of shack tenants in gardens or yards attached to formal housing (Plate 9.8), conditions are cramped and there are grossly inadequate washing and toilet facilities (Grant, 1995; Auret, 1995). Essentially, this is the privatisation of housing, the benefits of which usually filter upwards through a hierarchy of landlords and owners.

State responses to the continued housing crisis have been strongly influenced by the neoliberal trends in international development and have shifted away from the more direct subsidies of aided self-help to the formulation of partnerships between the national and local states, together with a variety of local agencies, such as non-governmental and community-based organisations (NGOs and CBOs). The focus for these partnerships is on facilitating the access of households to land or credit through the removal of existing

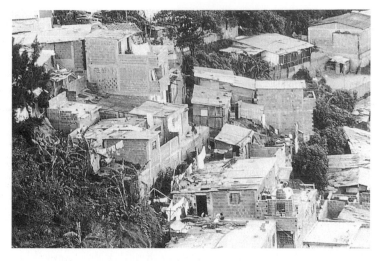

Plate 9.7 Self-help housing in Caracas, Venezuela soon consolidates
(photos: Rob Potter)

constraints – helping the poor to help themselves. However, the poorest and most needy are often not capable of the organised and sustained collective action required to improve their housing, health care or education. Enablement programmes are often used by the authorities as an excuse to abandon many of their social responsibilities, privatising them to the NGOs and CBOs. In the 1990s, therefore, the pursuit of adequate shelter as a human right has proceeded in theory rather than in practice. Reliance on market forces is simply resulting in a mounting but hidden problem that is liable to create social and political tensions for many years to come. This situation has many resonances of the modes of production argu-

ment presented in Chapter 3 (Potter, 1994; Potter and Conway, 1997).

The brown agenda

For many people, urban sustainability in developing countries equates only to environmental issues. Moreover, this environmental agenda tends to prioritise those issues which are of greatest concern to the West, e.g. global warming and the rapid use of finite resources. The result has been a Western-led series of programmes in many developing countries, programmes which do not correlate with, or respond

Plate 9.8 'Shack farming': backyard shack housing in Harare, Zimbabwe (Photo: Rob Potter).

to, many of the real concerns and priorities of the residents of cities in the developing world. Indeed, the environmental problems facing such cities vary enormously according to the local combination of contributing factors. In broad terms, these would encompass the following:

- the nature of the urbanisation process itself – the rate, scale and degree of concentration in growth
- the ecosystem within which the settlement is located
- the level and nature of the development process, which affects the ability of the family and the state to respond to problems
- the development priorities of the state

Within the development process, urban environmental problems usually emanate from two principal sources. The first is irresponsible or poorly managed economic growth. Despite the arguments of development economists, the market has responded poorly to environmental problems created by the philosophy of 'grow now and clear up later' unless coerced by the enforcement of regulatory legislation. In general, such controls have been weak, often because urban managers themselves are frequent beneficiaries of uncontrolled development. The second major contributor is poverty and vulnerability, which forces low-income households to survive as best they can, leaving the environment to look after itself. This does not mean that the poor are

unaware of the environmental consequences of their actions, rather that they have other priorities. The implication is that improvements to the urban environment must be strongly linked to poverty alleviation as well as the regulation of industry.

This combination of contributory factors to the brown agenda also varies spatially and in terms of scale – from the household scale to the regional and global scales. In general, the concerns of the household, workplace or community are more immediate and relate primarily to health and to equality of access to basic services (Table 9.15, overleaf). At the regional and global levels the problems are more long-term in nature and are linked to the impact of resource use on future generations – the major concerns of the West. Between these sets of concerns lies the city itself, combining all these issues in a complex situation that requires careful management to ensure sustainable urbanisation.

Household and community matters   It is estimated that some 600 million urban residents live in conditions that continually threaten their health. For most families, simply trying to feed themselves takes up most of their income, so there is little money for shelter or health care. Many are forced to live in squatter settlements or tenements that exhibit a range of environmental problems. The second United Nations Centre for Human Settlements (Habitat) Report (1996) high-

Table 9.15 Spatial dimensions of the brown agenda.

	Principal service infrastructure	Problem issues
Household/workplace	Shelter	Substandard housing
	Water provision	Lack of water, expensive
	Toilets	No sanitation
	Solid waste	No storage
	Ventilation	Air pollution
Community	Piped water	Inadequate reticulation
	Sewerage system	Human waste pollution
	Drainage	Flooding
	Waste collection	Dumping
	Streets (safety)	Congestion, noise
City	Industry	Accidents, hazards, air pollution
	Transport	Congestion, noise, air pollution
	Waste treatment	Inadequate, seepage
	Landfill	Unmonitored, toxic, seepage
	Energy	Unequal access
	Geomorphology	Natural hazards
Region	Ecology	Pollution, deforestation, degradation
	Water sources	Pollution, overuse
	Energy sources	Overextended, pollution

Source: Bartone (1994).

Table 9.16 Water charges in selected cities.[a]

	Average piped tariff (US$ M^3)	Private vendor tariff (US$ M^3)	Private/public
Jakarta	0.363	1.848	5.1
Bandung	0.268	6.161	23.0
Manila	0.232	1.873	8.1
Calcutta	0.049	2.099	42.8
Madras	0.046	0.875	19.0
Karachi	0.047	1.747	37.2
Ho Chi Minh City	0.045	1.511	33.6

[a] Data from various sources.

lights four particular problems: water, sewerage, over-crowding and air pollution.

Of all basic needs, access to clean water is probably the most important, and yet some 170 million urban residents lack access to potable water *near* (not in) their homes (World Bank, 1992). For example, in Indonesia only one-third of the urban population has access to safe drinking-water. Moreover, those with such access are usually the better off; the poor, who can least afford it, are forced to buy their water from vendors at much higher prices (Table 9.16). Little wonder that they often resort to contaminated water with disastrous consequences for their health. With rising populations some cities have been forced to overexploit their aquifer resources, so cities such as Bangkok and Mexico City have experienced widespread subsidence.

Closely linked to water provision is the problem of waste removal through sewerage systems (Pernia, 1992). Again, in most developing countries, because of increasing populations, this situation is worsening; during the 1980s alone, the number of urban residents without access to adequate sanitation increased by 25 per cent (World Bank, 1992). Human waste, therefore, often lies untreated around the household, increasing health risks, and is eventually washed into waterways, lakes or seas, polluting aquifers and poisoning aquatic resources (Stren, White and Whitney, 1992). The health problems created by poor water and sanitary conditions are often exacerbated by poor diets and by overcrowding and poor ventilation, which intensifies the transfer of respiratory infections, especially where biomass fuels are used. Those who are more involved in domestic activities (women and children) are therefore more prone to tuberculosis or bronchitis, still major killers in the cities of the Third World (Satterthwaite, 1997).

The city environment   The problems experienced in and around the household can be compounded by city-wide issues that often reflect the particular setting of the settlement. For example, many cities are located in hazard-vulnerable zones and it is usually the poor who are forced to live in the most marginal areas such as steep slopes or flood-prone lowlands (Main and Williams, 1994). Usually the impact of natural disasters is intensified by poor urban management in allowing such areas to be settled without providing adequate safeguards. In Rio de Janeiro in 1988, the floods and landslides which followed torrential rain were partially caused by neglected, blocked or inadequate drainage systems in the favelas (World Bank, 1993). In the same way, the increased incidence of landslides and mud-slides in Caracas has been attributed to the growth of informal rancho areas (Jiminez-Dias, 1994; Potter, 1996).

Humans also contribute to these problems as a result of inadequate supervision of economic growth; governments are reluctant to enforce what few regulatory controls they have, for fear of discouraging investment. As a result, industrial air and water pollution from uncontrolled discharges increasingly contaminate Third World cities. For example, industrial discharges have increased twelvefold over the last fourteen years in the newer Asian industrialising countries of Thailand, Indonesia and the Philippines. The most infamous example of such pollution remains the Union Carbide plant in Bhopal, where poisonous gases killed some 3,300 people and seriously injured another 150,000 (Figure 9.13, overleaf). Most were from poor households living adjacent to the plant (Gupta, 1988).

Increasing vehicle ownership and extensive use of fossil fuels are also contributing to air pollution in developing countries. In Bangkok, 26 million workdays are lost annually through respiratory problems; the incidence of lung cancer in Chinese cities is up to seven times greater than in the country as a whole. However, we must put this into a global perspective,

since many East European cities have much worse urban air pollution levels and the three leading global producers of carbon emissions are the United States, Canada and Australia.

Solid waste disposal compounds these problems for most cities (Plate 9.5). City-wide collection services are a rarity, and where they exist, they are often confined to the wealthier districts. Sometimes private garbage removal services do exist, not necessarily provided by companies but by groups of people traditionally associated with such activities. Again, however, the fees necessary for such services restrict them to those households who can afford them. In poor districts the rubbish is simply dumped in open spaces, where it is occasionally removed, usually for uncontrolled incineration. Ironically, it is the poor themselves who recycle solid waste, saving and selling bottles, cans or paper (Plate 9.5). Sometimes families even live on the city garbage dumps, as in the famous Smokey Mountain site in Manila, where some 20,000 scavengers live and work.

Regional impacts   The regional impact of cities has now spread far beyond their immediate hinterland. Food, fuel and material goods are drawn into cities from all over the nation and the world, affecting the lives of many. The area that the city affects by its waste output has also grown. Indeed, the wealthier the individual or the city, the more distance they can afford to put between themselves and their waste. This is clearly illustrated by the export of toxic and noxious wastes from developed countries to the more povertystricken parts of the Third World. It is useful to divide the regional impact of cities into two zones: (1) the immediate peri-urban area around the city where the urban footprint looms large and heavy; (2) the broader region beyond this.

Peri-urban zones exhibit two broad areas of concern (Sattherthwaite 1997):

Unplanned and uncontrolled urban sprawl. Often in the form of squatter settlements and illegal small-scale industries beyond the city boundaries, these areas can also contain large-scale municipal uses such as power stations or sports stadia.
Liquid waste disposal. Untreated sewage and industrial effluent enter rivers, lakes or aquifers, making peri-urban areas concentrations of intense contamination.

The impact of these processes is worsened by the fact that the peri-urban area is often a zone of major

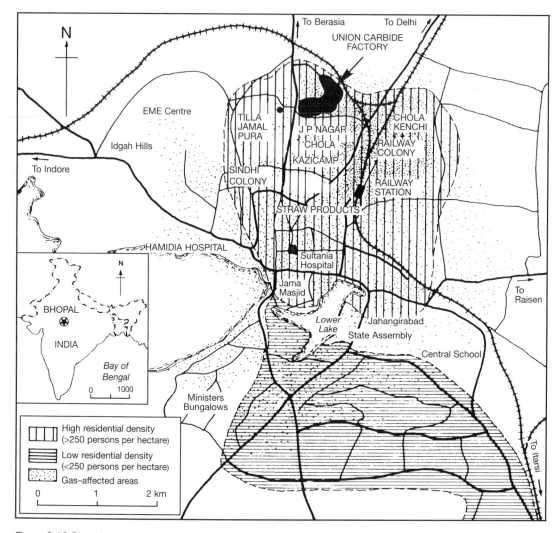

Figure 9.13 Bhopal: gas escape at Union Carbide. Adapted from Gupta (1988).

importance for recent migrants, offering land for shelter and agriculture and perhaps for woodfuel. The destruction and pollution of the peri-urban area is thus viewed rather differently by urban and peri-urban residents.

Farther afield in terms of regional impact, the urban footprint is becoming more marked every year, breaking the traditional links between the city and its hinterland. Thus the immediate regions around African capital cities are now producing exotic vegetables, flowers and other crops for export rather than to feed the city population, thereby exacerbating the nutritional problems of the poor (Smith, 1998). On the other hand, meeting the energy needs of the city has

often created other environmental problems such as denudation of biomass or expansion of coal mining and associated waste tipping. There are many other examples of regional urban footprints; the growing demand for cement and bricks lead respectively to quarrying and pollution, or to loss of valuable soil resources. Water demands too can affect regions way beyond the city, particularly for cities in semi-arid areas. Bulawayo in Zimbabwe is seeking funds to draw on the waters of the Zambesi. If this succeeds, what will be the impact on the fragile ecosystems of this part of southern Africa? The Three Gorges Dam on the Yangtse too is partly designed to provide energy for urban industry but will have enormous environmental effects on the 600km of the new lake between Chongqing and Ichang.

Urban management for sustainable urbanisation

The above discussion of the brown agenda reveals some of the main issues related to urban management for sustainable urbanisation, such as false priorities, lack of adequate legislation, self-interest, poor knowledge and training and a susceptibility to external influences, both benign and malign. Lack of public awareness of appropriate policy responses and the fact that protest is usually fragmented means the authorities discover that opposition is easy to contain. Indeed, most pressure for change in urban management towards greater responsiveness to problems of sustainability has come from external sources. Both the World Bank (1991) and the United Nations (UNDP, 1991) became increasingly nervous about potentially restive urban populations in the early 1990s and they have recommended new management strategies, as noted earlier in this chapter.

Such approaches have mirrored the emergence of neoliberal development strategies in general, with an emphasis on market-led solutions and limited interference by the state, with many urban management policies being substantially shaped by national governments in conjunction with their external advisers. Municipal governments thus have to work within limits set by agencies beyond their control, although even within the city, there are often more specific management problems too. Many would claim that overall the shift to the market economy has simply resulted in the transfer of responsibilities from the state to the poor, largely by removing the constraints on letting them help themselves. Certainly the poor have taken advantage of such moves in the development of their coping mechanisms but these are essentially small in scale and focused on the household. Collective and larger-scale responses are difficult without proper knowledge, training and funding. Recycling waste is possible for the poor, constructing a sewerage system is not.

Increasingly, intermediaries have become involved in order to 'improve local capacity'. The private sector, the embodiment of the market, has been slow to respond to the needs of sustainability. Although there are some examples of the private sector meeting some basic needs in large Latin American and Pacific Asian cities (A. Gilbert, 1992), they are far from universally replicable and, unsurprisingly, tend not to affect those in greatest need of assistance. Indeed, the role of the private sector in water vending has increased the exploitation of the poor and has to be countermanded by the state (Choguill, 1994).

Despite early pessimism, community participation and cooperation emerged on a substantial scale, perhaps due to growing competition for resources by increasingly diverse ethnic and social groups. In this context the role of NGOs and CBOs as facilitators of urban community development has been equivocal. Many NGOs are themselves large global organisations driven by Western agendas and funding sources. Indeed, many NGOs have been criticised for assisting in the privatisation of resource provision and the retreat of the state. In this context the underprivileged have little option but to engage in their own protests and civil action for improved access to urban resources. For some analysts, such as Escobar (Chapter 1), these movements could form the basis of new development strategies, but often they are deliberately limited in their objective and they are consciously non-political. If and when they achieve some of their aims, such social movements tend to fade away.

There is no doubt that the appropriate level at which to tackle problems related to urban sustainability is the local state, the city itself. At present, however, most but not all urban management is poorly informed, poorly motivated and poorly organised. A decentralisation of power, funding and responsibilities from the national to the local state would be a start, but it must also be accompanied by greater democratisation at the level of the city itself. Those most affected by the inadequacies and inequalities of unsustainable urban development – in short, those whose coping mechanisms sustain the unsustainable – must become part of the process of policy formulation and enactment. At present, there is little sign of this occurring on a widespread scale, and with urban populations and poverty continuing to grow, the problems of urban sustainability for a whole range of different cities within the Third World are likely to get worse, rather than better.

CHAPTER 10

Rural spaces

Rural spaces in development

Although the world's population is becoming increasingly urbanised, in many parts of the developing world it is the rural areas which will continue to accommodate the majority of people for the foreseeable future. Not only will these rural areas have to deliver more food to expanding populations in the next decades, but they will also be expected to safeguard many aspects of the 'global commons' such as biodiversity (Morse and Stocking, 1995). Currently, however, these areas are the locus for some of the most insecure livelihoods anywhere in the world.

This chapter examines how rural areas have been both included and excluded by development thinking, and it considers some of the key successes and limitations of the interventions that have been made. The diversity and flexibility of rural livelihoods within the developing world are emphasised throughout, and it is suggested that these features are precisely those which have been neglected in past development efforts, and on which many of the more successful initiatives of the 1990s are being built.

In and out of development thinking

As explained in Chapter 3, early development models often emphasised the importance of agricultural innovation and improved rural productivity for the release of capital and surplus labour, which could then be used in emerging urban and industrial activities. Based largely on the historical experience of Western Europe, the interdependence of the rural-agricultural and urban-industrial sectors and the transformation of a country's economy from 'one that is dominantly rural and agricultural to one that is dominantly urban, industrial, and service-oriented in composition' (Mellor, 1990: 70), were central to the concept of development itself. For example, in his classic text *The Theory of Economic Growth,* Arthur Lewis argues that

> industrialisation is dependent upon agricultural improvement: it is not profitable to produce a growing volume of manufactures unless agricultural production is growing simultaneously. This is also why industrial and agricultural revolutions always go together, and why economies in which agriculture is stagnant do not show industrial development. (Lewis, 1955: 433)

Rostow's influential economic growth model also recognised the significance of increasing agricultural productivity as a precondition for his all-important 'take-off' phase, at which point there is a steady increase in industrial investment, and economy and society are transformed, 'in such a way that a steady rate of growth can be, thereafter, regularly sustained' (Rostow, 1960: 8–9).

The colonial encounter in much of the developing world was centred principally on the raw materials, labour supply and opportunities which the rural areas offered for both export production and satisfying the food requirements of growing urban populations. Post-independence, national development plans have frequently attached great importance to rural and agricultural development, and substantial domestic and foreign financial resources have been directed to rural areas. However, it has been argued that both during and since the colonial period, governments have been guilty of perpetuating both a state and process of 'urban bias', in which urban areas have been consistently favoured relative to rural areas (Lipton, 1977). The suggestion has been that in reality national politicians, keen to maintain a hold on power, have generally been much more concerned to keep their urban populations contented, since these communities are invariably better educated, more articulate, organised in trade unions and other groupings, and therefore likely to be a much greater potential threat to economic and political stability than the less educated and less

Table 10.1 Population and employment in selected countries.

	Population (millions) in mid 1995[a]	Rural population as percentage of total population, 1995[a]	Agriculture as percentage of GDP, 1994[b]	Percentage of labour force in agriculture, 1990[a]
Low-income economies				
Haiti	7.2	68	44	68
India	929.4	73	30	64
Nigeria	111.3	61	43	43
Middle-income economies				
Egypt	57.8	55	20	43
Indonesia	193.3	66	17	57
Venezuela	21.7	7	5	12
Upper middle-income economies				
Brazil	159.2	22	13	23
South Africa	41.5	49	5	14
Malaysia	20.1	46	14	27
High-income economies				
Republic of Korea	44.9	19	7	18
United Kingdom	58.5	10	2	2
United States	263.1	24	2	3

[a] Data from World Bank (1997).
[b] Data from UNDP (1997).

well-organised rural poor. In the absence of massive state subsidies, a policy of ensuring cheap urban food supplies has usually meant maintaining low prices for rural producers.

Urban bias can also be detected in the lack of knowledge among key national decision makers and development institutions about rural dwellers and their needs. Rural communities and their people are often regarded as being 'out of the way' and 'off the beaten track' as far as development planners are concerned. As Chambers (1983) forcefully argues: 'In Third World countries as elsewhere, academics, bureaucrats, foreigners and journalists are all drawn to towns or based in them. All are victims, though usually willing victims, of the urban trap' (Chambers, 1983: 7). Further biases in the investigation process may contribute to an incomplete and inaccurate understanding of rural needs. For example, 'dry season bias' results from only visiting rural areas during the dry season, when travel is usually easier. However, in tropical countries with well-defined wet and dry seasons, it is during the rainy season that most crops are grown, people have to work for long hours in the fields, and disease and malnutrition are more common. 'Tarmac

bias' refers to the reality that many visitors to rural areas travel only on good roads and rarely venture into remote areas, thus failing to make contact with what are often the poorest communities. A further factor in generating inaccurate perceptions is 'person bias', where visitors only speak to influential community leaders, who are invariably men. The views of women and 'ordinary' community members are therefore rarely heard (Chambers, 1983).

Rural realities

A broad sense of the reality of rural living in the developing world is shown in Table 10.1. Although there are some variations in the overall pattern, it can be seen that the proportion of the total population living in rural areas is generally higher in lower-income countries. The importance in poorer countries of the agricultural sector both as a source of employment and within overall national production is also evident.

Table 10.2 (overleaf) confirms the divergent patterns of urban and rural poverty that were noted in

Table 10.2 Rural–urban gaps in the 1990s.

	Rural–urban disparity[a] in access[b]		
	Health (1985–95)	Water (1990–95)	Sanitation (1990–95)
Bangladesh	–	98	40
Nigeria	73	41	75
Uganda	42	68	55
India	80	93	20
Peru	–	32	43
Mexico	75	68	24
Republic of Korea	100	76	100

[a] Figures in the four columns for rural–urban disparity are expressed in terms of the urban average, which is indexed to equal 100. Thus, the smaller the figure shown, the bigger the rural–urban disparity; the closer to 100, the smaller the disparity. Rural–urban parity would equal 100.
[b] The following definitions of access are from UNDP (1991: 193–6) Health: ability to reach appropriate health services on foot or by the local means of transport in not more than one hour. Water: reasonable access to safe (i.e. uncontaminated) water supply. Sanitation: access to sanitary means of excreta and waste disposal.
Source: UNDP (1996).

Table 10.3 Infant mortality rate and urban–rural residence for selected countries.

Country	Infant mortality rate		
	Rural	Urban	Rural/urban
Sub-Saharan Africa			
Ivory Coast	121	70	1.7
Ghana	87	67	1.3
Kenya	59	57	1.0
Asia			
India	105	57	1.8
Indonesia	74	57	1.3
Philippines	55	42	1.3
Thailand	43	28	1.5
Latin America			
Guatemala	85	65	1.3
Mexico	79	29	2.7
Panama	28	22	1.3
Peru	101	54	1.9

Source: Allen and Thomas (1992).

Chapter 9. Using disparities in access to basic services as utilised by UNDP, the figures are expressed in terms of the urban average (indexed to equal 1). The relative poverty of rural areas is clear in all cases and extreme disparities shown by the lower figures.

Figure 10.1 shows that in some Asian countries such as Bhutan, Nepal and Pakistan, rural people are much less well-served in terms of provision of safe water supplies. Such disparities have a major bearing on the health of the population. The higher infant mortality rates found in rural areas compared with urban centres, as shown in Table 10.3, are a further expression of the generally more impoverished conditions characteristic of rural areas in developing countries.

In many African countries there is evidence that rural people are now relatively worse off than they were at independence. Three-quarters of African countries produced less food per person at the end of the 1980s than at the beginning of the decade (Pinstrup-Andersen, 1994). In Latin America too, it seems there has been little progress in reducing the overall level of poverty in the past twenty-five years. As the International Food Policy Research Institute reported in 1995: 'Forty-five per cent of the Latin American population

are poor, and the number of poor has increased from 120 million in 1970 to 200 million today. Thirty-five per cent of people in rural areas and 15 per cent of those in urban areas live in extreme poverty' (IFPRI, 1995a: 1). Although, as we will see later, South Asia has benefited from crop yield increases associated with the introduction of Green Revolution technology, population is growing rapidly and more than 50 per cent of the malnourished children in all developing countries are today found in South Asia (IFPRI, 1995b: 2).

Agrarian structures and landholding

Agriculture is fundamental to economy and society in the rural areas of the developing world. In turn, patterns of landholding in poor agrarian economies are key determinants of 'agrarian structure', which concerns the different ways in which land and labour are combined in varying forms of production, as well as the social relations, such as class, which 'structure' the processes of production and reproduction. Fundamentally, an analysis of agrarian structure attempts to answer the questions: 'Who owns what, does what, gets what and what do they do with it?' (Bernstein, 1992a: 24). Not only do patterns of land ownership provide a context in which land and labour

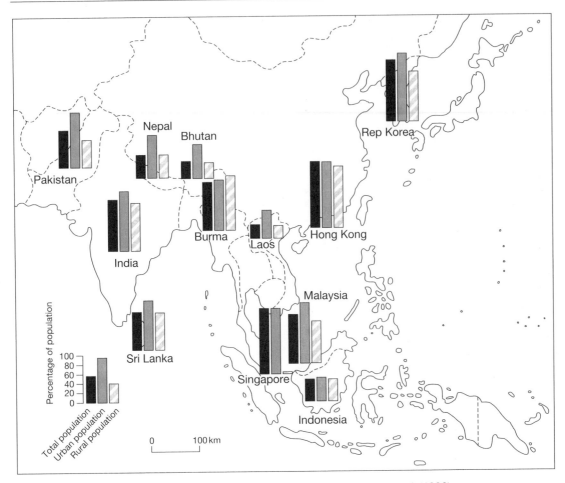

Figure 10.1 Safe drinking water: access in Asia, 1990. Adapted from Dickenson *et al.* (1996).

are combined in the production process, but land is also often the primary means through which rural inhabitants of the developing world define their personal, social and political identities. As such, land ownership is a major correlate of political and social prestige in these areas. Furthermore, since food is the major product of the land, such patterns have clear implications for the relative and absolute well-being of the population (Ghose, 1983) (Plate 10.1, overleaf).

Agrarian structures are neither static nor uniform across space. They reflect varied historical experience in rural areas, encompassing factors such as environment, culture and political economy. It should be recognised that the fortunes of particular groups of rural people are also often influenced by processes operating well beyond the countryside, in wider regional, national and international settings. For example, colonialism and the penetration of capitalism

have had a significant impact on agrarian structures in many parts of the developing world. Such processes have taken diverse forms, were variously imposed, taken up or rejected over time and space, and have left distinctive legacies within contemporary agrarian structures. The considerable inequality in landholding which exists in some countries and regions is clearly shown in Table 10.4 (overleaf).

Latin America   In much of Latin America, land ownership is increasingly concentrated in the hands of a small minority, within what is known as the latifundio system. This system includes very large plantation estates, cattle ranches and the haciendas persisting from the colonial era. The majority of the population of these countries have access to very small family farms, or minifundios, and/or are subject to exploitative landlord–tenant systems. Typically, the share of

Plate 10.1 Banana cutting in Dominica, West Indies
(photo: Philip Wolmuth, Panos Pictures)

Table 10.4 Inequality in the distribution of landholdings
in selected countries.

Country	Year	Gini coefficient[a]
Very high inequality (Gini above 0.75)		
Paraguay	1981	0.94
Brazil	1980	0.86
Panama	1981	0.84
Uruguay	1980	0.84
Saudi Arabia	1983	0.83
Madagascar	1984	0.80
Kenya	1981	0.77
High inequality (0.51 to 0.75)		
Colombia	1984	0.70
Dominican Republic	1981	0.70
Ecuador	1987	0.69
Grenada	1981	0.69
Chile	1987	0.64
Honduras	1981	0.64
Yemen	1982	0.64
Sri Lanka	1982	0.62
Peru	1984	0.61
Nepal	1982	0.60
Uganda	1984	0.59
Turkey	1980	0.58
Jordan	1983	0.57
Pakistan	1980	0.54
Philippines	1981	0.53
Medium inequality (0.40 to 0.50)		
Bahrain	1980	0.50
Bangladesh	1980	0.50
Morocco	1982	0.47
Togo	1983	0.45
Ghana	1984	0.44
Low inequality (below 0.40)		
Malawi	1981	0.36
Mauritania	1981	0.36
Egypt	1984	0.35
Niger	1981	0.32
Republic of Korea	1980	0.30

[a] The Gini coefficient is a measure of inequality in
distribution. It ranges from zero to 1; the closer the
value to 1, the greater the inequality.
Source: UNDP (1993).

total agricultural land accommodated by the minifundios is very small. In Guatemala, farms under 5
hectares constituted 90 per cent of all farms in 1979,
but occupied only 16 per cent of the land area. In Brazil
in 1980, these figures were 37 per cent and 1 per cent
respectively (Bernstein, 1992b).

The polarisation of landholding in contemporary
Latin America is deeply rooted in the colonial history
of the region, although subsequent policies in the agricultural sector and processes of commercialisation
during the nineteenth and twentieth centuries have
reinforced and extended the latifundio system. In
the seventeenth century, Spanish colonists were given
rights to expropriated land, originally in return for
military service, and were able to levy tribute from
indigenous communities in the form of labour or goods.
With the capital acquired, large landed estates, or
haciendas, were established, on which peasants typically cultivated land allocated to them, in return for
paying rent to the landlord in cash or a share of their
crops, or alternatively worked on the landlord's farm

in return for a small subsistence plot. The expansion
of this system depended on acquiring further labour
and appropriating land from Indians, a process which
was often unpopular and resisted by indigenous communities. The scale and nature of expansion varied
according to local factors such as population density,
the strength of peasant organisations, the extent of
land fragmentation and the demand for cash crop

production. For those people who resisted such incorporation, the alternative minifundio system often proved inadequate for subsistence production, since plots were often too small or located in areas which were marginal for agricultural production. Many small farmers had no alternative but to seek out other sources of income, often through wage employment. At the time of Guatemala's independence from Spain in 1821, it was estimated that one-third of the population depended upon support from migrant labour, such was the pattern of land concentration in the hands of a few wealthy landowners (Colchester, 1991).

Asia Landholding in Asia is heavily concentrated, as in Latin America, but due to land scarcity, the size of farms is much smaller and the cultivation of land is more decentralised through systems of tenancy and sharecropping. The significance of access to land and tenancy in determining human welfare is shown in Table 10.5, where it can be seen that landless households in the Philippines and Indonesia own less than the average size of landholding or are tenants, and are among the poorest households.

Much of Asia has a relatively recent colonial history, and the legacy of that period is evident in present agrarian structures, particularly with respect to the introduction and reinforcement of private property ownership. Before British rule in India, property rights had rested with the community, with village heads allocating individual rights to land, supervising the management of common resources such as meadows and rangelands, and collecting taxes in kind. Under Crown rule from 1868, a system of private property rights in land and taxation payable in cash was introduced. For example, in the Zamindari system of landholding, which was particularly widespread in northern India, de facto landlords were created as intermediaries between the tenants (the real landowners) and the British Administration. These 'landlords' was required to pass on to the government treasury an agreed percentage of land rents collected from tenants. In return, they were effectively given private property rights through their assigned privilege to extract rents from tenants at whatever level they thought was feasible (Shariff, 1987).

This system of 'parasitic landlordism' (Bernstein, 1992c) enabled the British to control land in the colony and to extract rents from the peasants who worked it without intervening directly in the production process. The system was parasitic in the sense that rents were not reinvested to enhance the productivity of farming, and unlike the communal land system, landlords had no duties to the tenant, such as providing

Table 10.5 Characterising the poor in Indonesia and the Philippines.

	Indonesia	Philippines
Rural-based	✓	✓
Dependent on farming	✓	✓
Concentrated in peripheral regions	✓	✓
Smaller than average landholdings	✓	✓
More likely to be tenants or landless agricultural wage labourers	✓	✓
Larger than average family	✓	✓
Young household head		✓
Household head with a low level of education	✓	✓
Low level of access of social services such as health and education facilities	✓	✓
High level of underemployment		✓

Source: Rigg (1997).

assistance in times of hardship. Although peasant cash crop production grew, indebtedness and landlessness also rose and peasants found themselves subordinated to a newly created class of overlords – the landlords and moneylenders. At India's independence in 1947, 50 per cent of the land area was held by approximately 4 per cent of the rural population. Meanwhile, some 27 per cent of the population were landless and a further 53 per cent had farms of less than 5 acres (Ghose, 1983).

Africa In broad terms, African agrarian structures are more widely characterised by communal systems of landownership, a lower concentration of landholding and less widespread tenancy and leasing arrangements than in Asia or Latin America. At a continental scale, and in comparison with other major world regions, land is relatively abundant in Africa, and it was labour rather than land shortages which was generally perceived throughout history to be the more critical constraint on agricultural development. Colonialism occurred much later in Africa, and generally speaking, peasant farmers were neither incorporated into a system of private property rights nor had their land expropriated. However, there are some exceptions to this rule, particularly in French colonial countries such as Cameroon, Madagascar, Guinea and Ivory Coast, where large-scale European settlement occurred (Plate 10.2, overleaf). The so-called White Highlands in British-ruled Kenya are another example of this. In

Plate 10.2 Harvest of pineapples on cash crop plantation, Dabou, Ivory Coast
(photo: Ron Giling, Panos Pictures)

southern Africa, the expropriation of land for European agriculture forced African farmers into 'reserve' areas. These reserves were often more marginal for cultivation, and together with policies such as compulsory labour recruitment, the active blocking of opportunities for cash cropping by African farmers and many more indirect means such as policies on soil conservation, African labour was released into European agriculture, mining and infrastructural development projects (Elliott, 1991). The legacies of settler colonialism are evident within contemporary agrarian structures in Zimbabwe, where on independence in 1980, approximately half of the country's agricultural land was in private freehold ownership of around 7,000 European farmers. The majority of the rural population were resident in the former reserve or 'communal' areas, with usufruct rights to land held in trust by the community (Elliott, 1995). Figure 10.2 illustrates the dualist system of land tenure in Zimbabwe in the early 1980s, and also shows the locations of land resettlement schemes established since independence; they are discussed more fully below. Plate 10.3 (page 258) is an extract from an aerial photo taken at the boundary between an 'African' and a 'European' farming area. The contrast in land use and population density between the two sectors is very evident.

Colonial interests in increasing cash crop production in Africa were generally undertaken without fundamental change to the system of landholding. Although peasant producers remained in control of their lands, the 'colonial triad' of taxation, export commodity production and monetisation, impacted significantly on the relations of production throughout the continent, as shown in Figure 10.3 (page 259). On the basis of work within Hausa communities in northern Nigeria, both Scott (1976) and Watts (1984) have suggested that although some groups during this period were able to benefit from increased agricultural production and sales, or engaged in trading and moneylending enterprises, many more people became increasingly vulnerable under these processes to long-standing hazards such as drought and market fluctuations. Land and labour became goods to be bought and sold, and there was a decline in the 'moral economy' – the reciprocal sharing of tasks and goods – thereby removing options and flexibility in times of hardship.

The legacy of colonial policies on agrarian structures in Africa is therefore extremely varied. Rarely in Africa today is there a neat distinction between imposed systems of individual freehold ownership and customary African tenure, as has often been suggested. Rather, as Siddle and Swindell propose,

> diverse and parallel systems of tenure and rights to farm, which include communal usufructary systems, loaning, pledging, different forms of labour renting associated with squatters and share contractors,

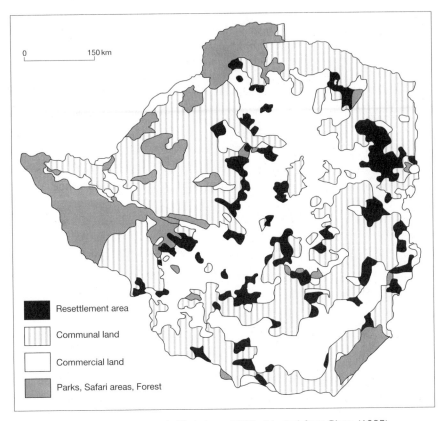

Figure 10.2 Resettlement areas in Zimbabwe, 1983. Adapted from Binns (1995).

fixed rents, leasing freehold purchase and land nationalization. These different forms of landholding and farming are rooted in different relations of production underwritten by religion, kinship and political authority, as well as varying with ecological circumstance. (Siddle and Swindell, 1990: 72)

There is further evidence that land in Africa is not always plentiful. Many recent case studies indicate that access to land – notably the best quality land, land surrounding urban areas and lands in areas of population expansion – is becoming difficult for some groups to obtain, particularly women, pastoralists and poorer households (Bassett, 1993: 4).

Principal systems of production

Indigenous food production systems across the developing world are characterised by their diversity and dynamism. Food production strategies result from complex combinations of different factors, managed with

considerable expertise by indigenous communities (Box 10.1). Longitudinal research into household coping strategies, particularly during times of hardship, has shown how even the concept of what constitutes 'food' may change at such times, with communities utilising their considerable environmental knowledge to seek out wild foods to supplement their diets (Mortimore, 1989). Non-agricultural income-generating activities can also play a greater or lesser role in the overall livelihood system for particular rural households and at certain times. For example, Chuta and Liedholm (1990) suggest that, on the basis of surveys in eighteen developing countries, one-fifth or more of the rural labour force may be primarily engaged in non-farm activities.

In contrast to the general situation in industrialised countries, households in the developing world are regularly the unit of production as well as consumption (Moock, 1986). Such households generally produce a high proportion of their subsistence requirements, and indeed the majority of such production is usually for subsistence purposes, rather than for cash income.

Plate 10.3 Communal/Commercial boundary. Marondera district, Zimbabwe.
(*source*: Surveyor General's Office, government of Zimbabwe).

Rural households themselves are incredibly varied and dynamic. Typically, a developing world household is often much larger than the 'nuclear household' of the developed world, and includes parents, children, grandparents, and quite often also unmarried aunts and uncles, as well as distant relatives. In polygamous societies, families can be very large indeed, since each wife would typically live with her own children and other relatives. Movement between households for reasons such as employment, marriage, religious duty or in response to crisis is also common in the developing world (Chapter 8). Between and within particular households, social relations based on age and gender may be complex and varied. Crehan (1992) observes:

> How [younger and older] women, men and children gain access to resources [who owns what?] is likely to be significantly different, as are their roles in production [who does what?] and their rights over what is produced [who gets what?]. (Crehan, 1992: 91)

Agricultural production

In the space available here, some generalisation of the diverse agricultural systems in the developing

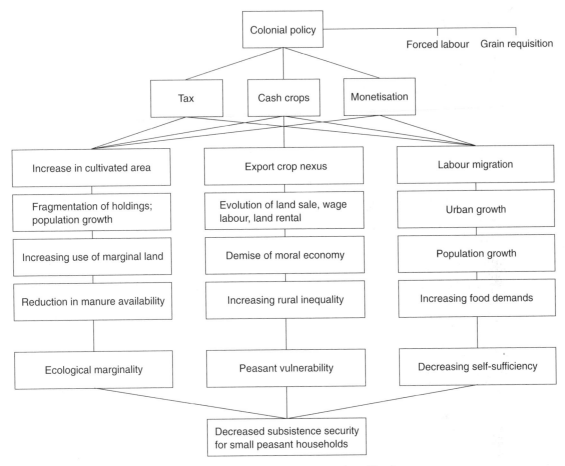

Figure 10.3 The effects of colonial penetration on food security in northern Nigeria.
Adapted from Watts (1984).

BOX 10.1   Farming and fishing on the Kenyan shores of Lake Victoria

Although Kenya controls only 4,100 square kilometres of Lake Victoria, 6 per cent of its surface area, the lake produces a massive 85 per cent of Kenya's fish supply, valued at US$80 million in 1994 (see figure) (Geheb, 1995; Geheb and Binns, 1997). In recent years, however, the ecology and economy of the lake have come under serious threat, and fishermen have been forced to devote more of their time to other activities, such as farming, in an effort to satisfy household food requirements and supplement income.

The people living around the Kenyan shores of Lake Victoria are mainly of the Luo ethnic group, whose ancestors migrated south from Sudan, arriving in what is now Kenya in the late fifteenth or early sixteenth centuries. The livelihood system of Luo lakeside communities around the Winam Gulf has traditionally involved three important elements: fishing, farming and livestock herding. Each of these elements varies in importance at different times during an annual cycle, whereas longer-term variations depend upon such factors as the availability of fish or periods of drought which affect farming.

In the precolonial period, Luo artisanal fishing activities were characterised by the use of low-technology gears, such as reed traps, papyrus beach seines and sisal long-lines, which served to control

Box 10.1 continued

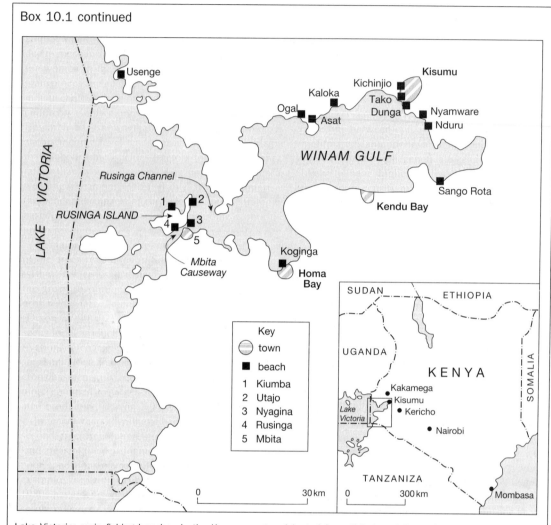

Lake Victoria: main fishing beaches in the Kenyan sector. Adapted from Geheb and Binns (1997).

the level of exploitation of fish resources. Furthermore, lakeside Luo clans and subclans each had well-defined territories, extending into the lake as far as their boats could travel. Territorial rules were strictly enforced, forbidding members of one clan from fishing in their neighbours' waters. The precolonial period was therefore characterised by control of access through territorial rules and regulations that closely regulated the fishing 'effort', and by a shared responsibility for the clan's territorial waters and maintenance of fish reserves.

In 1901 the railway from Mombasa and Nairobi reached Kisumu on the shores of Lake Victoria, opening up a vast potential market for fish. An acceleration of the fishing effort was prompted by the introduction of hut tax in 1900 and poll tax in 1910, so the fishermen had to earn money from something. Further expansion was enabled by the introduction of the flax gill-net in 1905 and the beach seine soon afterwards. Between 1949 and 1953 the number of fishermen increased from 35,000 to around 60,000. During this same period, however, there was a steady decrease in overall catch tonnage, such that the British colonial authorities introduced regulatory measures, including restrictions on net mesh-size. As the fishing intensity continued to increase, a number of exotic fish species were introduced. In 1954 the Nile perch, a voracious pre-

Box 10.1 continued

dator, was introduced into the Ugandan sector of the lake, together with six exotic tilapia species. The introduction of the Nile perch had a drastic effect on the ecology of the lake, not least in reducing the species diversity with the loss of between two and three hundred *Haplochromis* species.

Following Kenya's independence in 1963 there was a further rapid growth in fishing activity, but costs of hired labour, boats and nets escalated. The fishery has increasingly attracted outside interests with higher investment potential, such that in the 1990s there are many absentee boat-owners, many of whom are Nairobi-based business people and politicians. Probably as many as 50 per cent of the boats in the Kenyan sector of Lake Victoria are now owned by people who do not themselves work on them. The local communities have therefore lost much of the control they once had over the fishery, and fish yields have suffered. Pollution levels are also rising due to fertiliser and pesticide use on neighbouring agricultural land, which is contributing to eutrophication and the increased incidence of oxygen-using algal blooms. In 1997, parts of the Winam Gulf, particularly close to Kisumu, were suffering badly from water hyacinth, a floating mass of vegetation which makes any fishing activities virtually impossible. It is the combination of such economic and ecological pressures which have encouraged fishermen to invest their energies and seek additional income elsewhere.

A survey undertaken in 1994–95 found that 94 per cent of fishermen interviewed also farmed, and most respondents agreed that farming is becoming a greater necessity for fishermen, particularly since the early 1990s (Geheb and Binns, 1997). Respondents gave three main reasons to explain the increased

attention to farming: (1) fishing yields and incomes are declining; (2) after almost four years of irregular and poor rainfall, many fishermen were turning back to the land after better crop yields and rising food prices in 1994; (3) farming is generally regarded as being more reliable and more easily monitored, whereas fishing is increasingly perceived as a 'hit and miss' activity. Many Luo feel that the land is likely to offer better security than the lake, given the problem of declining fish yields, the greater incidence of boat thefts and the rising prices of fishing gear.

This illustrates the adaptability of rural communities and the importance of diversification into more than one activity in an unpredictable environment and resource base under pressure. Future development strategies will need to consider ways to ensure the sustainability of both land and water resources and their associated food production systems, if nutritional security is to be assured for lakeside communities. The essence of a solution may well be to build upon the diversification of activities which have long been an integral feature of Luo communities. With regard to achieving a sustainable future for the lake, there is an urgent need to improve the efficiency of regulations controlling the exploitation of fish stocks. Possibly the most appropriate solution is to strengthen traditional Luo institutions, which seemed to provide effective controls during the precolonial period. Simultaneously, the various fishing cooperatives located on key beaches around the Winam Gulf might also be strengthened to provide investment loans, security for boats and gear, and reliable marketing links. What is clear, however, is that well-informed action is urgently needed, both to preserve the ecology of the lake and to protect the livelihoods of many thousands of people.

world is unavoidable. Table 10.6 (overleaf) contrasts traditional subsistence production with modern, commercial agriculture. This type of analysis should not discount the significance of cash crop production within small-scale livelihood systems, as discussed below. Indeed, the same crop may be grown as a food staple, and any surplus sold subsequently for cash, such as in the case of rice in Asia or maize in many parts of sub-Saharan Africa.

The insecurity of rural livelihoods and the need to minimise risk are features of many small-scale food production systems, and can be key factors affecting

the willingness to innovate. Many 'outside' commentators have been slow to appreciate the rationality of indigenous cultivation practices. Innovations such as new crops and production methods like irrigation and the application of fertilisers and pesticides are frequently costly, unavailable and unreliable, and require a major departure from well-tested methods, which are often carefully adapted to the particularities of the environment. Farmers with low-level technology and limited financial resources cannot afford to take such risks. Furthermore, indigenous techniques, such as the burning of debris after clearing and the intercropping

Table 10.6 Differences between subsistence and commercial farmers.

	Traditional, subsistence	Modern, commercial
Proportion of output sold off the farm	Low	High
Destination of foods	Local direct consumption and some processed locally	High proportion processed and to food manufacturers
Origin of inputs		
Power	Draught animals, human labour	Petroleum, electricity
Plant nutrients	Legumes, ash, bones, manure	Chemical fertilisers
Pest control	Crop rotations, intercropping	Insecticides, fungicides, break crops
Weed control	Rotations, hoeing, use of plough	Herbicides
Implements and tools	Hoe, plough, sickle, scythe.	Machinery, often self-propelled combine harvesters
Seed	From own harvest	Purchased from seed merchants
Livestock feeds	Grass and fodder crops grown on farm or common land	Purchased from compound feed mixers
Economic aims	Prime aim to provide family food	Profit maximisation
	Land and labour main inputs; few capital inputs	Capital and land major inputs; labour a declining input
	Diversity of crops grown	Specialised production
	Aims at maximising gross output and yield per acre	Aims at maximising output per head and minimising production costs
	Prime aim avoidance of risk; reluctant to innovate	Innovation

Source: Grigg (1995).

of a number of crop types in the same plot, were once strongly criticised by Western observers as being environmentally destructive, but are now viewed much more positively and seen as being environmentally more sustainable (Hill, 1972). Intercropping is a valuable risk-aversion strategy; if one crop suffers from drought or pest attack, there will be others to supply household food needs. In addition, the fertility of the soil is likely to be maintained for a longer period if a variety of crops are grown, and total crop yields from intercropped plots are actually often greater than those where a single crop is grown (Richards, 1985, 1986).

An important factor affecting the nature of land use and farming systems is the size and density of population. Traditionally, in areas of sparse population within the tropics, extensive farming systems such as shifting cultivation and rotational bush fallow have been commonly practised (Ruthenberg, 1976). Both types of system depend on long fallow periods, sometimes of twenty years or more, to restore soil fertility after relatively short periods of cultivation, perhaps only one or two years. However, fallowing practices are often

complex and can vary over relatively short distances due to factors such as population density, the physical environment, land tenure arrangements and the management practices of individual farmers (Gleave, 1996). Typically within such fallowing systems, households clear and cultivate land around settlements, moving from one plot to another when falling yields indicate declining soil fertility. Whereas settlements are usually fixed under rotational bush fallowing systems, under true shifting cultivation, settlements will also move when all the land in the surrounding area has been cultivated. Plots under such fallowing systems are often irregularly shaped and unfenced clearings in the forest or bush, and their size varies according to the availability of land and labour.

Much has been written on population pressure leading to a reduction in fallow periods and environmental degradation, but in reality there is a complex relationship between population size and land use (Chapter 6). Ester Boserup has been prominent in suggesting an argument in which population pressure frequently induces rural communities to become more efficient, adapting their environment, engaging in more intensive production and, in time, increasing productivity (Boserup, 1993). A 25-year study of Machakos District in Kenya has given much support to the Boserup hypothesis, revealing progressive intensification of land use, increased productivity and careful environmental management in the face of steadily growing population (Tiffen, Mortimore and Gichuki, 1994). With increased population and land intensification, plot boundaries generally become more clearly defined, often with fences, trees or bushes to demarcate one household's land from another's. As fallow periods become shorter, fertility can no longer be maintained without additions to the soil, which usually take the form of animal manure or compost made from kitchen waste and other organic material. Sometimes crop rotations and successions are used to maintain fertility, with root crops such as sweet potatoes or cassava being planted after a harvest of grain crops such as rice, maize, millet or sorghum. Another response to increasing population pressure is to cultivate hitherto unused land, perhaps by constructing terraces or irrigation systems. Plot size may decline even further, and more intensive, even permanent cultivation develops.

In low-technology agricultural systems the most important input is human labour. Where most family labour is unpaid, the size and composition of the household labour force, together with access to communal labour, can have a crucial impact on agricultural productivity and the nutritional status of both individuals and households (Moock, 1986). The availability of farm labour can be a problem where there is a predominance of either very young or old members whose contribution to farm work is generally less effective than that of able-bodied young adults. A further factor affecting the quality of household labour is ill-health. Debilitating diseases such as malaria, onchocerciasis (river blindness) and schistosomiasis (bilharzia) can have a marked effect on labour availability and efficiency. Since the 1980s, the spread of AIDS has had a significant effect on certain communities in rural Africa, depriving households of valuable labour resources (Chapter 5).

The degree of monetisation can also be an important factor in determining household labour supply and performance even in subsistence production. This will determine the extent to which households require cash income to purchase inputs, and therefore they must sell their own labour in other activities and/ or pay for labour at certain times. All such factors

can influence the ability of households to undertake productive tasks at the required point in the cultivation cycle, or to fulfil essential 'reproductive' tasks such as fuelwood or water collection.

Capital is one factor of production that is frequently in short supply among rural cultivators in developing countries. For many producers, traditional tools such as hoes, cutlasses and axes are little changed from those used by their ancestors, but they are often well adapted to the local environment and the nature of the work undertaken. In contrast, more sophisticated tools and machinery, such as tractors and ploughs, may be quite inappropriate and they also require fuel and regular maintenance. Simple irrigation technologies, such as shadoofs and other water-lifting devices, are used to raise water from wells, streams and rivers onto neighbouring fields. There is often little incentive to invest time and effort into improving land under extensive rotation systems with only short periods of cultivation. However, with more permanent farming systems, capital investment is common in such forms as terracing, soil erosion control, irrigation infrastructure, fencing, farm buildings and tree planting. In the absence of mechanised implements on many farms, such improvements can represent hours of work by the household or community.

In contrast to small-scale agricultural production for largely subsistence purposes, commercial or plantation agriculture, as shown in Table 10.6, is characterised by labour-intensive production methods, high investments in technologies, a limited range of crops grown, expatriate ownership and management, often by a multinational company, and a hierarchical relationship between management and labour force. Plantations are frequently 'total institutions', with workers living on the plantations and being heavily dependent on them for their livelihood (Laing and Pigott, 1987; Tiffen and Mortimore, 1990). As we have already seen, it was the colonial encounter which introduced many new cash crops, primarily directed at export, and forced the commodification of land and labour in rural areas of the developing world. Such crops included cotton, sugar, coffee, cocoa, tea, sisal and groundnuts. The earliest plantations were probably established on the Canary Islands, though they became more widespread in parts of the Americas from about 1550. Coffee became an important plantation crop in Brazil from the late eighteenth century, with cocoa joining it in the nineteenth century. In West Africa whereas the French colonialists established plantations for growing coffee, cocoa and bananas in Ivory Coast and bananas in Guinea, the Americans grew rubber in Liberia, and from 1884 the Germans set up rubber, oil palm and banana plantations in Cameroon. Meanwhile, in East Africa, White settler farmers established estates in countries such as Kenya to grow tea, coffee, pyrethrum, tobacco, cotton and sugar. The rapid development of the car industry in the late nineteenth century prompted the widespread establishment of rubber estates by Europeans, particularly in peninsular Malaysia.

The specific impacts of plantation agricultural production on the wider economy and society in any particular location can be complex; this was discussed in Chapter 7 with reference to enhanced cash crop production as an element of structural adjustment programmes. Here is Young's summary:

> Expansion of cash crop production may be beneficial for food security if land ownership is relatively equitable and if exports are diverse, but clearly a rush to promote cash crops for export may just intensify food insecurity, increase economic and political marginalisation, and increase the likelihood of environmental catastrophe. (Young, 1996: 72)

But cash crops, whether destined for domestic consumption or export, are by no means always grown on estates and plantations. Indeed, in the case of Ghana, Hill (1963) has shown that it was mainly small farmers who planned and developed cocoa production from the early 1900s, reinvesting their wealth in the expansion of cocoa lands and also in infrastructure, such as roads and bridges. A major concern over cash crop cultivation on smallholder farms relates to problems of food security and the nutritional status of households that may be neglecting food production to cultivate crops for the market. Longhurst (1988) suggested that the children of cash crop farmers do not enjoy better nutritional status than children of non-growers. Studies of cocoa production in Nigeria and Mexico, sisal in Brazil, sugar cane in Kenya and coffee in Papua New Guinea all revealed negative effects of cash crop production on family food consumption and/or nutritional status. In determining the nutritional effects of cash crop production among smallholder farmers, a critical factor was the extent to which women controlled production and marketing. Although the picture is complex and varies between and within rural communities, it is likely that whoever controls these aspects will have the greatest influence on how cash is spent. Increased women's wealth is likely to translate into improved household nutrition, particularly children's nutrition.

Pastoralism

According to Sandford 'Pastoralists are people who derive most of their income or sustenance from keeping domestic livestock in conditions where most of the feed that their livestock eat is natural forage rather than cultivated fodders and pastures' (Plate 10.4) (Sandford, 1983: 1). A distinction is often made between settled ranching and various forms of livestock herding that involve migration; these migratory forms range from pure nomadism, through seminomadism to transhumance, involving seasonal movements from permanent settlements (Galaty and Johnson, 1990). Rural livelihood systems based on livestock production are most commonly found in arid and semi-arid areas, where crop cultivation is restricted by water availability. Pastoralists utilise marginal environments where resources are widely dispersed and land may be too dry, rocky or steep for cultivation.

It was estimated in the early 1980s that some 30–40 million people were engaged in 'animal-based' economies in the world's arid and semi-arid areas, of which '50–60 per cent were found in Africa, 25–30 per cent in Asia, 15 per cent in all of America, and less than 1 per cent in Australia' (Sandford, 1983: 2). The most important countries in terms of numbers of pastoralists are Sudan, Somalia, Chad, Ethiopia, Kenya, Mali, Mauritania, India and China. Some of the main pastoral groups in the developing world include the Masai, Sumburu and Karamojong of East Africa, the Fulani and Tuareg in the savanna–sahel zone of West Africa, the Bedouin of the Middle East and the Kazakhs and Mongols of Central Asia.

Whereas many rural households may own some cattle or small livestock such as chickens, sheep and goats, to provide draught power and supplement diets, livestock are central to pastoralist production, as is a degree of mobility within their lifestyles. Among pastoral communities, cattle, camels, sheep and goats provide the primary sustenance base for the household, providing milk, yoghurt and meat at certain times, although meat may not be eaten on a regular basis. Social status is also often linked closely to the size and quality of animal herds. A completely nomadic lifestyle is now quite rare, and it is more common for pastoralists to make seasonal migrations, returning to a permanent base which may be occupied throughout the year by women and children. The precise timing and extent of migrations may be governed by factors such as altitude, the farming calendar, the prevalence of tsetse infestation, and the progression and severity of the dry season (Stock, 1995).

Plate 10.4 Cattle being taken to Ballyera market, Niger (photo: Mark Edwards, Still Pictures)

Pastoralists, like cultivators, have been frequently misunderstood by academic writers and government officials, who criticise herders for keeping livestock purely for prestige purposes, for allowing overgrazing of rangeland, for transmitting human and animal diseases, for avoiding tax payments and for disregarding international boundaries. Another common misunderstanding is that pastoralists do not grow crops. On the contrary, a large proportion of pastoralists do actually engage in cultivation to a greater or lesser extent, and some groups attach equal importance to farming and are therefore more accurately described as 'agro-pastoralists'. The Fulani of northern Nigeria are frequently subdivided into four groups. The Bororo are true nomads, shifting camp regularly and moving with their herds in search of pasture and water. Seminomadic Fulani have permanent homesteads with adjacent farms, where a variety of crops are grown. The elders stay in the homesteads for most of the year, to be joined by the returning young herdsmen at the start of the rainy season when the fields must be prepared for planting. A third group, the semisettled Fulani, regard cultivation and livestock rearing as equally important. They have fixed settlements and migrate over shorter distances, mainly at the height of the dry season. A fourth group, the settled stock-owners, graze or corral their herds close to the villages, and young men take animals out daily into the surrounding area. These stock-owners often supplement their income by tending the livestock owned by wealthy town-based people, who increasingly invest in livestock as a symbol of wealth and status. All four groups of Fulani show a common attachment to their animals and are respected for their skills in rearing and managing livestock (Binns, 1994b: 105).

One key problem relating to the development and sustainability of pastoralism in arid and semi-arid areas of developing countries, is that much conventional range management has adopted what has been termed an 'equilibrium view' (Scoones, 1995). For example, the concept of 'carrying capacity' has been used to indicate the human and animal population limit which cannot be exceeded without setting in motion the process of land degradation. This concept, however, is largely based on relatively sophisticated beef ranching systems in places such as Australia and the United States; whereas in the dryland areas of Africa and Asia, equilibrium conditions just do not apply. These areas are characterised instead by unpredictable climates and considerable spatial heterogeneity, where non-equilibrium dynamics apply much of the time. Scoones comments:

> Stocking rate [numbers/area/time] adjustments have always concentrated on the changing of animal numbers, rather than seeking management options that manipulated area or time. Pastoralists operating in the non-equilibrium environments of the drylands use the range of strategies, but flexible movement

and spatial and temporal adjustments are the key to success. (Scoones, 1995: 26)

Where the traditional flexibility and mobility of pastoral systems have been constrained by government policies directed at sedentarisation, or the construction of large dams and irrigation systems across traditional grazing areas and migratory routes, tensions between pastoralists and cultivators have arisen, as in northern Nigeria (Binns and Mortimore, 1989).

Another issue concerning the behaviour of pastoralists is what has been called the 'tragedy of the commons', a situation where grazing land is held communally, but livestock are owned by individuals or households. It is argued that, because of these different types of ownership, individuals have an incentive to graze as many animals as they can without concern for the sustainable management of the communal grazing resources. As a result, it is suggested that overstocking leads to overgrazing, which in turn can lead to land degradation and even desertification (Hardin, 1968). Sandford (1983), however, is concerned about the assumption that the 'tragedy of the commons' scenario is widespread, whereas in reality there is much variation from one region and pastoral group to another:

> Not only are the argument's assumptions often not fulfilled in practice in particular cases, but its importance has been exaggerated to the point where it sometimes appears as the only factor to be considered in deciding on land-tenure policy in pastoral areas; and a number of unjustifiable conclusions have been drawn from it. (Sandford, 1983: 119)

There is evidence that many pastoral communities do in fact control the intensity of grazing on communally owned land, and in arid and semi-arid regions a move to individual ownership of land would probably cut across the need for flexibility and mobility, both natural adaptive responses to environments with inadequate and unreliable water and grazing resources. Since pastoralism in many cases takes place in marginal areas, where little cultivation is possible without the introduction of costly irrigation schemes, it might be argued that livestock herding represents the best possible use of the land. However, in spite of this, many Third World governments would prefer pastoralists to settle and take up commercial ranching. Government policies towards the rural and livestock sectors frequently place a strong emphasis on the merits of sedentarisation, although this runs counter to the lifestyles of many pastoral peoples.

Fishing

Fish provides an important source of protein for many people in Third World countries and particularly those living in rural areas. However, its importance as an economic activity, source of employment and as a contribution to diets and household nutrition is frequently undervalued. Fishing activities may be broadly classified by location into marine fishing and inland, freshwater fishing; they are also classified by method into artisanal (or traditional) and modern. In reality, however, there is a vast and complex array of different fishing methods, ranging from line and spear fishing to basket traps, and a great variety of nets (Plate 10.5). In the West African region alone there is a great diversity of fishing activities, with the Fante and Ewe fishermen of Ghana using mainly wooden seagoing boats and beachseine nets off the Atlantic coast, whereas the Bozo and Somono peoples use canoes, basket traps and seine, gill and cast nets in the inland Niger Delta of Mali (Binns, 1992). Other important wetland areas, such as the Hadejia-Jama'are floodplain of north-eastern Nigeria, are also valuable sources of fish and the home of fishing communities (D.H.L. Thomas, 1996).

Malaysia produced 603,000 tonnes of fish in 1990 using fishing methods ranging from modern trawlers to the use of traditional craft propelled by oars and sails. As in Africa, a wide range of traditional nets and traps are used. In shallow coastal waters up to 7 fathoms deep, lines of fixed stakes are used, known as *kelongs*, which are constructed of nibong palms. In the 1950s this method contributed some 30 per cent of total fish catches in peninsular Malaysia, but is now much less common. Purse seines and trawl nets now account for over 60 per cent of total fish catches, and have displaced many traditional items such as the fixed *belat* trap, constructed from bamboo and placed across streams and rivers.

Fishermen typically have a good knowledge of the breeding cycles and movements of fish and they often migrate in search of better catches. The Somono fishermen of Mali migrate in the early part of the rainy season following the *Alestes* migration, and also in the dry season when water levels in the inland Niger Delta fall. Considerable numbers of Fante fishermen have moved westwards from Ghana and have established communities on the Gambian coast, from where they catch mainly shark and ray, which are then dried, salted and bagged before being sent by ship to Ghana, where there is considerable demand. The marketing

Plate 10.5 Fishing boats landing their catches, Dunga Beach, Kisumu, Kenya
(photo: Tony Binns)

and processing of fish involves large numbers of people in Africa – many of them women – who buy fish from fishermen, and sell them in the local market-place or to long-distance traders. In hot climates where there is no refrigeration, fish are often fried or smoked over wood or charcoal to preserve them.

Fish catches have declined in some areas as the resource base has come under steadily increasing pressure. The small, yet important, Kenyan sector of Lake Victoria is experiencing such pressure and there is an urgent need to bring some order and control to the management and exploitation of fish stocks (Box 10.1). There is much potential for increasing fish production in many developing countries. The provision of outboard motors for boats, improved nets and improvements in the handling and processing of fish catches, all can have a significant impact on the size and quality of production. Among fishing communities, large amounts of capital can be tied up in fishing boats and tackle, and the availability of credit at reasonable rates can be helpful to fishermen wanting to repair or buy new boats and nets (Geheb and Binns, 1997).

In south-west India, the depletion of forests has resulted in a shortage of wood for building traditional fishing boats. Artisanal fishing is also being threatened by more sophisticated fishing methods such as bottom trawling and purse seining. In 1981 the UK-based

Intermediate Technology Group, in cooperation with another UK company, the South Indian Federation of Fishermen Societies and the Fishermen's Welfare Society, produced a new type of boat. It is light, stable, powered by sail and oars and has a life of 6–10 years. With increasing competition from trawlers, fish catches in inshore waters declined and many artisanal fishermen considered using outboard motors on the new-style boats. By 1984 new and stronger plywood boats with outboard motors were being built. The use of motors increased the fishing range and considerably reduced the crew's energy expenditure. Furthermore, fishermen using the new motorised boats caught ten times more fish than those using traditional sail-powered boats. Several boatyards have since been established in Kerala State and, having acquired the necessary constructional skills, some workers have set up their own small businesses to build similar plywood boats (Kurien, 1988).

In an effort to improve the protein intake of poor rural people, much interest has been shown in recent years in the development of 'aquaculture' or 'fish farming', with the establishment of fish hatcheries and the stocking of ponds and lakes. The value of such practices has long been recognised in China and parts of Southeast Asia. Many rural Chinese communities have long-established and ecologically sustainable

systems of fish ponds, together with large flocks of
ducks and geese. The possibility of introducing such
systems into other rural areas of the developing world
requires careful evaluation and could play a significant role in raising nutritional levels.

Forestry

Trees, woodlands and forests are multi-purpose
resources that provide varied functions in society
(Chapter 6) and differing roles in rural livelihoods
across the developing world. At all scales, patterns of
forest resource use and management reflect the diverse
ecologies of forests over space, the numerous different
combinations of forest type, the changing value placed
on particular forest products and services, and the
many interventions over time which have aimed to
secure those resources for human development.

Approximately 60 per cent of the world's 'closed
forests', conventionally defined as where tree crowns
cover more than 20 per cent of the land area, are in
the humid tropics, (Mather and Chapman, 1995). Such
forests have been the locus of indigenous livelihoods
built upon a close association with the forest and
dating back thousands of years. It is estimated that as
many as 15–20 million people dependent on hunting,
gathering and shifting cultivation techniques, lived in
the Amazon Basin in the sixteenth century. Indeed,
shifting cultivation systems, by definition, depend on
the manipulation of the dynamics of forest ecologies:
'All tropical regions provide examples of the way in
which farmers, as part of their traditional systems of
shifting cultivation, incorporate trees from the original
forest stand into their fields, for ground protection or
for useful products' (Weidelt, 1993: 39).

For many more rural people in the developing
world generally, it is the resources of open woodlands and scrub vegetation which play a particularly
significant role in farming systems and livelihoods.
The generally high dependence of communities on
biomass sources of energy, as noted in Chapter 6, in
itself demands a close association with local woodland ecologies to secure fuelwood for cooking, heating and lighting. In addition to firewood, woodlands
may also provide a whole host of products, including

> building timber; wood for kraal fences, tools, transport
> and construction (boats, scotchcarts, sledges, etc.);
> edible leaves, pods, nuts and fruits; honey; natural
> fibres; fodder; medicines; utensils; and a whole range
> of other items. (Munslow *et al.*, 1988: 45)

Securing such products often requires extensive local
environmental knowledge and varied degrees of management. In arid zones, particularly where populations
are sparse and livestock herding is the dominant livelihood system, woodland management may be based
on the selective harvesting of natural vegetation, such
as is characteristic of large parts of sub-Saharan Africa. A knowledge of local ecologies and the regen-
erative capacity of the natural vegetation is used to
secure fodder for livestock, and to provide shade and
rubbing poles for pest control. At the other end of a
continuum of management, in areas of high population density and plentiful rainfall, trees of various
indigenous and exotic species are planted and managed carefully and intensively for their varied products. In many parts of Southeast Asia, multi-storey,
home or kitchen gardens are host to varied tree and
plant species, often with distinct horizontal levels: food
crops at ground level, coffee bushes and medicinal
plants in the next zone, through fruit, fuel and fodder
species at higher levels (Christanty, 1986).

The term *agroforestry* refers to the deliberate growth
and management of woody perennials in conjunction
with annual crops and/or animals. Table 10.7 (pages
270–1) shows aspects of the extent and diversity of
agroforestry systems in the developing world. Although
the concept is not new, agroforestry experienced a
resurgence of interest in the 1980s, as the relative
failure of large agriculture and forestry initiatives in
the developing world started to become evident. Conventional approaches to forest management have generally depended on excluding people from access to
forest resources, such as through the establishment of
forest plantations for industrial purposes or through
the creation of forest reserves, particularly for watershed management. In Peru and India this 'scientific
perspective' has led to massive afforestation programmes. These programmes may meet industrial timber demands but they are often at the expense of local
use of grazing and agricultural lands and the loss of
varied 'non-timber forestry products', so valued by
local communities (Dankelman and Davidson, 1988).

Similarly, attempts to address the 'fuelwood problem', particularly in Africa and India, have often misunderstood the role of the woody biomass component
in farming systems. For example, extension efforts
aimed at the introduction of fast-growing species within
communal woodlots quickly encountered a number
of problems, including the question of responsibility
for management of these 'mini-plantations', and the
paucity of species such as eucalypts relative to indigenous species in providing the multiple products so

valued and needed. Furthermore, the wood-burning properties of introduced species rarely matched those demanded by women. Conventional forestry approaches misunderstood how species selection and use would vary, depending on what was being cooked, the socio-economic characteristics of the household and the multiple functions of the hearth.

Tree management systems among rural households of the developing world are under constant change. At a national level, many countries of the developing world do not have the forest resources that they used to. The closed forests of Central America are estimated to be 82 per cent of what existed formerly and Cuba has only 2 per cent of its original forests remaining (Lohmann, 1993: 21). On average, the countries of the developing world are losing their forest resources at a rate of 1.1 per cent per annum (UNDP, 1996). Table 10.8 (page 272) shows the annual rates of deforestation in the 1980s for selected countries.

The forces of deforestation are many and complex, as discussed in Chapter 6. However, discussions of deforestation are now being informed by improved understanding of the role of trees and woodlands in farming systems, and of the diverse management systems and institutions which have developed to secure their multiple products. For example, the 'islands' of forest immediately surrounding villages in West Africa and the Amazon were widely assumed until recently to be the last vestiges of original forest, and they were taken as evidence of ecosystems on the verge of collapse. However, research in Guinea (Fairhead and Leach, 1995) and among the Kayapo Indians in the Amazon (Anderson, 1990) has suggested that much of what is now forest in Guinea was once cultivated, and the secondary forest has been created by people as part of a complex cultivation system to provide various products and services within rural livelihood systems.

In Latin America the importance of local environmental knowledge in enabling the sustainable management of forests is evidenced by the deforestation that is occurring as people are forced into the forests, by the increasing concentration of land ownership, to secure resources for agriculture, but without the necessary knowledge or interest in forest ecologies and dynamics (Lohmann, 1993). Similarly, state-sponsored programmes of transmigration in Indonesia, although alleviating land pressure in the more populated islands, have led in some cases to deleterious impacts on forest environments in receiving areas, as migrants have lacked sufficient skills for forest management (Secrett, 1986).

An improved understanding of the role of trees in farming systems of the developing world is now also being used to modify conventional forest management practices. As well as the resurgence of interest in agroforestry, programmes of 'social forestry' are now increasingly recognised as a promising approach to forestry development. The concept of social forestry may encompass a broad spectrum of activities, with the common aim of managing more explicitly the varied social values of trees in community development. Programmes are seen particularly as a means of addressing women's concerns for food production and energy management, and for potentially meeting the needs of the poorest households, which have been left out of conventional forest management programmes.

Approaches to agricultural and rural development

Strategies designed to promote development in rural areas and among rural peoples have taken many forms. Since agriculture provides both a direct and indirect source of income in the rural sector, development efforts have focused frequently, though not exclusively, on raising agricultural production. This section will consider some key ideas and practices relating to agricultural and rural development in developing countries.

Rural development initiatives may be undertaken for different reasons and at different times by various agencies or institutions, including community organisations and government departments, as identified in Chapter 7. Poverty alleviation, the desire to secure political patronage or to effect social transformation, are just some of many possible motivations for developing rural areas. As was considered earlier in this chapter, any intervention in prevailing agrarian structures has implications not only for the material conditions of rural life, but also for personal and social identity, for power relations within and between households and for traditions and customs associated with land, and its use and significance within society (Shipton and Goteen, 1992). Such complexities have, unfortunately, often been overlooked in rural development strategies and may actually be an important factor in explaining the continuing prevalence of widespread rural poverty in the developing world.

The specific nature of rural development strategies has also been shaped by prevailing development

Table 10.7 Examples of agroforestry systems in the developing world.

Major system	Subsystem and practices	South Pacific examples	Southeast Asian examples
AGROSILVICULTURAL SYSTEMS	Improved fallow (in shifting cultivation areas)		Forest villages of Thailand; various fruit trees and plantation crops used as 'fallow' species in Indonesia
	The Taungya system	(e.g. Taro with *Cedrella* and *Anthocephalus* trees)	Widely practised; forest villages of Thailand are improved forms
	Tree Gardens	Involving fruit trees	Dominated by fruit trees
	Hedgerow intercropping (alley cropping)		Extensive use of *Sesbania grandiflora, Leucaena leucocephala* and *Calliandra callothyrsus*
	Multi-purpose trees and shrubs on farmlands	Mainly fruit or nut trees (e.g. *Canarium, Pometia, Barringtonia, Pandanus, Artocarpus altilis*)	Dominated by fruit trees; also *Acacia mearnsii* cropping system, Indonesia
	Crop combination with plantation crops	Plantation crops and other multipurpose trees (e.g. *Casuarina* and coffee in the highlands of PNG; also *Cliricidia* and *Leucaena* with cacao)	Plantation crops and fruit trees; smallholder systems of crop combinations with plantation crops; plantation crops with spice trees
	Agroforestry fuelwood production	Multi-purpose fuelwood trees around settlements	Several examples in different ecological regions
	Shelterbelts, windbreaks, soil conservation hedges	*Casuarina oligodon* in the highlands as shelterbelts and soil improvers	Terrace stabilisation in steep slopes
SILVOPASTORAL SYSTEMS	Protein bank (Cut-and-carry fodder production)	Rare	Very common, especially in highlands
	Living fence of fodder trees and hedges	Occasional	*Leucaena, Calliandra*, etc., used extensively
	Trees and shrubs on pastures	Cattle under coconuts, pines and *Eucalyptus deglupta*	Grazing under coconuts and other plantations
AGROSILVOPASTORAL SYSTEMS	Woody hedges for browse, mulch, green manure, soil conservation, etc.	Various forms; *Casuarina oligodon* widely used to provide mulch and compost	Various forms
	Home gardens (involving a large number of herbaceous and woody plants with or without animals)	Several types of home gardens and kitchen gardens	Very common; Java home gardens often quoted as examples; involving several fruit trees
OTHER SYSTEMS	Agrosilvofishery (aquaforestry)		Silviculture in mangrove areas; trees on bunds of fish-breeding ponds
	Various forms of shifting cultivation	Common	e.g. swidden farming
	Apiculture with trees	Common	Common

Source: Gholz (1987).

South Asian examples	Middle East and Mediterranean examples	East and Central African examples	West African examples	American Tropics examples
Improvements to shifting cultivation; several approaches (e.g. in the north-eastern parts of India)		Improvements to shifting cultivation (e.g. gum gardens of the Sudan)	*Acioa barterii*, *Anthonontha macrophylla*, *Gliricidia sepium*, etc., tried as fallow species	Several forms
Several forms, several names		The 'Shamba' system	Several forms	Several forms
In all ecological regions	The Dehesa system; 'Parc arboree'			e.g., 'Paraiso Woodlot' of Paraguay
Several experimental approaches (e.g. conservation farming in Sri Lanka)		The corridor system of Zaire	Experimental systems on alley cropping with *Leucaena* and other woody perennial species	Experimental
Several forms both in lowlands and highlands (e.g. hill farming in Nepal; 'Khejri'-based system in the dry parts of India)	The oasis system; crop combinations with the Carob tree; the Dehesa system; irrigated systems; olive trees and cereals	Various forms; the Chagga system of Tanzania highlands; the Nyabisindu system of Rwanda	*Acacia albida*-based food production systems in dry areas; *Butyrospermum* and *Parkia* systems; 'Parc arboree'	Various forms in all ecological regions
Integrated production systems in smallholdings; shade trees in plantations; other crop mixtures including various tree spices	Irrigated systems; olive trees and cereals	Integrated production; shade trees in commercial plantations; mixed systems in the highlands	Plantation crop mixtures; smallholder production systems	Plantation crop mixtures; shade trees in commercial plantations; mixed systems in smallholdings; spice trees; babassu palm-based systems
Various forms including some forms of social forestry		Various forms	Common in the dry regions	Several forms in the dry regions
Use of *Casuarina* spp. as shelterbelts; several windbreaks	Tree spices for erosion control	The Nyabisindu system of Rwanda	Various forms	Live fences, windbreaks especially in highlands
Multi-purpose fodder trees on or around farmlands especially in highlands		Very common	Very common	Very common
Sesbania, *Euphorbia*, *Syzigium*, etc., common		Very common in all ecological regions		Very common in the highlands
Several tree species used very widely	Very common in the dry regions; the Dehesa system	The *Acacia* dominated system in the arid parts of Kenya, Somalia and Ethiopia	Cattle under oil palm; cattle and sheep under coconut	Common in humid as well as dry regions (e.g. grazing under plantation crops in Brazil)
Various forms especially in lowlands		Common; variants of the 'Shamba' system	Very common	Especially in hilly regions
Common in all ecological regions; usually involving fruit trees	The oasis system	Various forms (e.g. the Chagga homegardens; the Nyabisindu system)	Compound farms of humid lowlands	Very common in the thickly populated areas
Occasional				
Very common; various names		Very common	Very common in the lowlands	Very common in all ecological regions
Common	Common	Common	Common	

Table 10.8 Annual rates of deforestation in selected countries of the developing world.

Country	Annual rate of deforestation, 1980–89 (%)
Brazil	0.7
Indonesia	0.8
Malaysia	1.5
Pakistan	0.4
Nepal	4.0
Rwanda	2.3
Zimbabwe	0.4
Ghana	0.8
India	2.3
Colombia	1.7
Equador	2.3
Peru	0.4
Papua New Guinea	0.1

Source: UNDP (1996).

ideologies. For example, in the 1950s and 1960s it was generally assumed that the benefits of urban, industry-led development would spontaneously 'trickle down' without active planning in the rural sector. Where specific agricultural development schemes were introduced, they were often characterised by 'top-down' planning, inadequate environmental knowledge and heavy mechanisation. Such an example was the ill-fated East African Groundnut Scheme in the 1950s, which involved ploughing up vast areas of semi-arid Tanganyika (now Tanzania) using inappropriate machinery to grow groundnuts to supply Britain with vegetable oils (Binns, 1994b: 97–98).

In contrast, during the 1970s many governments felt that substantial state intervention was required to achieve greater equity and poverty alleviation in rural areas. Organisations such as the World Bank stressed the need for 'integrated rural development' schemes which, in addition to raising agricultural productivity and improving nutrition, also emphasised the importance of improving rural health care, education, transport and marketing. But in reality, many so-called integrated schemes actually showed little evidence of integration, such that rural education and health care were often neglected in favour of raising agricultural output through the use of high-yielding varieties, fertilisers, pesticides and mechanisation (Airey, Binns and Mitchell, 1979; Binns and Funnell, 1983). Also during the 1970s, and following the publication of Schumacher's influential book *Small is Beautiful*, a lengthy debate ensued concerning the issue of transferring technology from rich to poor

countries and the need for 'appropriate' technology (Schumacher, 1974).

By the 1980s, however, as considered in Chapter 3, development ideologies reflected a concern for accountability and efficiency, which was translated into limiting the use of state resources. International donors and governments themselves began to look for alternative institutions to deliver services and foster development in rural areas, particularly within the private and non-governmental sectors, as discussed in Chapter 7. NGOs were considered as frequently more successful in promoting local participation, democracy and empowerment, as well as being more cost-effective. In consequence, rural development projects during the 1980s were generally smaller, involving a better understanding of rural communities; they were also more responsive to the perceived aspirations and constraints of local farmers.

Land reform

At one time or another, but especially since 1960, virtually every country in the world has passed land reform laws. . . . Yet in spite of decades of land reform activities, land ownership remains extremely skewed, concentration of land ownership is almost universally increasing, the mass of landless is growing rapidly, and the extent of rural poverty and malnutrition has reached horrendous proportions. (de Janvry, 1984: 263)

Land reform programmes may take many different forms, but generally involve tenancy reforms and/or changes in the distribution and scale of land ownership. In practice, therefore, land reform may involve such measures as the elimination of certain kinds of rent or cropping arrangements, the creation of new kinds of farm, such as cooperatives or state farms, or the expropriation and redistribution of lands, including through the implementation of ceilings on ownership and resettlement programmes. The objectives of land reform may combine social, political and economic intentions, including enhancing social stability, increasing political participation and patronage, widening economic opportunity and promoting more efficient use of land and labour.

The political significance of land reform is confirmed by the number of countries which initiated programmes following their gaining independence or other major political events. In cases such as post-revolutionary Ethiopia in 1974, or China in 1958, the radical nature of land reform is confirmed by the new forms of societal organisation created. In Cuba, land reform was high on the agenda of the revolutionaries in the late 1950s, as shown in Box 10.2. Under colonial rule in Africa, the dispossession of lands and oppression of the agricultural workforce were less prevalent than in Latin America or Asia. Radical land reform programmes following independence were most likely in settler-dominated regimes such as Zimbabwe (Box 10.2).

Programmes of land reform have also been undertaken with more reformist intentions, such as to prevent the accumulation of large landholdings or to enhance security of tenure. In South Korea, during a major programme of land reform initiated after the Japanese colonial period in 1949, those landlords holding more than 3 hectares had to turn over lands to tenants in return for compensation from the state. In Taiwan, during a programme of land reform in the

---

BOX 10.2   Some experiences of land reform

Zimbabwe

On independence in 1980 the new majority government in Zimbabwe quickly committed itself to a programme of land reform. The objectives included overcoming the dualist agricultural history of the country, satisfying the political demands of the peasantry and fostering the proposed socialist transformation of society. The resettlement programme aimed to provide 162,000 landless families and persons displaced by the war with land in newly serviced resettlement villages on former European farms. Several 'models' for resettlement were implemented, including cooperative farms run on a collective basis, and the more widespread, individual family farms with common grazing resources (Elliott, 1995). However, the Lancaster House Agreement between Britain and Zimbabwe served to protect private property interests in the country for a further ten years and was an important factor in limiting the pace of land reform. By the beginning of the 1990s, only 52,000 families had been moved to 3.3 million hectares of land, and since the government had no control as to where lands became available for purchase, scheme areas tended to be in fragmented rather than contiguous blocks, and they were often located in more marginal areas. The cooperative model for resettlement has been particularly limited. In 1991 the uptake on those schemes was only 42 per cent of planned capacity (Elliott, 1995), and there were problems of management, infrastructural developments and finance, with the British government refusing to provide aid to this model. Although there is now legislation to enable the compulsory purchase of lands for the programme on behalf of the Zimbabwean government, escalating costs, population increase and

changing class interests, e.g. the rise of black middle-class interests, are among a variety of factors that restrict the impact of land reform on Zimbabwe's economy and society.

Cuba

At the end of the revolution in Cuba in 1959, 28 sugar companies controlled 20 per cent of the farmlands of the country and 40 cattle ranches controlled a further 10 per cent of the cultivated area (Macewan, 1992: 162). Approximately 70 per cent of the land area was held by fewer than 10 per cent of the farming community. Agrarian reform laws passed in 1959 and 1963, together with other social and political reforms, are widely accepted to have destroyed these inherited forms of rural inequality and to have provided the basis for Cuba's economic transformation into the 1990s. The main elements of land reform were the staged expropriation of large farming units to state control and the enlargement of the small-scale private sector through legislation which gave land ownership to all tenants, sharecroppers and squatters cultivating up to 67 hectares at the time of the revolution. The state farm sector provided the basis for socialist agricultural development, and the large units enabled the effective provision of services in health and education. With a greater number of peasant farmers after the revolution, the new government also undertook measures to organise the role of the enlarged private agricultural sector within the economy. For example, new institutions were created to provide credit, to initiate the formation of cooperatives and to purchase products from the peasantry.

Box 10.2 continued

India

Major elements of Indian land reform were initiated during the first five-year plan from 1951. According to Shariff (1987), they included four important aspects:

- Measures for fixing rents and enhancing security.
- A ceiling on landholding, and the subsequent redistribution to landless farmers, with some exemptions being made for well-managed estates where break-up would lead to losses in production, particularly of important cash crops such as rubber and sugar.
- The development of cooperative village management and farming to consolidate holdings and prevent fragmentation.
- The abolition of intermediaries, through the vesting of rights in states, thereby raising incentives among individual cultivators to invest in management and improvement.

In practice, implementation of Indian land reform policy lay with individual states, and they pursued particular elements with highly varying degrees of commitment. By the late 1980s, intermediary tenures encompassed within the Zamindaria system had largely been abolished throughout the country. Furthermore, all but three states had fixed rents at a maximum of one-quarter to one-fifth of gross produce. However, progress in conferring ownership rights on cultivating tenants and in redistributing ceiling surplus lands was more limited. Bernstein (1992c) suggests that comprehensive transfer of land was obstructed by the power of small-scale landlords who were well represented at the state level and who had close interconnections with trading and moneylending, through caste and class relations at the village level. Despite this, significant changes in Indian agrarian structures have occurred since independence, particularly as incentives have increased for landlords to invest in commercial agricultural production as an alternative to trade, moneylending or renting activities.

1950s, rents were cut to a maximum of 37.5 per cent of crop value from a figure in excess of 70 per cent (Barke and O'Hare, 1991). Land reform in the Philippines was effected through the conversion of share tenancies to fixed rents in the 1970s. In India, the exploitative tenancy arrangements inherited from the British colonial period, were identified within the first five-year plan of 1950, as fundamental barriers to raising agricultural production and as the source of much social injustice in the rural sector. A fourfold programme of land reform was subsequently introduced in 1951 (Box 10.2).

The specific impacts of land reform are hard to identify in many cases since they often arise from diverse programmes, frequently implemented in phases and in conjunction with wider agrarian reform measures. Many such programmes remain incomplete, and there is often a lack of data monitoring their performance. They often demand considerable financial investment, but this may not be forthcoming, particularly as international backing for land reform has declined. Large amounts of aid from the United States are generally considered to have been a critical factor in the relative success of South Korea's land reform programme, whereas underfunding is deemed a principal reason for the failure of land redistribution efforts

in the Philippines (Dixon, 1990). Table 10.9 shows that in Zimbabwe the cost of land purchases rose from Z$20 per hectare at the outset of the resettlement programme in 1980, to over Z$200 per hectare a decade later. In 1997, while visiting the United Kingdom for the Commonwealth Heads of Government Meeting, the president of Zimbabwe, Robert Mugabe, suggested that Britain should compensate White farmers in his country, since up to 13 million hectares of land expropriated by White settler farmers in the early twentieth century had been scheduled for the future resettlement programme. Mugabe argued, 'If the British government wants us to compensate its children, it must give us the money, or it does the compensation itself' (Raath, 1997).

Corruption and considerable vested interests in land among senior politicians and businessmen may further prevent the accurate registration of titles and the transfer of lands. In Kenya a vigorous programme of land reform was pursued from independence in 1963, building upon colonial schemes for land consolidation. Studies in the early 1970s, however, noted the prevalence of landowning MPs, ambassadors, senior civil servants and other government officials as a substantial political barrier to radical land reform:

Table 10.9 Annual land acquisition and expenditure within the Zimbabwean resettlement programme.

Year	Land (000 hectares)	Purchase price (Z$000)	Average price (Z$ per hectare)
1979–80	87	1 798	20.59
1980–81	223	3 517	15.76
1981–82	900	18 803	20.89
1982–83	935	22 009	23.42
1983–84	160	4 536	28.37
1984–85	75	2 967	39.53
1985–86	86	4 445	51.57
1986–87	134	3 898	29.20
1987–88	20	874	43.02
1988–89	64	2 807	43.92
1989–90	62	10 669	170.76
1990–91	34	8 060	234.23
Total	2 786	84 386	

Source: Zimbabwe Ministry of Lands, Agriculture and Rural Resettlement (1992) Second report of settler households in normal model A resettlement schemes.

The situation in Kenya today is no longer the same as it was in the early 1960s, when an alliance of African politicians and European businessmen in Kenya, in conjunction with the colonial authorities, bypassed the European settlers in taking the decision to buy out the latter. Now there is no colonial authority, no business class with an equally strong political voice, and the African leadership own much of the land. (Hunt, 1984: 288–9)

Political conflict and upheaval are regularly generated by land reform programmes, not solely between the landed elite and popular interests, but also between such groups as cultivators and pastoralists. In Eritrea there is no provision within the current Land Proclamation to protect the rights of pastoralist claims to land (Fullerton-Joireman, 1996).

Land reform is evidently not in itself a panacea for rural development, and it is now generally recognised that such reforms need to be undertaken in conjuction with other measures, such as ensuring effective social institutions at the local level and raising land productivity. Furthermore, particular attention is required to examine the sociopolitical, economic and environmental contexts in countries and regions where land reform programmes are being introduced as a component of future rural development strategies.

The green revolution

Many more rural development strategies have focused explicitly on raising agricultural productivity through technocratic packages without any fundamental reform of agrarian structure. The *green revolution* is the term widely used to refer to the application of Western technologies to raising agricultural production in the developing world. This encompasses the breakthroughs in the 1960s in plant genetics, which produced high-yielding varieties (HYVs) of grains such as rice and wheat, and the associated technological package required for their production, including fertilisers and insecticides (Dixon, 1990). The largest impacts have been in Asia, where in the early 1980s more than 75 per cent of the wheat plantings and 30 per cent of the rice plantings were HYVs (Barke and O'Hare, 1991: 107).

Part of the attraction for planners of the green revolution lay in the assumed scale-neutrality of the technologies and the power of the market to encourage and disseminate improvements in well-being. It was assumed that the biochemical technologies of seeds and fertilisers would be equally viable at all scales of operation, whether on small or large farms. It was therefore thought that the yields and incomes of all farmers could be enhanced without raising rural inequalities. In practice, however, the green revolution has had extremely uneven regional and social impacts.

In India the green revolution is considered fairly unambiguously to have delivered national food self-sufficiency (Bernstein, 1992c). However, per capita grain production actually fell in eleven of the fifteen major states to 1985, and was correlated strongly with the distribution of irrigation which enabled multiple cropping (Bernstein, 1992c). Social inequality was also aggravated in circumstances where rising landlessness forced increasing numbers of women into wage employment to ensure household survival. However, due to mechanisation, some employment opportunities for women in the traditional areas of harvesting and grain processing declined. In sharp contrast, for women in households which owned land, their burden in agriculture was often increased with double cropping. Research has shown that investment in technology at this income level tends to be in areas which save male labour time, such as tractors for land clearance, rather than women's time taken up in tasks such as weeding (Pearson, 1992). Table 10.10 summarises the gendered impacts of rural development strategies based on technocratic packages, including within the green revolution.

A major factor in explaining the limited impact of the green revolution in South America and Africa, is that 'the crops that have formed the mainstay of the green revolution, rice and wheat, are simply not grown by large numbers of third world farmers' (Dixon, 1990: 92). In much of Africa, as we have seen, food production is dominated by rain-fed cultivation of coarse grains such as maize, millet and sorghum. Research into improved varieties of these crops has generally been more limited and the impact more confined to larger-scale, export-oriented, irrigated production (Dixon, 1990).

In recent years there have been efforts to develop a second green revolution, more suited to the vast dryland areas of the developing world and to the resource-poor conditions of many farmers who live there. For example, conventional plant breeding schemes in the 1980s have developed high-yielding strains of millet, sorghum, cowpeas and cassava, which have achieved some success in raising production in parts of India and Africa, without requiring increased farm inputs (Barke and O'Hare, 1991). However, it is through biotechnology, encompassing genetic engineering and also many more diverse and complex processes and technologies, that the most profound changes are predicted. Although the products of these technologies have not as yet been employed on a large scale, biotechnology is believed to offer the potential for raising food security in the developing world, as well as delivering many other human goals globally, ranging from detoxification of hazardous wastes, to a possible cure for AIDS (Conway, 1997; Morse, 1995).

Genetic engineering has the ability to break the natural barriers for genetic transfer between species and as such enables the more precise and speedy transfer of genes than within conventional breeding programmes (Morse, 1995). Biotechnologies also widen the sources of genes available, through the transfer between very different organisms and potentially enable the creation of custom-made genes which are not available in nature. The morals, costs and dangers of biotechnology, as well as its advantages, are stimulating much contemporary debate (World Resources Institute, 1994; van der Gaag, 1997). As van der Gaag comments, 'The line between evil and good, between dream and nightmare is an unbelievably fine one' (van der Gaag, 1997: 8). Although genetically engineered crops are likely to have higher yields, Vandana Shiva warns that food insecurity may be aggravated by genetically engineered food as 'more and more peasants will see their crops substituted through biotechnology' (van der Gaag, 1997: 8).

Further reappraisal of the green revolution is also being undertaken in the 1990s through social science research, which is attempting to separate the impacts of technological change *per se* from the parallel social and economic changes that are under way. For example, in the case of the impact of mechanisation, consideration is now being given as to whether 'blame' should be apportioned to the technology or to the social and economic environment in which the technology is being adopted. Rigg argues:

> The difficulty lies in knowing whether labour shortages, brought on by the increase in non-farm employment, have induced mechanisation, or whether mechanisation has displaced labour, thereby forcing agricultural workers to look for employment in the non-farm sector.
> (Rigg, 1997: 244)

Evidently, evaluations of agricultural modernisation packages such as the green revolution need to look beyond changes in the agricultural economy, for 'there is more to rural life than agriculture and this "more" is getting larger and more significant' (Rigg, 1997: 197). The causes and effects of rural diversification are themselves diverse and can rarely be understood merely in terms of agricultural modernisation. In conclusion, it seems that further research is needed to examine the full implications of introducing green revolution technology at different points in space and time.

Irrigation

Irrigation has played an important role in the development of human civilisation and in enabling rural production in the developing world (Chapter 6).

Table 10.10 Rural development: technology and its impacts on women.

Property ownership	Employment	Decision making	Status	Level of living and nutrition	Education
GREEN REVOLUTION: NEW SEEDS, BREEDS AND AGROCHEMICALS					
May lose usufruct rights as land is used more intensively. Land owned by women is often physically marginal and not suitable for optimum applications of new inputs	Women exclude themselves from use of chemicals because of threat to their reproductive role. New crops may not need traditional labour inputs of women. Women generally displaced from the better-paid, permanent jobs	Decline. Training in new methods in agriculture limited to men. Use of new technology and crops generally subsumed by men. Women farmers equally innovative when given opportunity	Increase in family income may allow women to concentrate on reproductive activities. In patriarchal society this increases status of male head of household	New crops may be less acceptable in family diet and nutritionally inferior because of chemicals	Increase in additional disposable income of family may be used for children's education
MECHANISATION					
Women operate smaller farms in general and so may not find it economic to invest in new implements	Women usually excluded from use of mechanical equipment. Women farmers have difficulty obtaining male labourers	Decline	Decline because of reduced role on farm and downgrading of female skills	New implements not used for subsistence production	Growth of interest in mechanical training but limited to males
COMMERCIALISATION OF AGRICULTURE AND CHANGES IN CROP PATTERNS					
Female-operated farms tend to concentrate on subsistence crops and crops for local market. Tend to remain at small scale	Decline because technical inputs substituted for female labour	Decline because less involved in major crop production activity	Decline	Decline because cash crops take over land traditionally used for subsistence production by women. Males allocate more income to developing enterprise and for personal gratification than to family maintenance	Increased time available for education
POST-HARVEST TECHNOLOGY					
New equipment owned by men	Women's traditional food processing skills no longer in demand. May employ young women in unskilled jobs in agro-industries	Decline because ownership of equipment and skills passed to men	Decline because female skills downgraded	Decline because loss of women's independent income from food processing activities. New product may be nutritionally inferior. Women deprived of use of waste products for animal feed and so lose important part of traditional family diet	Increased time available for education

Source: Momsen (1991).

Table 10.11 Proportion of arable land under irrigation for selected countries in 1993.

Country	Irrigated land as a proportion of arable land (%)
Nigeria	3.2
India	28.9
China	53.6
Indonesia	24.3
Zimbabwe	7.0
Philippines	28.6
Equador	34.1
Mexico	26.3
Brazil	6.7

Source: UNDP (1996).

Rural development interventions based on irrigation technologies may be undertaken with varied objectives, such as to control flooding or to increase agricultural production through the timely application of water, which can lead to extended cropping seasons or double cropping. The extension of irrigation in the developing world has occurred most rapidly from the 1960s, and has been closely associated with the ideology of development as modernisation. Irrigation has been achieved mainly through the construction of large-scale reticulation systems using big dams and barrages on permanent rivers (Heathcote, 1983). Large-scale dam construction provided a means for subsidising industrial production and, where associated with irrigation, a technocratic route to rural development by raising agricultural production.

South Asia and Southeast Asia account for over half the developing world's irrigated area, with China, India and Indonesia all having a significant proportion of their agricultural land under some form of irrigation (Table 10.11, Plate 10.6). The spread of irrigation in Africa, however, has been much slower and is concentrated in countries such as Sudan and Egypt, where agriculture is virtually wholly under irrigation (Barrow, 1987).

One of the world's largest irrigation developments is the Gezira scheme in eastern Sudan, covering almost one million hectares and involving 100,000 tenants. Project planning for the Gezira started in the early twentieth century and was developed in stages from the 1920s using water from the Blue Nile. Before this development, small farmers used the banks of the Nile for the cultivation of date palms, vegetables and a type of millet known as dura, which is the staple food crop. Water was raised from the river using

Plate 10.6 Irrigation of crops by hand in a part of Bangladesh (photo: Mark Edwards, Still Pictures)

ox-powered *saskias*. The British colonial concern in the Gezira scheme was to expand cotton production, and a centrally planned rotation of cotton, wheat, sorghum, groundnuts and fallow was introduced onto the scheme. In the early years, the Gezira programme was quite successful in producing export and food crops, certainly in relation to other similar schemes, such as the inland Niger Delta programme in Mali. However, cropping intensities on Gezira fell in the 1970s and 1980s, as maintenance of irrigation structures and the clearance of silt from channels became less effective, leading to water supply problems and the withdrawal of some lands from cultivation, often in mid-season (Fadl, 1990). Furthermore, problems of pesticide pollution and salinisation threaten the future productivity of the scheme (Stock, 1995).

Programmes for the expansion and rehabilitation of irrigation in Southeast Asia have been closely related to the perceived opportunities provided by the green revolution. The Muda irrigation scheme in north-west

Malaysia was completed in 1974 with World Bank support, and aimed to increase rice production in the region by upgrading canal, drain and farm road density, together with fertiliser subsidies and price supports. Although rice production did rise, income distribution between groups involved in the programme had worsened by the end of the 1980s. The number of tenants declined as landowners perceived the productivity gains to be made and demanded their lands back, although increased mechanisation reduced the demand for hired labour, and the poorest tenants and small owner-operators experienced losses in income (Ghee, 1989).

With respect to irrigation schemes in sub-Saharan Africa, it has been suggested that 'the larger the projects are, and the higher the level of their technology, the poorer is their performance' (*The Courier*, 1990: 84). Large-scale irrigation schemes are certainly costly to construct, maintain and operate. The three largest schemes in Nigeria had a development cost of approximately US$25,000 per hectare of land developed (Stock, 1995: 168). Costs also fall to certain individuals who are not compensated fully, e.g. when their lands are submerged through the construction of dams. There are also various unquantified social costs for farmers involved in these schemes, such as their loss of autonomy. Large irrigation schemes are usually operated by semi-state institutions in conjunction with peasant organisations, but farmers are required to adhere to specific crops and cropping calendars, schedules of maintenance, and to use designated sources of credit and marketing outlets. Further hidden costs of large-scale irrigation schemes fall on downstream farmers who operate small-scale irrigation techniques, based on a sensitive knowledge of local flows and ecologies. Fish catches may also decline downstream of large dams and irrigation schemes (Adams, 1985). The environmental costs of irrigation developments and the associated dams include the disturbance of water regimes in rivers and floodplains, the loss of habitat, and frequently a decline in water quality and health (*New Internationalist*, 1995).

Funding for large irrigation schemes is declining, in part due to their poor performance up to now. However, it seems likely that future irrigation projects will be even more expensive, if only because the most easily constructed projects are already built. Interestingly, more than half of World Bank spending on irrigation has been on the rehabilitation and upgrading of old irrigation schemes (World Bank, 1996a). Since the mid 1980s, many development institutions have been increasingly involved in small-scale and

less formal irrigation, which accounts for over half of the total irrigated area in developing countries (FAO, 1987). Small-scale systems based on indigenous practices such as 'flood cropping', stream diversion and simple lift irrigation, are now widely held to offer a more acceptable strategy for agricultural development (Binns and Funnell, 1989; Kimmage, 1991). Although smaller may be cheaper, easier and potentially more sustainable, Adams and Anderson (1988) warn that small-scale schemes may also fall prey to the same basic problems as large-scale systems. In particular, they point to the general failure within irrigation schemes to frame development initiatives in the reality of historical experience and 'the operation of the prevailing social systems and economic networks that sustain local production' (Adams and Anderson, 1988: 535).

Dam construction, specifically, is becoming increasingly privatised in the developing world (Stewart, 1995) and it is suggested that 'only in Asia are governments determined to continue to build dams, with or without international aid' (F. Pearce, 1997: 353). The Narmada dams in India are going ahead despite the withdrawal of World Bank funding. And in 1993 work began on the world's biggest dam, the Three Gorges in China:

> It will flood 350 miles of river canyon and create an inland sea some 392 miles long. It will displace more than a million people and no-one knows if it will be able to withstand earthquakes. . . . But foreign companies don't seem to mind. Caterpillar of the US and CAE Electronics Ltd have both signed contracts for work on Three Gorges and are actively pursuing more. Management information systems will be supplied by Agra Industries of Calgary for $25 million. Merrill Lynch, the US investment giant, is involved in the financing. . . . A consortium of 25 Brazilian corporations has formed the 'Three Gorges Brazilian Joint Venture'. Bridon of Doncaster in England has also jumped on the bandwagon.
> (Stewart, 1995: 17)

Although, in recent decades, thousands of lives have been lost through flooding in China, the prospective impact of the Three Gorges dam on the security of rural livelihoods is as yet untested, and the flash floods created by the dam may be much worse than any flood yet experienced. The disruption of livelihoods for those people being displaced through construction is more certain, suggesting that rural development may not be the primary factor supporting this particular development in China.

Rural non-farm activities

Obtaining sufficient food to satisfy domestic requirements is the key aim of most, if not all, rural households in developing countries; nevertheless, rural people do engage in a wide range of other activities besides food production. Traditional crafts are particularly important (Plate 10.7), and it is being increasingly appreciated that in many rural communities of the developing world there is a wealth of knowledge and skills about non-agricultural activities such as textile production, baking, wood- and metalworking, pottery and various forms of construction. Furthermore, in the light of the general failure of large-scale industrialisation policies to alleviate underemployment and poverty in many countries, an interest has emerged particularly among NGOs in fostering rural industry as a strategy towards rural development and as a means for supplying urban and/or tourist markets.

In contrast to many developing countries, rural industries are a common feature in China, partly due to the widespread richness and nature of minerals and ores and also as a result of the country's past development strategies. A central feature of the Great Leap Forward between 1958 and 1961 was the mobilisation of rural peasants 'to convert surplus labour into capital with the aim of producing more iron and steel and of creating greater agricultural and light industrial production' (Endicott, 1988: 51). All communities were encouraged to establish 'five small industries': agricultural machinery, coal mining, hydroelectricity, chemical fertilisers and building materials. Unfortunately, with the diversion of labour into industrial growth, agricultural production suffered, and in the winter of 1960 there was widespread starvation across China. A decade later, the Walking on Two Legs initiative emphasised a more balanced approach to rural development; it involved 'being open to the new and the modern while not overlooking the tried and trusted' (Endicott, 1988: 81). Endicott reports how, in a rural Sichuan community from 1976, industrial enterprises multiplied and flourished, notably with the arrival of a young and dynamic commune party secretary, but also assisted by state loans and grants, technical advice and a three-year tax holiday. By 1980 the original seven enterprises had increased to twenty-three, including an oilseed rape processing workshop, a Chinese medicine pharmaceutical plant, a coal briquette factory and an electrometallurgy factory producing calcium carbide and ferrosilicon. These developments very much reflected Mao's pronouncement that 'the peasants can become workers right where they are' (Endicott, 1988: 87).

Plate 10.7 Cottage industry: block printing, Manikgarj, Bangladesh (photo: Cooper & Hammond, Panos Pictures)

Elsewhere in the developing world, the significance of rural industries relative to urban industries can be considerable. In Sierra Leone it has been estimated that 86 per cent of total industrial sector employment and 95 per cent of industrial establishments were located in rural areas. Textile and clothing production accounted for 53 per cent of such employment. Rural enterprises are typically small, family-owned and family-run; 59 per cent of rural industrial firms in Honduras are run by one person and 95 per cent of firms have fewer than five workers (Chuta and Liedholm, 1990). In some countries, traditional skills have been strengthened by NGOs; in rural Gambia, blacksmiths are being encouraged to make simple farm implements and donkey carts. In countries where there is a growing tourism industry, there is considerable potential for producing craft items using local skills and resources, to satisfy the demands of overseas visitors. This can also lead to a valuable injection of foreign exchange into poor communities. Governments can assist the development of rural industries by reducing import duty on commodities such as sewing machines, and by providing them with extension services and credit assistance. The Grameen Bank in Bangladesh has been particularly successful in delivering credit to poor rural households (Box 10.3).

Research into the dynamics of rural livelihood systems in Indonesia, including the introduction of technological packages within agriculture, has suggested that increasing and diversified opportunities for employment and income in the non-farm economy have enabled the poor to increase their standard of living, in some cases compensating for the predicted land concentration and impoverishment of some groups. It was found that off-farm, local opportunities in informal activities such as trading, home-based shops, food manufacturing and taxi services, but also employment (part-time, seasonal or full-time) in factory work and the public sector, were growing in importance. The location of large factory enterprises in rural areas of Java and the central and northern regions of Thailand enabled people to remain in rural areas and commute

---

BOX 10.3 The Grameen Bank and poverty alleviation in rural Bangladesh

Rural poor people frequently lack funds and equipment to innovate and raise their productivity. In some cases they are deeply in debt to local business people and moneylenders, with no prospect of breaking out of a cycle of persistent indebtedness. And although many governments and agencies recognise the importance of providing easily available and relatively cheap credit as a means towards reducing rural poverty and promoting development, most rural financial institutions in the developing world have had a rather mediocre record in developing countries. A notable exception is the Grameen Bank in Bangladesh, which works with some 2 million poor people in over 34,000 villages and has achieved a recovery rate from its loans of around 98 per cent.

Bangladesh is one of the world's most densely settled states, with some 832 people per square kilometre, a figure only exceeded by small territories and island states such as Hong Kong, Singapore, Gibraltar and Bermuda (World Bank, 1997). Bangladesh is also one of the world's poorest countries, ranked 13 from the bottom of the World Bank's table of gross national product per capita, with only US$240 per capita in 1995, and placed 144 out of 175 countries in the UNDP Human Development Index, with an HDI of 0.368 in 1994 (UNDP, 1997; World Bank, 1997). In 1990, 64 per cent of the labour force was employed in agriculture and in 1995, 82 per cent of the population was classified as rural (World Bank, 1997). With such extreme pressure on the land, the incidence of landlessness is high. It was estimated in 1983–84 that about 46 per cent of all rural households owned less than 0.5 acres of land and were therefore regarded as 'functionally landless', with an inadequate income accruing from their land (Hossain, 1988).

The Grameen Bank originated in 1976 as an experimental research project led by economist Professor Muhammad Yunus at Chittagong University. The initial aim of the project was to tackle the serious problem of indebtedness among the rural poor in Bangladesh, through providing loans to households which owned less than 0.5 acres of land. The word *Grameen* means 'village' and the project was concerned with assessing whether if the poor received financial help at reasonable terms and conditions, they could then generate productive self-employment without external assistance. Some households are paying back small loans with an interest rate as high as 10 per cent per month, which further compounds their difficulties. Professor Yunus managed to persuade a local branch of a commercial bank to provide credit at an interest rate of 13 per cent a year, provided he could guarantee recovery of the loans. This pilot project was successful and in June 1979 the Grameen Bank Project was launched in five districts of Bangladesh, with support from rural branches of commercial banks and the agricultural development bank, and with financial assistance from the state bank of Bangladesh and the International Fund for Agricultural Development. Within a year, 24 branches had been set up, though expansion was constrained by some reluctance from participating banks. A government ordinance in September 1983 transformed the project into the Grameen Bank, a specialised financial institution for the rural poor, in which government has a 60 per cent share and the borrowers 40 per cent.

As Todd comments, 'The "essential Grameen" . . . is an exclusive focus on the poor, with preference to poor women, simple loan procedures administered in the village, small loans repaid weekly and used for

Box 10.3 continued

any income-generating activity chosen by the woman herself, collective responsibility through groups, bolstered by compulsory group savings, strict credit discipline and close supervision through weekly meetings and home visits' (Todd, 1996: 7).

Each branch of the Grameen Bank covers between 15 and 20 villages, and members of households owning less than 0.5 acres of cultivated land, or with assets with a value equivalent to less than 1.0 acres of medium quality land, are eligible to receive a loan. The loan and interest must be repaid in 52 weekly instalments. The bank's default rate of less than 3 per cent is most impressive when compared with default rates of 40–60 per cent in other rural credit programmes in developing countries. Since many poor rural dwellers are illiterate, bank workers often visit their homes to assist with loan arrangements. One of the guiding mottos of the bank is: 'Take the bank to the people, not the people to the bank.'

In order to receive loans, the poor have to organise themselves into groups and associations and must be prepared to interact with each other. Several groups from each village constitute a 'centre' and weekly meetings of all centre members are held with the bank representative in attendance. All business is conducted in a transparent manner in front of the members, and an important incentive for regular repayment is the assurance of a new and bigger loan at the end of the repayment cycle.

The Grameen Bank is no longer merely an agency for disbursing loans, but it has generated a wider 'culture' of development, involving many other aspects of life among the rural poor in Bangladesh. In 1984 a social development programme called Sixteen Decisions was introduced to promote discipline, unity, hard work and improve living standards (see table).

The success of the Grameen Bank indicates the importance of credit as a starting point for the introduction of a wide-ranging economic and social development programme and poverty alleviation. A study undertaken in 1985 found that those people who had joined the bank had about 50 per cent higher income than target groups in villages where there were no bank groups (Hossain, 1988). A large number of countries have shown interest in the success of the Grameen Bank, such that the 'Grameen model' is now being replicated in India, Nepal and Vietnam (Todd, 1996).

The Sixteen Decisions of the Grameen Bank.

1 The four principles of Grameen Bank – discipline, unity, courage and hard work – we shall follow and advance in all walks of our lives.
2 We shall bring prosperity to our families.
3 We shall not live in dilapidated houses. We shall repair our houses and work towards constructing new houses as soon as possible.
4 We shall grow vegetables all the year round. We shall eat plenty of them and sell the surplus.
5 During the planting seasons, we shall plant as many seedlings as possible.
6 We shall plan to keep our families small. We shall minimise our expenditures. We shall look after our health.
7 We shall educate our children and ensure that they can earn enough to pay for their education.
8 We shall always keep our children and the environment clean.
9 We shall build and use pit latrines.
10 We shall drink tubewell water. If it is not available, we shall boil water or use alum.
11 We shall not take any dowry in our sons' weddings, neither shall we give any dowry in our daughters' weddings. We shall keep the centre free from the curse of dowry. We shall not practise child marriage.
12 We shall not inflict any injustice on anyone, neither shall we allow anyone to do so.
13 For higher income we shall collectively undertake bigger investments.
14 We shall always be ready to help each other. If anyone is in difficulty, we shall all help.
15 If we come to know of any breach of discipline in any centre, we shall all go there and help restore discipline.
16 We shall introduce physical exercise in all our centres. We shall take part in all social activities collectively.

Note: Formulated in a national workshop of 100 female centre chiefs in March 1984, these sixteen decisions might be called the social development constitution of Grameen Bank. All Grameen Bank members are expected to carry out these decisions.
Source: Hossain (1988).

daily for employment. As Rigg comments, 'In large part, the emergence of occupational diversity has put off the displacement of the rural poor from their land and has, perversely, helped to sustain "traditional" rural economies' (Rigg, 1997: 189). Often, however, it is the younger members of the household who engage in alternative activities. This may impact in economic terms, as the most productive labour is lost to agriculture; but perhaps most significant are its social and cultural effects. Rigg observes:

> The slack season is a period for courting, festivals and fairs, religious devotions, and the telling of tales. It is the season when households and villages collectively renew their bonds and identities. If young men and women are absent from such festivities, this creates the conditions in which the activities either become meaningless, or their meanings and roles change.
> (Rigg, 1997: 241)

In spite of such observations, there seems no doubt that the diversification of rural economies can have some benefits as an element of rural development strategies, perhaps most particularly in reducing the dependence of households and communities on the success of agricultural production. For example, where crop yields are affected by low rainfall, pests or disease, income generated by non-farm activities can reduce the vulnerabilty of rural households and enable them to survive until the next harvest. In fact, such activities have often been an important part of the repertoire of coping strategies which communities have developed to ensure their survival from one year to the next.

Conclusion

In his book *Challenging the professions: frontiers for rural development*, Robert Chambers (1993) identifies the fundamental limitations of previous ideologies of rural development as being:

> a planner's core, centre-outwards, top-down view of rural development. They start with the economies, not the people; with the macro and not the micro; with the view from the office, not the view from the field. And in consequence, their prescriptions tend to be uniform, standard and for universal application. (Chambers, 1993: 110)

The brief overview here of the green revolution indicated that a standard package was indeed required by farmers, including HYV seeds, fertilisers, pestic-

ides and irrigation systems which, as far as possible, matched those of the experimental stations where the technologies were developed. The technologies were generally developed on research stations, not on farms, and required substantial government investment to ensure adoption at the local level. However, this package was frequently not modified sufficiently for successful implementation in the very contrasting regional and local environments within the developing world. Furthermore, there are concerns as to the sustainability of such energy-intensive forms of agricultural production (Chapters 6 and 7).

Many writers would agree that Chambers' work has been central in promoting a new ideology of rural development, based on putting poor rural people first and 'reversing' the centralising and simplifying tendencies of many past rural development efforts. By definition, there are no blueprints for the rural development interventions which flow from this emerging ideology; instead, poor rural people provide 'starting points which are at once dispersed, diverse and complicating' (Chambers, 1993: 120). Guiding principles in this fresh approach to rural development include emphasis on providing 'baskets of choices' rather than uniform packages, in combining traditional and Western science in the generation and dissemination of technologies, on participatory methods of enquiry and action, and on empowering communities through appropriate interactions with accountable and democratic state and non-governmental organisations (Chambers, 1993, 1997; Chambers, Pacey and Thrupp, 1989).

The 'reversals' required throughout research and development confirm that rural poverty alleviation, the health of ecosystems and the sustainability of rural livelihoods depend on much more than just raising agricultural production. They constitute substantial challenges in practice and an obvious focus for sceptics. However, an 'increasing minority' of rural development programmes now demonstrate greater empathy with the needs of poor rural households and eschew some of the principles of 'farmer first'. In addition, the core assumptions, guiding concepts and styles of investigation are under vigorous review and modification in an attempt to move 'beyond farmer first' and to further the original work and evident success of this new approach (Scoones and Thompson, 1994).

It seems as if considerable progress has been made in understanding and responding to the needs of rural households and communities; however, despite some success stories, the experience on the ground in many developing countries has been one of remarkably little progress in alleviating rural poverty. In particular, sub-Saharan Africa still has a long, steep hill to climb out of rural poverty. Hopefully in time, a greater use of participatory methodologies and the introduction of more appropriate rural development strategies will be translated into a genuine improvement in rural livelihoods. But the reality is still too often one of urban bias and political expediency, such that the rural poor continue to be well down the 'development agenda' in many developing countries.

References

Actionaid (1994) *Kyuso Rural Development Area Plan and Budget*. Nairobi: Actionaid

Actionaid (1995) *Listening to Smaller Voices: Children in an environment of change*. Chard: Actionaid

Adams, W.M. (1985) The downstream impacts of dam construction: a case study from Nigeria. *Transactions of the Institute of British Geographers*, **10**, 292–302

Adams, W.M. (1990) *Green development: environment and sustainability in the third world*. London: Routledge

Adams, W.M. (1996) Irrigation, erosion and famine: visions of environmental change in Marakwet, Kenya, in Leach, M. and Mearns, R. (eds) *The lie of the land: challenging received wisdom on the African environment*. Oxford: International African Institute

Adams, W.M. and Anderson, D.M. (1988) Irrigation before development: indigenous and induced change in agricultural water management in East Africa. *African Affairs*, **87**, 519–35

Aeroe, A. (1992) The role of small towns in regional development in Southeast Africa, in Baker, J. and Pedersen, P.O. (eds) *The Rural–Urban Interface in Africa*. Uppsala: Nordic Institute for African Studies, 51–65

Aikman, D. (1986) *Pacific Rim: area of change, area of opportunity*. Boston: Little Brown

Airey, A., Binns, T. and Mitchell, P.K. (1979) To integrate or . . . ? Agricultural development in Sierra Leone. *IDS Bulletin*, **10**(4), 20–7

Allen, J. (1995) Global worlds, in Allen, J. and Massey, D. (eds) *Geographical Worlds*. Oxford: Oxford University Press and Open University, 105–44

Allen, J. and Hamnett, C. (1995) Uneven worlds, in Allen, J. and Hamnett, C. (eds) *A Shrinking World*. Oxford: Oxford University Press, 233–54

Allen, T. and Thomas, A. (eds) (1992) *Poverty and development in the 1990s*. Oxford: Oxford University Press

Alonso, W. (1968) Urban and regional imbalances in economic development. *Economic Development and Cultural Change*, **17**, 1–14

Alonso, W. (1971) The economics of urban size. *Papers of the Regional Science Association*, **26**, 67–83

Anderson, A. (1990) (ed) *Alternatives to deforestation: steps towards sustainable use of the Amazon rainforest*. New York: Columbia University Press

Apter, D. (1987) *Rethinking Development: Modernization, Dependency and Postmodern Politics*. Newbury Park CA: Sage

Armstrong, W. and McGee, T.G. (1985) *Theatres of Accumulation: Studies in Asian and Latin American Urbanization*. London: Methuen

Armstrong, W. and McGee, T.G. (1985) *Theatres of Accumulation*. London: Methuen

Augelli, J.P. and West, R.C. (1976) *Middle America: Its Land and Peoples*. Englewood Cliffs NJ: Prentice Hall

Auret, D. (1995) *Urban Housing: A National Crisis*. Gweru: Mambo Press

Austin-Broos, D.J. (1995) Gay nights and Kingston Town: representations of Kingston, Jamaica, in Watson, S. and Gibson, K. (eds) *Postmodern Cities and Spaces*. Oxford: Blackwell, 149–64

Auty, R. (1979) World within worlds. *Area*, **11**, 232–35

Auty, R. (1993) *Sustaining Development in Mineral Economies: the Resource-Curse Thesis*. London: Routledge

Auty, R. (1994) *Patterns of Development*. London: Methuen

Baez, A.L. (1996) Learning from experience in the Monteverde Cloud Forest, Costa Rica, in Price, M.F. (ed) *People and tourism in fragile environments*. Chichester: John Wiley, 109–22

Bairoch, P. (1975) *The Economic Development of the Third World Since 1900*. London: Methuen

Baker, J. and Pedersen, P.O. (eds) (1992) *The Rural–Urban Interface in Africa*. Uppsala: Nordic Institute for African Studies

Baran, P. (1973) *The Political Economy of Growth*. Harmondsworth: Penguin

Baran, P. and Sweezy, x (1968) *Monopoly Capitalism*. Harmondsworth: Penguin

Barff, R. and Austen, J. (1993) 'It's gotta be da shoes': domestic manufacturing, international subcontracting, and the production of athletic footwear. *Environment and Planning A*, **25**, 1103–14

Barke, M. and O'Hare, G. (1991) *The Third World*, 2nd edn. Harlow: Oliver & Boyd

Barratt-Brown, M. (1974) *The Economics of Imperialism*. Harmondsworth: Penguin

Barrett, H. and Browne, A. (1995) Gender, environment and development in Sub-Saharan Africa, in Binns, T. (ed) *People and Environment in Africa*. Chichester: John Wiley, 31–8

Barrett, H.R., Binns, T., Browne, A.W., Ilbery, B.W. and Jackson, G.H. (1997) Prospects for horticultural exports under trade liberalisation in adjusting African economies. Unpublished report to the Overseas Development Administration, London

Barrow, C. (1987) *Water resources and agricultural development in the tropics*. London: Longman

Barrow, C.J. (1995) *Developing the environment: problems and management*. London: Longman

Bartone, C. (1994) *Towards Environmental Strategies for Cities*, Urban Management Policy Paper 18. Washington DC: World Bank

Bassett, T.J. (1993) The land question and agricultural transformation in sub-Saharan Africa, in Bassett, T.J. and Crummey, D.E. (eds) *Land in African agrarian systems*. Madison WI: University of Wisconsin Press, 3–34

Bauer, P.T. (1975) Western guilt and Third World poverty. *Quadrant*, **20**(4), 13–22

Becker, C.M. and Morrison, A.R. (1997) Public policy and rural–urban migration, in Gugler, J. (ed) *Cities in the Developing World*. Oxford: Oxford University Press, 88–105

Beckford, G. (1972) *Persistent Poverty: Underdevelopment in Plantation Economies of the Third World*. New York: Oxford University Press

Bell, M. (1980) Imperialism: an introduction, in Peet, R. (ed) *An Introduction to Marxist Theories of Underdevelopment*, Monograph HG14, RSPACS. Canberra: Australian National University, 39–50

Bernstein, H. (1992a) Agrarian structures and change: India, in Bernstein, H., Crow, B. and Johnson, H. (eds) *Rural livelihoods: crises and responses*. Oxford: Oxford University Press, 51–64

Bernstein, H. (1992b) Agrarian structures and change: Latin America, in Bernstein, H., Crow, B. and Johnson, H. (eds) *Rural livelihoods: crises and responses*. Oxford: Oxford University Press, 27–50

Bernstein, H. (1992c) Poverty and the poor, in Bernstein, H., Crow, B. and Johnson, H. (eds) *Rural livelihoods: crises and responses*. Oxford: Oxford University Press, 13–26

Berry, B.J.L. (1961) City size distributions and economic development. *Economic Development and Cultural Change*, **9**, 573–87

Berry, B.J.L. (1972) Hierarchical diffusion: the basis of development filtering and spread in a system of growth centres, in Hansen, N.M. (ed) *Growth Centres in Regional Economic Development*. New York: Free Press

Binns, T. (1992) Traditional agriculture, pastoralism and fishing, in Gleave, M.B. (ed) *Tropical African development*. London: Longman, 153–91

Binns, T. (1994a) Ghana: West Africa's latest success story? *Teaching Geography*, **19**(4), 147–53

Binns, T. (1994b) *Tropical Africa*. London: Routledge

Binns, T. (ed) (1995) *People and environment in Africa*. Chichester: John Wiley

Binns, T. and Funnell, D.C. (1983) Geography and integrated rural development. *Geografiska Annaler B*, **65**(1), 57–63

Binns, T. and Funnell, D.C. (1989) Irrigation and rural development in Morocco. *Land Use Policy*, **6**(1), 43–52

Binns, T. and Mortimore, M.J. (1989) Ecology, time and development in Kano State, Nigeria, in Swindell, K., Baba, J.M. and Mortimore, M.J. (eds) *Inequality and Development: Case Studies from the Third World*. London: Macmillan, 359–80

Biswas, A.K. (1992) Water for Third World Development. *Water Resources Development*, **8**(1), 3–9

Biswas, A.K. (1993) Management of international waters. *International Journal of Water Resources Development*, **9**(2), 167–89

Biswas, A.K. and Biswas, A. (1985) The global environment. *Resources Policy*, **11**(1), 25–42

Black, R. (1996) Refugees and environmental change: the case of the Forest Region of Guinea. Unpublished Project CFCE Report No. 2, University of Sussex, Brighton

Black, R. (1997) Refugees, land cover, and environmental change in the Senegal River Valley. *Geojournal*, **41**(1), 55–67

Blaikie, P. (1985) *The political economy of soil erosion in developing countries*. London: Longman

Blaikie, P. and Brookfield, H. (eds) (1987) *Land Degradation and Society*. London: Methuen

Blaut, J. (1993) *The Colonizers' Model of the World*. London: Guildford

Bogert, C. (1995) Midlife Crisis, *Newsweek*, 30 October, 12–18

Borchert, J.R. (1967) American metropolitan evolution. *Geographical Review*, **57**, 301–23

Bongaarts, J. (1994) Demographic transition, in Eblen, R.A. and Eblen, W.R. (eds) *Encyclopedia of the Environment*. Boston: Houghton-Mifflin, 132

Bongaarts, J. (1995) Global and regional population projections to 2025, in Islam, N. (ed) *Population and food in the early twenty-first century: meeting future food demand of an increasing population*. Washington DC: International Food Policy Research Institute, 7–16

Booth, D. (1985) Marxism and development sociology: interpreting the impasse. *World Development*, **13**, 761–87

Booth, D. (1993) Development research: from impasse to new agenda, in Schurmann, F. (ed) *Beyond the Impasse: New Directions in Development Theory*. London: Zed Books

Booth, K. (1997) Exporting ethics in place of arms. *The Times Higher*, 7 November, 118

Boserup, E. (1965) *The conditions of agricultural growth: the economics of agricultural change under population pressure*. London: Allen & Unwin

Boserup, E. (1993) *The Conditions of Agricultural Growth*. London: Earthscan (first published in 1965)

Botes, L.J. (1996) Promoting community participation in development initiatives. Paper presented to the Development Studies Association Annual Conference, University of Reading

Boyden, J. and Holden P. (1991) *Children of the Cities*. London: Zed Books

Brandt, W. (1980) *North–South: a programme for survival*. London: Pan

Brandt, W. (1983) *Common Crisis. North–South: co-operation for world recovery*. London: Pan

Brazier, C. (1994) Winds of change. *New Internationalist*, **262**, 4–7

Brierley, J. (1989) A review of development strategies and programmes of the People's Revolutionary Government in Grenada, 1979–83. *Geographical Journal*, **151**, 40–52

Brierley, J.S. (1985a) Idle land in Grenada: a review of its causes and the PRG's approach to reducing the problem. *Canadian Geographer*, **29**, 298–309

Brierley, J.S. (1985b) The agricultural strategies and programmes of the People's Revolutionary Government in Grenada, 1979–1983, in *Conference of Latin American Geographers Yearbook*, 55–61

Brohman, J. (1996) *Popular Development: Rethinking the Theory and Practice of Development*. Oxford: Blackwell

Brookfield, H. (1975) *Interdependent Development*. London: Methuen

Brookfield, H. (1978) Third World Development. *Progress in Human Geography*, **2**(1), 121–32

Brown, L.R. (ed) (1996a) *Vital signs, 1996/1997: the trends that are shaping our future*. London: Earthscan

Brown, L.R. (1996b) *The potential impact of AIDS on population and economic growth rates*, Food, Agriculture and the Environment, Discussion Paper 15. Washington DC: International Food Policy Research Institute

Browne, A.W. and Barrett, H.R. (1995) *Children and AIDS in Africa*, African Studies Centre Paper 2. Coventry: Coventry University

Bryant, R.L. and Bailey, S. (1997) *Third world political ecology*. London: Routledge

Buchanan, K. (1964) Profiles of the Third World. *Pacific Viewpoint*, **5**(2), 97–126

Buckley, R. (1994) *NAFTA and GATT: the impact of free trade*, Understanding global issues series, 94/2. Cheltenham: European Schoolbooks

Buckley, R. (1995) *The United Nations: overseeing the new world order*, Understanding global issues series, 93/6. Cheltenham: European Schoolbooks

Buckley, R. (ed) (1996) *Fairer Global Trade: the challenge for the WTO, Understanding Global Issues 96/6*. Cheltenham: Understanding Global Issues Ltd

Burgess, R. (1990) The State and Self-Help Building in Pereira, Colombia. Unpublished PhD thesis, University of London

Burgess, R. (1992) Helping some to help themselves: Third World housing policies and development strategies, in Mathéy, K. (ed) *Beyond Self-Help Housing*. London: Mansell, 75–91

Burgess, R., Carmona, K. and Kolstree, T.C. (1997) *The Challenge of Sustainable Cities*. London: Zed Books

Campbell, D.J., Zinyama, L.M. and Matiza, T. (1991) Coping with food deficits in rural Zimbabwe: the sequential adoption of indigenous strategies. *Research in Rural Sociology and Development*, **5**, 73–85

Cardoso, F.H. (1976) The consumption of dependency theory in the United States, in *Proceedings of the Third Scandinavian Research Conference on Latin America*, Bergen

Castells, M. (1977) *The Urban Question: A Marxist Approach*. London: Edward Arnold

Castells, M. (1978) Urban social movements and the struggle for democracy. *International Journal of Urban and Regional Research*, **1**, 133–46

Castells, M. (1983) *The city and the grassroots*. London: Edward Arnold

Cater, E. (1992) Must tourism destroy its resource base? in Mannion, A.M. and Bowlby, S.R. (eds) *Environmental issues in the 1990s*. London: John Wiley, 309–24

Chambers, R. (1983) *Rural Development: Putting the Last First*. London: Longman

Chambers, R. (1993) *Challenging the Professions: frontiers for rural development*. London: Intermediate Technology Publications

Chambers, R. (1997) *Whose Reality Counts?* London: Intermediate Technology Publications

Chambers, R., Pacey, A. and Thrupp, L.A. (1989) *Farmer first*. London: Intermediate Technology Publications

Chandra, R. (1992) *Industrialization and Development in the Third World*. London and New York: Routledge

Chant, S. (1996) *Gender, Uneven Development and Housing*. New York: UNDP

Chatterjee, P. (1994) Riders of the apocalypse. *New Internationalist*, **262**, 10–11

Choguill, C. (1994) Crisis, chaos, crunch: planning for urban growth in the developing world. *Urban Studies*, **31**, 935–45

Christaller, W. (1933) *Die zentralen Onte in Suddeutschland*. Doctoral thesis translated by Baskin, C.W. (1966) *Central Places in Southern Germany*. Eaglewood Cliffs: Prentice Hall

Christanty, L. (1986) Traditional agroforestry in West Java: the *pekarangan* (home garden) and *kebuntalun* (annual perennial rotation) cropping systems, in Marten, G.G. (ed) *Traditional agroforestry in Southeast Asia: a human ecology perspective*. Boulder CO: Westview, 132–58

Chuta, E. and Liedholm, C. (1990) Rural small-scale industry: empirical evidence and policy issues, in Eicher, C.K. and Staatz, J.M. (eds) *Agricultural development in the Third World*, 2nd edn. Baltimore MD: Johns Hopkins University Press, 327–41

Clapham, C. (1985) *Third World Politics*. London: Croom Helm

Clayton, A. and Potter, R.B. (1996) Industrial development and foreign direct investment in Barbados. *Geography*, **81**, 176–80

Clayton, K. (1995) The threat of global warming, in O'Riordan, T. (ed) *Environmental science for environmental management*. London: Longman, 110–31

Cliff, A.D. and Smallman-Raynor, M.R. (1992) The AIDS pandemic: global geographical patterns and local spatial processes. *Geographical Journal*, **158**(2), 182–98

Clifford, M. (1994) Social engineers. *Far Eastern Economic Review*, 14 April, 56–58

Cochrane, A. (1995) Global worlds and worlds of difference, in Anderson, J., Brook, C. and Cochrane, A. (eds) *A Global World?* Oxford: Oxford University Press and Open University, 249–80

Colchester, M. (1991) Guatemala: the clamour for land and the fate of the forests. *The Ecologist*, **21**(4), 177–85

Colchester, M. and Lohmann, L. (eds) (1993) *The struggle for land and the fate of the forests*. London: Zed Books

Conroy, C. and Litvinoff, M. (1988) *The Greening of Aid: sustainable livelihoods in practice*. London: Earthscan

Conway, G. (1997) *The doubly green revolution: food for all in the 21st century*. London: Penguin

Conway, G. and Barbier, E. (1995) Pricing policy and sustainability in Indonesia, in Kirkby, J., O'Keefe, P. and Timberlake, L. (eds) *The Earthscan reader in sustainable development*. London: Earthscan, 151–7

Cooke, P. (1990) Modern urban theory in question, *Transactions of the Institute of British Geographers, New Series*, **15**, 331–43

Corbridge, S. (1986) *Capitalist World Development*. London: Macmillan

Corbridge, S. (1992) Third World development. *Progress in Human Geography*, **16**(54), 584–95

Corbridge, S. (1993a) Colonialism, post-colonialism and the Third World, in Taylor, P. (ed) *Political Geography of the Twentieth Century*. London: Belhaven, 173–205

Corbridge, S. (1993b) Marxisms, modernities and moralities: development praxis and the claims of distant strangers. *Environment and Planning D*, **11**, 449–72

Corbridge, S. (ed) (1995) *Development Studies: A Reader*. London: Edward Arnold

Courier, The (1990) *Irrigation*, **124**, 64–95

Courier, The (1996) Country report – Kenya. *The Courier*, **157**, 19–36

Courtenay, P.P. (ed) (1994) *Geography and Development*. Melbourne: Longman Cheshire

Cowan, M.P. and Shenton, R. (1995) The invention of development, in Crush, J. (ed) *Power of Development*. London: Routledge, 27–43

Cowan, M.P. and Shenton, R.W. (1996) *Doctrines of Development*. London: Routledge

Craig, G. and Mayo, M. (eds) (1995) *Community empowerment: a reader in participation and development*. London: Zed Books

Crehan, K. (1992) Rural households: making a living, in Bernstein, H., Crow, B. and Johnson, H. (eds) *Rural livelihoods: crises and responses*. Oxford: Oxford University Press, 87–112

Crook, C. (1991) Two pillars of wisdom, *The Economist*, 12 October, 3–4

Crook, N. (1997) *Principles of population and development*. Oxford: Oxford University Press

Crow, B. (1992) Rural livelihoods: action from above, in Bernstein, H., Crow, B. and Johnson, H. (eds) *Rural livelihoods: crises and responses*. Oxford: Oxford University Press, 251–74

Crush, J. (ed) (1995) *Power of Development*. London: Routledge

Crush, J. (1995) Imagining development, in Crush, J. (ed) *Power of Development*. London: Routledge, 1–26

Cuthbert, A. (1995) Under the volcano: postmodern space in Hong Kong, in Watson, S. and Gibson, K. (eds) *Postmodern Cities and Space*. Oxford: Blackwell, 138–48

Dandekar, H.C. (1997) Changing migration strategies in Deccan Maharashtra, India, 1885–1990, in Gugler, J. (ed) *Cities in the Developing World*. Oxford: Oxford University Press, 48–61

Dankelman, I. and Davidson, J. (1988) *Women and environment in the Third World: alliance for the future*. London: Earthscan

Dann, G.M.S. and Potter, R.B. (1994) Tourism and postmodernity in a Caribbean setting. *Cahiers du Tourisme, Series C*, **185**, 1–45

Dann, G.M.S. and Potter, R.B. (1997) Tourism in Barbados: rejuvenation or decline? in Lockhart, D.G. and Drakakis-Smith, D. (eds) *Island Tourism: Trends and Prospects*. London: Mansell, 205–28

Datta, G. and Meerman, J. (1980) *Household income and household income per capita in welfare comparisons*, World Bank Staff Working Paper 378. Washington DC: World Bank

Davin, D. (1996) Migration and rural women in China: a look at the gendered impact of large-scale migration. *Journal of International Development*, **8**(5), 655–65

Debray, R. (1974) *A Critique of Arms*. Paris: Seuil

de Albuquerque, K. (1996) Computer technologies and the Caribbean. *Caribbean Week*, **8**, 32–33.

de Janvry, A. (1984) The role of land reform in economic development, in Eicher, C. and Staatz, J.M. (eds) *Agricultural Development in the Third World*. Baltimore MD: John Hopkins University Press, 262–77

DFID (Department for International Development) (1997) *White Paper on Eliminating World Poverty: a challenge for the twenty first century*. London: Government Stationery Office

Devas, N. and Rakodi, C. (eds) (1993) *Managing Fast Growing Cities: New Approaches to Urban Planning and Management in the Developing World*. Harlow: Longman

Dey, J. (1981) Gambian women: unequal partners in rice development projects? *Journal of Development Studies*, **17**(3), 109–22

Dicken, P. (1992) *Global shift: the internationalization of economic activity. (second edition)* London: Paul Chapman

Dicken, P. (1993) The growth economies of Pacific Asia in their changing global context, in Dixon, C. and Drakakis-Smith, D. (eds) *Economic and social development in Pacific Asia*. London: Routledge, 22–42

Dicken, P. (1998) *Global Shift. (third edition)* London: Paul Chapman

Dickenson, J.P. (1994) Manufacturing industry in Latin America and the case of Brazil, in Courtenay, P.P. (ed.) *Geography and Development*, 165–191

Dickenson, J., Gould, B., Clarke, C., Mather, C., Prothero, M., Siddle, D., Smith, C. and Thomas-Hope, E. (1996) *A Geography of the Third World*, 2nd edn. London: Routledge

Dixon, C. (1990) *Rural development in the Third World*. London: Routledge

Dixon, C. (1998) *Thailand*. London: Routledge

Dixon, C. and Drakakis-Smith, D. (eds) (1997) *Uneven Development in Southeast Asia*. Aldershot: Ashgate

Dixon, C. and Heffernan, M. (eds) (1991) *Colonialism and Development in the Contemporary World*. London: Mansell

Donaghue, M.T. and Barff, R. (1990) Nike just did it: international subcontracting, flexibility and athletic footwear production. *Regional Studies*, **24**, 537–52

Dooge, J.C.I. (1992) *An agenda for science for environment and development in a changing world*. Cambridge: Cambridge University Press

Dos Santos, T. (1970) The structure of dependency. *American Economic Review*, **60**, 125–158

Dos Santos, T. (1977) Dependence relations and political development in Latin America: some considerations. *Ibero-Americana*, **7**

Dowdeswell, E. (undated) Editorial. *Our Planet*, **6**(5), 2

Drakakis-Smith, D. (1981) *Urbanization, Housing and the Development Process*. London: Croom Helm

Drakakis-Smith, D. (1983) Advance Australia fair: internal colonialism in the Antipodes, in Drakakis-Smith, D. and Wyn Williams, S. (eds) *Internal Colonialism: Essays Around a Theme*. Developing Areas Research Group, Institute of British Geographers, Monograph 3, 81–103

Drakakis-Smith, D. (1987) *The Third World City*. London: Methuen

Drakakis-Smith, D. (1989) Urban social movements and the built environment. *Antipode*, **21**(3), 207–31

Drakakis-Smith, D. (1990) Food for thought or thought about food: urban food distribution systems in the Third World, in Potter, R.B. and Salau, A.T. (eds) *Cities and Development*. London: Mansell, 100–20

Drakakis-Smith, D. (1991) Colonial urbanization in Africa and Asia: a structural review. *Cambria*, **16**, 123–50

Drakakis-Smith, D. (1992) *Pacific Asia*. London: Routledge

Drakakis-Smith, D. (1995) Third World cities: sustainable urban development I. *Urban Studies*, **32**, 659–77

Drakakis-Smith, D. (1996) Third World cities: sustainable urban development II. *Urban Studies*, **33**, 673–701

Drakakis-Smith, D. (1997a) Sustainable urbanisation in Vietnam, *Geoforum*, **28**(1), 21–38

Drakakis-Smith, D. (1997b) Third World cities: sustainable urban development III. *Urban Studies*, **34**(5/6), 797–823

Drakakis-Smith, D. and Dixon, C. (1997) Sustainable urbanisation in Vietnam. *Geoforum*, **28**(1), 21–38

Drakakis-Smith, D., Doherty, J. and Thrift, N. (1987) What is a socialist developing country? *Geography*, **72**(4), 333–5

Drakakis-Smith, D., Graham, E., Teo, P. and Ling, O.G. (1993) Singapore: reversing the demographic transition to meet labour needs. *Scottish Geographical Magazine*, **109**, 152–63

Dwyer, D.J. (1975) *People and Housing in Third World Cities*. London: Longman

Dwyer, D.J. (1977) Economic development: development for whom? *Geography*, **62**(4), 325–34

Economist Intelligence Unit (1996a) *Country Profile: Côte d'Ivoire*. London: EIU

Economist Intelligence Unit (1996b) *Country Profile: Kenya*. London: EIU

Eden, M.J. and Parry, J. (eds) (1996) *Land Degradation in the Tropics: Environment and Policy Issues*. London: Mansell

Edwards, M. and Hulme, D. (eds) (1992) *Making a difference: NGOs and development in a changing world*. London: Earthscan

Edwards, M. and Hulme, D. (eds) (1995) *Non-governmental organisations – performance and accountability: beyond the magic bullet*. London: Earthscan

Ehrlich, P.R. (1968) *The Population Bomb*. New York: Ballantine Books

Eicher, C.K. and Staatz, J.M. (eds) (1990) *Agricultural development in the Third World*, 2nd edn. Baltimore MD: Johns Hopkins University Press

Elliott, J.A. (1990) The mechanical conservation of soil in Zimbabwe, in Cosgrove, D. and Petts, G. (eds) *Water, Engineering and Landscape*. London: Belhaven, 115–28

Elliott, J.A. (1991) Environmental degradation, soil conservation and the colonial and post-colonial state in Rhodesia/Zimbabwe, in Dixon, C. and Heffernan, M. (eds) *Colonialism and development in the contemporary world*. London: Mansell, 72–91

Elliott, J.A. (1994) *An introduction to sustainable development: the developing world*. London: Routledge

Elliott, J.A. (1995) Government policies and the population–environment interface: land reform and distribution in Zimbabwe, in Binns, T. (ed) *People and environment in Africa*. Chichester: John Wiley, 225–30

Elsom, D. (1996) *Smog alert: managing urban air quality*. London: Earthscan

Endicott, S. (1988) *Red Earth: Revolution in a Sichuan Village*. London: I.B. Tauris

Escobar, A. (1995) *Encountering Development*. Princeton NJ: Princeton University Press

Estes, R. (1984) World social progress, 1969–1979. *Social Development Issues*, **8**, 8–28

Esteva, G. (1992) Development, in Sachs, W. (ed) *The Development Dictionary*. London: Zed Books, 6–25

Evans, R. (1993) Reforming the union. *Geographical Magazine*, February, 24–27

Eyre, J. and Dwyer, D.J. (1996) Ethnicity and uneven development in Malaysia, in Dwyer, D.J. and Drakakis-Smith, D. (eds) *Ethnicity and Development*. London: John Wiley, 181–94

Fadl, O.A.A. (1990) Gezira: the largest irrigation scheme in Africa, *The Courier*, Nov/Dec, 91–95

Fairhead, J. and Leach, M. (1995) Local agro-ecological management and forest–savanna transitions: the case of Kissidougou, Guinea, in Binns, T. (ed) *People and environment in Africa*. Chichester: John Wiley, 163–70

FAO (Food and Agriculture Organisation) (1980)

FAO (Food and Agriculture Organisation) (1987) *Consultation on irrigation in Africa*. Irrigation and Drainage Paper 42, Rome: FAO

Ferguson, J. (1990) *Grenada: Revolution in Reverse*. London: Latin American Bureau

Foley, G. (1991) *Global warming: who is taking the heat?* London: Panos

Frank, A.G. (1966) The development of underdevelopment, *Monthly Review*, September, 17–30

Frank, A.G. (1967) *Capitalism and Underdevelopment in Latin America*. New York: Monthly Review Press

Frank, A.G. (1980) North–South and East–West paradoxes in the Brandt Report. *Third World Quarterly*, **2**(4), 669–80

Friedland, W.H. (1994) The global fresh fruit and vegetable system: an industrial organization analysis, in McMichael, P. (ed) *The global restructuring of agro-food systems*. Ithaca NY: Cornell University Press, 173–89

Friedmann, J. (1966) *Regional Development Policy: A Case Study of Venezuela*. Cambridge MA: MIT Press

Friedmann, J. (1986) The world city hypothesis. *Development and Change*, **17**, 69–83

Friedmann, J. (1992) *Empowerment: the Politics of Alternative Development*. Oxford: Blackwell

Friedmann, J. (1995) Where we stand: a decade of world city research, in Knox, P.L. and Taylor, P.J. (eds) *World Cities in a World-System*. Cambridge: Cambridge University Press, 21–37

Friedmann, J. and Weaver, C. (1979) *Territory and Function: the Evolution of Regional Planning*. London: Edward Arnold

Friedmann, J. and Wulff, G. (1982) World city formation: an agenda for research and action. *International Journal of Urban and Regional Research*, **6**, 309–43

Friedmann, M. (1962) *Capitalism and Freedom*. Chicago IL: University of Chicago Press

Fullerton-Joireman, S. (1996) The minefield of land reform: comments on the Eritrean Land Proclamation. *African Affairs*, **95**, 269–85

Furtado, C. (1964) *Development and Underdevelopment*. Berkeley CA: University of California Press

Furtado, C. (1965) *Diagnosis of the Brazilian Crisis*. Berkeley CA: University of California Press

Furtado, C. (1969) *Economic Development in Latin America*. Cambridge: Cambridge University Press

Galaty, J.G. and Johnson, D.L. (1990) Introduction: pastoral systems in global perspective, in Galaty, J.G. and Johnson, D.L. (eds) *The world of pastoralism*. London: Belhaven, 1–31

Gale, D.J. and Goodrich, J.N. (eds) (1993) *Tourism Marketing and Management in the Caribbean*. London: Routledge

Geheb, K. (1995) Exploring people–environment relationships: the changing nature of the small-scale fishery in the Kenyan sector of Lake Victoria, in Binns, T. (ed) *People and Environment in Africa*. Chichester: John Wiley, 91–101

Geheb, K. and Binns, T. (1997) 'Fishing farmers' or 'farming fishermen'? The quest for household income and nutritional security on the Kenyan shores of Lake Victoria. *African Affairs*, **96**, 73–93

German, T. and Randel, J. (eds) (1993) *The reality of aid*. London: Actionaid

Getis, A., Getis, J. and Fellman, J. (1994) *Introduction to Geography*, 4th edn. Dubuque IA: William C. Brown

Ghee, L.T. (1989) Reconstituting the peasantry: changes in landholding structure in the Muda irrigation scheme, in Hart, G., Turton, A. and White, B. (eds) *Agrarian transformations: local processes and the state in Southeast Asia*. Berkeley CA: University of California Press, 193–212

Gholz, H.L. (ed) (1987) *Agroforestry: realities, possibilities and potentials*. Dordrecht: Martinus Nijhoff

Ghose, A.K. (ed) (1983) *Agrarian reform in contemporary developing countries*. London: Croom Helm

Gibbon, D. (ed) (1995) *Structural Adjustment and the Working Poor in Zimbabwe*. Uppsala: Nordic Institute for African Studies

Gilbert, A.G. (1992) Third World cities: housing, infrastructure and servicing. *Urban Studies*, **29**, 435–60

Gilbert, A.G. (1993) Third World cities: the changing national settlement system. *Urban Studies*, **30**, 721–40

Gilbert, A.G. (1994) Third World cities: poverty, employment, gender roles and the environment during a time of restructuring. *Urban Studies*, **31**, 605–33

Gilbert, A.G. and Gugler, J. (1982) *Cities, poverty and development: urbanization in the Third World*. Oxford: Oxford University Press

Gilbert, A.G. (1976) The arguments for very large cities reconsidered. *Urban Studies*, **13**, 27–34

Gilbert, A.G. (1977) The argument for very large cities reconsidered: a reply. *Urban Studies*, **14**, 225–7

Gilbert, A.G. (1996) *The Mega-City in Latin America*. Tokyo: United Nations University Press

Gilbert, A.G. and Goodman, D.E. (1976) Regional income disparities and economic development, in Gilbert, A.G. (ed) *Development Planning and Spatial Structure*. Chichester: John Wiley

Gilbert, A.G. and Gugler, J. (1992) *Cities, Poverty and Development: Urbanization in the Third World*, 2nd edn. Oxford: Oxford University Press

Girvan, N. (1973) The development of dependency economics in the Caribbean and Latin America: review and comparison. *Social and Economic Studies*, **22**, 1–33

Gleave, M.B. (1996) The length of the fallow period in tropical fallow farming systems: a discussion with evidence from Sierra Leone. *Geographical Journal*, **162**(1), 14–24

Gottmann, J. (1978) Megalopolitan systems around the world, in Bourne, L.S. and Symmons, J.W. (eds) *Systems of Cities*. Oxford: Oxford University Press, 53–60

Goudie, A. (1990) *The human impact on the natural environment*, 3rd edn. London: Blackwell

Gould, P. (1969) The structure of space preferences in Tanzania. *Area*, **1**, 29–35

Gould, P. (1970) Tanzania, 1920–63: the spatial impress of the modernisation process. *World Politics*, **22**, 149–70

Gould, P. and White, R. (1974) *Mental Maps*, Harmondsworth: Penguin

Gould, W.T.S. (1992) Urban development and the World Bank. *Third World Planning Review*, **14**, iii–vi

Gould W.T.S. (1993) *People and education in the Third World*. Harlow: Longman

Gould W.T.S. and Prothero, R.M. (1975) Space and time in African population mobility, in Kosinski, L.A. and Prothero, R.M. (eds) *People on the Move*. London: Methuen, 39–49

Graham, E. (1995) Singapore in the 1990s: can population policies reverse the demographic transition? *Applied Geography*, **15**, 219–32

Grainger, A. (1993) *Controlling tropical deforestation*. London: Earthscan

Grant, M.C. (1995) Movement patterns and the intermediate sized city. *Habitat International*, **19**, 357–70

Griffin, K. (1980) Economic development in a changing world. Annual Lecture of the Development Studies Association, University of Swansea

Griffiths, I.L.L. (1994) *The Atlas of African Affairs*, 2nd edn. London: Routledge

Grigg, D. (1995) *An introduction to agricultural geography*, 2nd edn. London: Routledge

Gugler, J. (ed) (1996) *The Urban Transformation of the Developing World*. Oxford: Oxford University Press

Gugler, J. (1997) Over-urbanization reconsidered, in Gugler, J. (ed) *Cities in the Developing World*. Oxford: Oxford University Press, 114–23

Gupta, A. (1988) *Ecology and Development in the Third World*. London: Methuen

Gutkind, P.C.W. (1969) Tradition, migration, urbanization, modernity and unemployment in Africa: the roots of instability. *Canadian Journal of African Studies*, **3**, 343–65

Habitat (1996) *An urbanising world: global report on human settlements, UN Centre for Human Settlements*. Oxford: Oxford University Press

Haddad, L. (1992) Introduction, in *Understanding how resources are allocated within households*. Washington DC: International Food Policy Research Institute

Hagerstrand, T. (1953) *Innovationsforloppet ur Korologisk Synpunkt*. Lund: University of Lund

Haggett, P. (1990) *Geography: A Modern Synthesis*. London: Harper and Row

Hall, S. (1995) New cultures for old, in Massey, D. and Jess, P. (eds) *A Place in the World?* Oxford: Oxford University Press and Open University, 175–213

Hancock, G. (1997) Transmigration in Indonesia: how millions are uprooted, in Rahnema, M. and Bawtree, V. (eds) *The Post-Development Reader*. London: Zed Books, 234–43 (Reprinted from Hancock, G. 1989, *Lords of Poverty*. London: Macmillan)

Hansen, N.M. (1981) Development from above: the centre-down development paradigm, in Stöhr, W.B. and Taylor, D.R.F. (eds) *Development from Above or Below? The Dialectics of Regional Development in Developing Countries*. Chichester: John Wiley

Hardin, G. (1968) The tragedy of the commons. *Science*, **162**, 1243–8

Hardoy, J.E., Cairncross, S. and Satterthwaite, D. (eds) (1990) *The poor die young: housing and health in Third World cities*. London: Earthscan

Harmsen, R. (1995) The Uruguay Round: a boon for the world economy. *Finance and Development*, March, 24–26

Harris, N. (1989) Aid and urbanization. *Cities*, **6**, 174–85

Harris, N. (1992) *Cities in the 1990s: The Challenge for Developing Countries*. London: UCL Press

Harrison, D. (ed) (1992) *Tourism and the less developed countries*. London: Belhaven

Harrison, P. and Palmer, R. (1986) *News Out of Africa: Biafra to Band Aid*. London: Hilary Shipman

Harriss, B. and Crow, B. (1992) Twentieth century free trade reform: food market deregulation in sub-Saharan Africa and south Asia, in Wuyts, M., Mackintosh, M. and Hewitt, T. (eds) *Development policy and public action*. Oxford: Oxford University Press, 199–227

Harriss, J. and Harriss, B. (1979) Development studies. *Progress in Human Geography*, **3**(4), 577–82

Harvey, D. (1973) *Social Justice and the City*. London: Edward Arnold

Harvey, D. (1989) *The Condition of Postmodernity*. Oxford: Blackwell

Heathcote, R.L. (1983) *The arid lands: their use and abuse*. London: Longman

Henning, R.O. (1941) The furrow makers of Kenya. *Geographical Magazine*, **12**, 268–79

Hentati, A. (undated) Taking effective action. *Our Planet*, **6**(5), 5–7

Hettne, B. (1990) *Development Theory and the Three Worlds*. London: Longman

Hettne, B. (1995) *Development Theory and the Three Worlds*, 2nd edn. Harlow: Longman

Hewitt, T., Johnson, H. and Wield, D. (eds) (1992) *Industrialization and Development*. Oxford: Oxford University Press and the Open University

Hiebert, M. (1993) Long shot? *Far Eastern Economic Review*, 14 October, 58

Hildyard, N. (1994) The big brother bank, *Geographical*, June, 26–8

Hill, P. (1963) *The migrant cocoa-farmers of Southern Ghana: A study in rural capitalism*. Cambridge: Cambridge University Press

Hill, P. (1972) *Rural Hausa: a village and a setting*. Cambridge: Cambridge University Press

Hill, R. O'Keefe, P. and Snape, C. (1995) Energy planning, in Kirkby, J., O'Keefe, P. and Timberlake, L. (eds) *The Earthscan reader in sustainable development*. London: Earthscan, 78–101

Hirschman, A.O. (1958) *The Strategy of Economic Development*. New Haven CT: Yale University Press

Hoch, I. (1972) Income and city size. *Urban Studies*, **9**, 299–328

Hodder, R. (1992) *The West Pacific Rim*. London: Belhaven

Horvath, R. (1988) National development paths 1965–1987: measuring a metaphor. Paper presented to the International Geographic Congress, Sydney University

Hossain, M. (1988) *Credit for alleviation of rural poverty: the Grameen Bank in Bangladesh*, Research Report 65. Washington DC: International Food Policy Research Institute

Hoyle, B.S. (1979) African socialism and urban development: the relocation of the Tanzanian capital. *Tijdschrift voor Economische en Sociale Geografie*, **70**, 207–16

Hoyle, B.S. (1993) The 'tyranny' of distance – transport and the development process, in Courtney, P.P. (ed) *Geography and Development*. Melbourne: Longman Cheshire, 117–43

Hudson, B. (1989) The Commonwealth Eastern Caribbean, in Potter, R.B. (ed) *Urbanization, Planning and Development in the Caribbean*. London and New York: Mansell

Hudson, B. (1991) Physical planning in the Grenada Revolution: achievement and legacy, *Third World Planning Review*, **13**, 179–90

Hudson, J.C. (1969) Diffusion in a central place system. *Geographical Analysis*, **1**, 45–58

Hughes, J.M.R. (1992) Use and abuse of wetlands, in Mannion, A.M. and Bowlby, S.R. (eds) *Environmental issues in the 1990s*. London: John Wiley, 211–26

Hunt, D. (1984) *The impending crisis in Kenya: the case for land reform*. London: Gower

Huntington, E (1945) *Mainsprings of civilisation*, New York: John Wiley & Sons

Hutton, W. (1993) Gatt's principles have been corrupted by free market nihilism, *The Guardian*, 16 November

ICPQL (Independent Commission on Population and Quality of Life) (1996) *Caring for the future*. Oxford: Oxford University Press

IFPRI (International Food Policy Research Institute) (1995a) *A 2020 Vision for Food, Agriculture, and the Environment in Latin America: A Synthesis*. Washington DC: IFPRI

IFPRI (International Food Policy Research Institute) (1995b) *A 2020 Vision for Food, Agriculture, and the Environment in South Asia: A Synthesis*. Washington DC: IFPRI

IPCC (Intergovernmental Panel on Climatic Change) (1990) *Climate change: the IPCC assessment*. Cambridge: Cambridge University Press

Ignatieff, M. (1995) Fall of a blue empire. *The Guardian*, 17 October

Independent, The (1998) The population bomb defused. *The Independent*, 12 January

Jaffee, S. (1994) *Exporting high value food commodities*. Washington DC: World Bank

Jain, P.S. (1996) Managing credit for the rural poor: lessons from the Grameen Bank. *World Development*, **24**(1), 79–89

Jamal, V. and Weeks, J. (1994) *Africa misunderstood: or whatever happened to the rural–urban gap?* Basingstoke: Macmillan

Jameson, F. (1984) Postmodernism, or the cultural logic of late capitalism. *New Left Review*, **146**, 53–92

Janelle, D.G. (1969) Spatial reorganization: a model and a concept. *Annals of the Association of American Geographers*, **59**, 348–64

Janelle, D.G. (1973) Measuring human extensibility in a shrinking world. *Journals of Geography*, **72**, 8–15

Jenkins, R. (1987) *Transnational Corporations and Uneven Development*. London: Methuen

Jenkins, R. (1992) Industrialization and the global economy, in Hewitt, T., Johnson, H. and Wield, D. (eds) *Industrialization and Development*. Oxford: Oxford University Press in association with the Open University

Jiminez-Diaz, V. (1994) The incidence and causes of slope failures in the barrios of Caracas, Venezuela, in Main, H. and Williams, S.W. (eds) *Environment and Housing in Third World Cities*. Chichester: Wiley

Johnson, B.L.C. (1983) *India: Resources and Development*, 2nd edn. London: Heinemann

Johnston, R.J. (1996) *Nature, state and economy: a political economy of the environment*, 2nd edn. Chichester: John Wiley

Jones, E. and Eyles, J. (1977) *An Introduction to Social Geography*. Oxford: Oxford University Press

Jones, G. and Hollier, G. (1997) *Resources, Society and Environmental Management*. London: Paul Chapman

Jones, J.P., Natter, W. and Schatzki, T.R. (1993) *Postmodern Contentions: Epochs, Politics, Space*. London: Guildford Press

Jowett, J. (1990) People: demographic patterns and policies, in Cannon, T. and Jenkins, A. (eds) *The geography of contemporary China: the impact of Deng Xiaoping's decade*. London: Routledge, 102–32

Kaarsholm, P. (ed) (1995) *From Post-Traditional to Post-Modern? Interpreting the Meaning of Modernity in Third World Urban Societies*, Occasional Paper 14, International Development Studies, Roskilde University

Kabbani, R. (1986) *Imperial Fictions*. London: Pandora

Kabeer, N. (1992) Beyond the threshold: intrahousehold relations and policy perspectives, in *Understanding how resources are allocated within households*. Washington DC: International Food Policy Research Institute, 51–52

Kats, G. (1992) Achieving sustainability in energy use in developing countries, in Holmberg, J. (ed) *Policies for a small planet*. London: Earthscan, 258–89

Keeling, D.J. (1995) Transport and the world city paradigm, in Knox, P.L. and Taylor, P.J. (eds) *World Cities in a World-System*. Cambridge: Cambridge University Press, 115–31

Kelly, M. and Granich, S. (1995) Global warming and development, in Morse, S. and Stocking, M. (eds) *People and environment*. London: UCL Press, 69–107

Kennedy, E. and Bouis, H.E. (1993) *Linkages between agriculture and nutrition: implications for policy and research*. Washington DC: International Food Policy Research Institute

Killick, A. (1990) Whither development economics? *Economics*, **26**(2), 62–69

Kimmage, K. (1991) Small-scale irrigation initiatives in Nigeria: the problems of equity and sustainability. *Applied Geography*, **11**(5), 5–20

King, A. (1976) *Colonial Urban Development*. London: Routledge & Kegan Paul

King, A. (1990) *Colonialism, Urbanism and the World Economy*. London: Routledge

King, A. (1994) *Urbanism, Colonialism and the World Economy*. London: Routledge

Kirkby, J., O'Keefe, P. and Timberlake, L. (eds) (1995) *The Earthscan reader in sustainable development*. London: Earthscan

Kirton, C.D. (1988) Public policy and private capital in the transition to socialism: Grenada 1979–85, *Social and Economic Studies*, **37**, 125–50

Knight, J.B. (1972) Rural–urban income comparisons and migration in Ghana. *Bulletin of the Oxford University Institute of Economics and Statistics*, **34**(2), 199–229

Knox, P.L. and Taylor, P.J. (eds) (1995) *World Cities in a World-System*. Cambridge: Cambridge University Press

Komin, S. (1991) Social dimensions of industrialization in Thailand, *Regional Development Dialogue*, **12**, 115–37

Korten, D. (1990) *Voluntary organisations and the challenge of sustainable development*, Briefing Paper 15, Australia Development Studies Network, Australian National University, Canberra

Korten, D.C. (1990) *Getting to the twenty-first century: voluntary action and the global agenda*. Conneticut: Kumarian Press

Kuhn, T. (1962) *The Structure of Scientific Revolutions*. Chicago IL: University of Chicago Press

Kurien, J. (1988) Kerala Fishing-Boat Project, South India, in Conroy, C. and Litvinoff, M. (eds) *The Greening of Aid*. London: Earthscan, 108–12

Laing, R. and Pigott, M. (1987) Meeting the health and housing needs of plantation workers. *IDS Bulletin*, **18**(2), 23–29

Lasuen, J.R. (1973) Urbanisation and development – the temporal interaction between geographical and sectoral clusters. *Urban Studies*, **10**, 163–88

Leach, M. (1991) Locating gendered experience: an anthropologist's view from a Sierra Leonean village. *IDS Bulletin*, **22**(1), 44–50

Lee, J. and Bulloch, J. (1990) Spirit of war moves on Mid-East waters. *The Independent on Sunday*, 13 May, 13

Leeming, F. (1993) *The changing geography of China*. Oxford: Blackwell

Lefevre, A. (1995) *Islam, Human Rights and Child Labour*. Copenhagen: Nordic Institute of Asian Studies

Leinbach, T.R. (1972) The spread of modernization in Malaya: 1895–1969. *Tijdschrift voor Economische en Sociale Geografie*, **63**, 262–77

Lewis, W.A. (1950) The industrialisation of the British West Indies. Caribbean Economic Review, **2**, 1–61

Lewis, W.A. (1955) *The Theory of Economic Growth*. London: George Allen & Unwin

Leys, C. (1996) *The Rise and Fall of Development Theory*. London: James Currey

Leyshon, A. (1995) Annihilating space? The speed-up of communications, in Allen, J. and Hamnett, C. (eds) *A Shrinking World?* Oxford: Oxford University Press and the Open University, 11–54

Linsky, A.S. (1965) Some generalizations concerning primate cities. *Annals of the Association of American Geographers*, **55**, 506–13

Lipton, M. (1977) Why poor people *stay poor: Urban bias in world development*. London: Temple Smith

Littlejohns, M. and Silber, L. (1997) UN prepares to bite the bullet on reforms. *The Financial Times*, 16 September

Lloyd-Evans, S. and Potter, R.B. (1996) Environmental impacts of urban development and the urban informal sector in the Caribbean, in Eden, M.J. and Parry, J. (eds) *Land Degradation in the Tropics*. London: Mansell, 245–60

Lockhart, D. (1993) Tourism to Fiji: crumbs off a rich man's table? *Geography*, **78**(3), 318–23

Lohmann, L. (1993) Against the myths, in Colchester, M. and Lohmann, L. (eds) *The struggle for land and the fate of the forests*. London: Zed Books, 16–34

Longhurst, R. (1988) Cash crops and food security. *IDS Bulletin*, **19**(2), 28–36

Lösch, A. (1940) *Die räumliche Ordnung der Wirtschaft*, Jena, translated by Woglom, W.H. and Stolpen, W.F., 1954, *The Economics of Location*. New Haven CT: Yale University Press

Lowder, S. (1986) Inside Third World Cities, Beckenham: Croom Helm

Lowenthal, D. (1960) *West Indian Societies*. Oxford: Oxford University Press

Lundqvist, J. (1981) Tanzania: socialist ideology, bureaucratic reality, and development from below, in Stöhr, W.B. and Taylor, D.R. (eds) *Development from Above or Below?* Chichester: John Wiley, 329–49

MacCannell, D. (1976) *The Tourist: A New Theory of the Leisure Class*. New York: Schocken

McCormick, J. (1995) *The global environment movement*, 2nd edn. Chichester: John Wiley

McElroy, J.L. and Albuquerque, K. (1986) The tourism demonstration effect on the Caribbean. *Journal of Travel Research*, **25**, 31–4

MacEwan, A. (1982) Revolution, agrarian reform and economic transformation in Cuba, in Jones, S., Joshi, P.C. and Murmis, M. (eds) *Rural poverty and agrarian reform*. New Delhi: Allied Publishers 162–82

McGee, T. (1979) Conservation and dissolution in the Third World city: the 'shanty town' as an element of conservation. *Development and Change*, **10**, 1–22

McGee, T. (1994) The future of urbanisation in developing countries: the case of Indonesia. *Third World Planning Review*, **16**, iii–xii

McGee, T.G. (1967) *South East Asian City*. London: Bell

McGee, T.G. (1989) 'Urbanisasi' or Kotadesasi: evolving patterns of urbanisation in Asia, in Costa, F.J. (ed.) *Urbanization in Asia*. Honolulu: University of Hawaii Press.

McGee, T.G. (1995) Eurocentralism and geography, in Crush, J. (ed) *Power of Development*. London: Routledge, 192–207

McGee, T.G. (1997) The problem of identifying elephants: globalization and the multiplicities of development. Paper presented at the Lectures in Human Geography Series, University of St Andrews

McGee, T.G. and Greenberg, L. (1992) The emergence of extended metropolitan regions in ASEAN. *ASEAN Economic Bulletin*, **1**(6), 5–12

McGee, T.G. and Robinson, I. (eds) (1995) *The Mega-Urban Regions of Southeast Asia*. Vancouver: UBC Press

McIlwaine, C. (1997) Fringes or frontiers? Gender and export-oriented development in the Philippines, in Dixon, C. and Drakakis-Smith, D. (eds) *Uneven Development in Southeast Asia*. Aldershot: Ashgate, 100–23

Mackenzie, F. (1992) Development from within? The struggle to survive, in Taylor, D.R. and Mackenzie, F. (eds) *Development from within: survival in rural Africa*. London: Routledge, 1–33

Mackintosh, M. (1992) Questioning the state, in Wuyts, M., Mackintosh, M. and Hewitt, T. (eds) *Development policy and public action*. Oxford: Oxford University Press, 61–89

MacLeod, S. and McGee, T. (1990) The last frontier: the emergence of the industrial palate in Hong Kong, in Drakakis-Smith, D. (ed) *Economic Growth and Urbanization in Developing Areas*. London: Routledge

McLuhan, M. (1962) *The Gutenburg Galaxy: The Making of Typographic Man*, London: Routledge and Kegan Paul

Main, H. and Williams, S.W. (eds) (1994) *Environment and Housing in Third World Cities*. London: John Wiley

Makuch, Z. (1996) The World Trade Organisation and the General Agreement on Tariffs and Trade, in Werksman, J. (ed) *Greening international institutions*. London: Earthscan, 94–116

Maltby, E. (1986) *Waterlogged wealth: Why waste the world's wet places?* London: Earthscan

Mannion, A.M. and Bowlby, S.R. (eds) (1992) *Environmental issues in the 1990s*. London: John Wiley

Manzo, K. (1995) Black consciousness and the quest for counter-modernist development, in Crush, J. (ed) *Power of Development*. London: Routledge, 228–52

Masselos, J. (1995) Postmodern Bombay: fractured discourses, in Watson, S. and Gibson, K. (eds) *Postmodern Cities and Spaces*. Oxford: Blackwell, 200–15

Massey, D. (1991) A global sense of place, *Marxism Today*, June, 24–29

Massey, D. (1995) Imaging the world, in Allen, J. and Massey, D. (eds) *Geographical Worlds*. London: Oxford University Press, 5–52

Massey, D. and Jess, P. (1995) *A Place in the World? Places, Cultures and Globalization*. Oxford: Oxford University Press and the Open University

Mather, A.S. and Chapman, K. (1995) *Environmental resources*. London: Longman

Mayhew, S. (1997) *A Dictionary of Geography*, 2nd edn. Oxford: Oxford University Press, 122

Meadows, D.H., Meadows, D.L., Randers, J. and Behrens, W.W. (1972) *The Limits to Growth*. London: Pan

Mehmet, O. (1995) *Westernising the Third World*. London: Routledge

Mehta, S.K. (1964) Some demographic and economic correlates of primate cities: a case for revaluation. *Demography*, **1**, 136–47

Meier, G.M. and Baldwin, R.E. (1957) *Economic Development: Theory, History, Policy*. New York: John Wiley

Meillassoux, C. (1972) From reproduction to production. *Economy and Society*, **1**, 93–105

Meillassoux, C. (1978) The social organization of the peasantry: the economic basis of kinship, in Seddon, D. (ed) *Relations of Production: Marxist Approaches to Economic Anthropology*. London: Frank Cass, 159–70

Mellor, J.W. (1990) Agriculture on the road to industrialization, in Eicher, C.K. and Staatz, J.M. (eds) *Agricultural development in the Third World*, 2nd edn. Baltimore MD: Johns Hopkins University Press, 70–88

Mera, K. (1973) On the urban agglomeration and economic efficiency. *Economic Development and Cultural Change*, **21**, 309–24

Mera, K. (1975) *Income Distribution and Regional Development*. Tokyo: University of Tokyo Press

Mera, K. (1978) The changing pattern of population distribution in Japan and its implications for developing countries, in Lo, F.C. and Salih, K. (eds) *Growth Pole Strategies and Regional Development Policy*. Oxford: Pergamon

Merriam, A. (1988) What does 'Third World' mean? in Norwine, J. and Gonzalez, A. (eds) *The Third World: States of Mind and Being*. London: Unwin-Hyman

Merrick, T. (1986) World population in transition. *Population Bulletin*, **41**(2), 1–51

Messkoub, M. (1992) Deprivation and structural adjustment, in Wuyts, M., Mackintosh, M. and Hewitt, T. (eds) *Development policy and public action*. Oxford: Oxford University Press, 175–98

Middleton, N., O'Keefe, P. and Moyo, S. (1993) *Tears of the crocodile: from Rio to reality in the developing world*. London: Pluto Press

Mijere, N. and Chilivumbo, A. (1987) Rural urban migration and urbanization in Zambia during the colonial and post-colonial periods, in Kaliperi, E. (ed) *Population, Growth and Environmental Degradation in Southern Africa*. New York: Reinner

Miller, D. (1992) The young and the restless in Trinidad: a case of the local and the global in mass consumption, in Silverstone, R. and Hirsch, E. (eds) *Consuming Technology*. London: Routledge, 163–82

Miller, D. (1994) *Modernity: An Ethnographic Approach: Dualism and Mass Consumption in Trinidad*. Oxford: Berg

Milner-Smith, R. and Potter, R.B. (1995) *Public knowledge of attitudes towards the Third World*. CEDAR Research Paper 13, Royal Holloway College, University of London

MoBbrucker, H. (1997) Amerindian migration in Peru and Mexico, in Gugler, J. (ed) *Cities in the Developing World*. Oxford: Oxford University Press, 74–87

Mohan, G. (1996) SAPs and Development in West Africa. *Geography*, **81**(4), 364–8

Momsen, J.H. (1991) *Women and development in the third world*. London: Routledge

Moock, J.L. (ed) (1986) *Understanding Africa's rural households and farming systems*. London: Westview Press

Morse, S. (1995) Biotechnology: a servant of development? in Morse, S. and Stocking, M. (eds) *People and environment*. London: UCL Press, 131–55

Morse, S. and Stocking, M. (eds) (1995) *People and environment*. London: UCL Press

Mortimore, M.J. (1989) *Adapting to drought: farmers, famines and desertification in West Africa*. Cambridge: Cambridge University Press

Mountjoy, A.B. (1976) Urbanization, the squatter and development in the Third World. *Tijdschrift voor Economische en Sociale Geografie*, **67**, 130–7

Mountjoy, A.B. (1980) Worlds without end, *Third World Quarterly*, **2**(4), 753–57

Munslow, B. and Ekoko, F. (1995) Is democracy necessary for sustainable development? *Democratisation*, **2**, 158–78

Munslow, B., Katerere, Y., Ferf, A. and O'Keefe, P. (1988) *The fuelwood trap: a study of the SADCC region*. London: Earthscan

Murray, M. (1995) The value of biodiversity, in Kirkby, J., O'Keefe, P. and Timberlake, L. (eds) *The Earthscan reader in sustainable development*. London: Earthscan, 17–29

Myint, x (1964) *The Economics of Developing Countries*. London: Hutchinson

Myrdal, G. (1957) *Economic Theory and Underdeveloped Areas*. London: Duckworth

New Internationalist, (1995) The Flood of protest, No. 273

Newsweek (1995) The UN turns, 30 October

Noonan, T. (1996) In the rough, *Far Eastern Economic Review*, 25 January, 38–9

Norwine, J. and Gonzalez, A. (1988) Introduction, in Norwine, J. and Gonzalez, A. (eds) *The Third World: States of Mind and Being*. London: Unwin-Hyman, 1–6

Oberai, A.S. (1993) *Population Growth, Employment and Poverty in Third World Mega-Cities: Analytical and Policy Issues*. Basingstoke: Macmillan and New York: St Martin's Press

O'Brien, R. (1991) *Global Financial Integration: the End of Geography*. London: Pinter

O'Connor, A. (1976) Third World or one world. *Area*, **8**, 269–71

O'Connor, A. (1983) *The African City*. London: Hutchinson

O'Connor, A. (1991) *Poverty in Africa: a geographical approach*. London: Belhaven

Ohlsson, L. (ed) (1995) *Hydropolitics*. London: Zed Books

Olthof, W. (1995) Wildlife resources and local development: experiences from Zimbabwe's Campfire programme, in van de Breemer, J.P.M., Drijver, C.A. and Venema, L.B. (eds) *Local resource management in Africa*. Chichester: John Wiley, 111–28

O'Riordan, T. (ed) (1995) *Environmental science for environmental management*. London: Longman

O'Tuathail, G. (1994) Critical geopolitics and development theory: intensifying the dialogue. *Transactions of the Institute of British Geographers, New Series*, **19**, 228–38

Oxfam (1984) *Behind the weather: Lessons to be Learned. Drought and Famine in Ethiopia*. Oxford: Oxfam

Oxfam (1993) *Africa: make or break. Action for recovery*. Oxford: Oxfam

Oxfam (1994) *The Coffee Chain Game*. Oxford: Oxfam

Parnwell, M. (1994) Rural industrialisation and sustainable development in Thailand, *Quarterly Environment Journal*, **2**, 24–29

Parnwell, M. and Turner, S. (1998) Sustaining the unsustainable: city and society in Southeast Asia. *Third World Planning Review*, **20**, 147–164

Parry, M. (1990) *Climate Change and World Agriculture*. London: Earthscan

Patullo, P. (1966) *Last Resort? Tourism in the Caribbean*, London: Mansell and the Latin American Bureau

Pearce, D. (1995) *Blueprint 4: capturing global environmental value*. London: Earthscan

Pearce, F. (1993) How green is your golf? *New Scientist*, 25 September, 30–5

Pearce, F. (1997) The biggest dam in the world, in Owen, L. and Unwin, T. (eds) *Environmental management: readings and case studies*. Oxford: Blackwell, 349–54

Pearson, R. (1992) Gender matters in development, in Allen, T. and Thomas, A. (eds) *Poverty and development in the 1990s*. Oxford: Oxford University Press, 291–313

Pederson, P.O. (1970) Innovation diffusion within and between national urban systems. *Geographical Analysis*, **2**, 203–54

Peet, R. and Watts, M. (eds) (1996) *Liberation Ecologies: Environment, development, social movements*. London: Routledge

Pepper, D. (1996) *Modern Environmentalism: An introduction*. London: Routledge

Perloff, H.S. and Wingo, L. (1961) Natural resource endowment and regional economic growth, in Spengler, J.J. (ed) *Natural Resources and Economic Growth*. Washington DC: Resources for the Future

Pernia, E. (1992) Southeast Asia, in R. Stren (ed) *Sustainable Cities: Urbanization and the Environment in International Perspective*. Oxford: Westview Press, 233–58

Perroux, F. (1950) Economic space: theory and applications. *Quarterly Journal of Economics*, **64**, 89–104

Perroux, F. (1955) Note sur la notion de 'pâle de croissance'. *Économie Appliquée*, **1/2**, 307–20

Phillips, D.R. (1990) *Health and Health Care in the Third World*. Harlow: Longman

Phillips, D.R. and Yeh, G.O. (1990) Foreign investment and trade: impact on spatial structure of the economy, in Cannon, T. and Jenkins, A. (eds) *The geography of contemporary China*. London: Routledge, 224–48

Pinches, M. (1994) Urbanisation in Asia: development, contradiction and conflict, in Jayasuriya, L. and Lee, M. (eds) *Social Dimensions of Development*. Sydney: Paradigm Press

Pinstrup-Andersen, P. (1994) *World food trends and future food security*. Washington DC: International Food Policy Research Institute

Pletsch, C. (1981) The three worlds or the division of social scientific labour 1950–1975. *Comparative Studies in Society and History*, **23**, 565–90

Pleumarom, A. (1992) Course and effect: golf tourism in Thailand. *The Ecologist*, **22**(3), 104–10

Porteous, D. (1995) in Crush, J. (ed) *Power of Development*, London: Routledge

Porter, G. (1995) Scenes from childhood, in Crush, J. (ed) *Power of Development*. London: Routledge, 63–86

Porter, G. (1996) SAPs and road transport deterioration in West Africa. *Geography*, **81**(4), 368–71

Portes, A., Castells, M. and Benton, L. (eds) (1991) *The Informal Economy: Studies in Advanced and Less Developed Countries*. Baltimore MD: Johns Hopkins University Press

Portes, A., Dore-Cabral, C. and Landolt, P. (1997) *The Urban Caribbean: Transition of the New Global Economy*. Baltimore MD: Johns Hopkins University Press

Potter, R.B. (1981) Industrial development and urban planning in Barbados. *Geography*, **66**, 225–8

Potter, R.B. (1983) Tourism and development: the case of Barbados, West Indies. *Geography*, **68**, 46–50

Potter, R.B. (1985) *Urbanisation and Planning in the Third World: Spatial Perceptions and Public Participation*. London: Croom Helm and New York: St Martin's Press

Potter, R.B. (1990) Cities, convergence, divergence and Third World development, in Potter, R.B. and Salau, A.T. (eds) *Cities and Development in the Third World*. London: Mansell

Potter, R.B. (1992a) *Housing Conditions in Barbados: A Geographical Analysis*. Mona, Kingston, Jamaica: Institute of Social and Economic Research, University of the West Indies

Potter, R.B. (1992b) *Urbanisation in the Third World*. Oxford: Oxford University Press

Potter, R.B. (1993a) Basic needs and development in the small island states of the Eastern Caribbean, in Lockhart, D. and Drakakis-Smith, D. (eds) *Small Island Development*. London: Routledge

Potter, R.B. (1993b) Little England and little geography: reflections on Third World teaching and research. *Area*, **25**, 291–4

Potter, R.B. (1993c) Urbanization in the Caribbean and trends of global convergence–divergence. *Geographical Journal*, **159**, 1–21

Potter, R.B. (1994) *Low-Income Housing and the State in the Eastern Caribbean*. Barbados: University of the West Indies Press

Potter, R.B. (1995a) Urbanisation and development in the Caribbean. *Geography*, **80**, 334–41

Potter, R.B. (1995b) Whither the real Barbados? *Caribbean Week*, **7**(4), 64–7

Potter, R.B. (1996) Environmental impacts of urban-industrial development in the tropics: an overview, in Eden, M. and Parry, J.T. (eds) *Land Degradation in the Tropics*, London: Pinter

Potter, R.B. (1997) Third World urbanisation in a global context. *Geography Review*, **10**, 2–6

Potter, R.B. and Conway, D. (eds) (1997) *Self-Help Housing, the Poor and the State in the Caribbean*. Knoxville TN: Tennessee University Press and Barbados: University of the West Indies Press

Potter, R.B. and Dann, G.M.S. (1994) Some observations concerning postmodernity and sustainable development in the Caribbean. *Caribbean Geography*, **5**, 92–107

Potter, R.B. and Dann, G. (1996) Globalization, postmodernity and development in the Commonwealth Caribbean, in Yeung, Yue-man (ed) *Global Change and the Commonwealth*. Hong Kong: Hong Kong Institute of Asia-Pacific Studies, Chinese University of Hong Kong, 103–29

Potter, R.B. and Lloyd-Evans, S. (1998) *The City in the Developing World*. Harlow: Longman

Potter, R.B. and Unwin, T. (1988) Developing areas research in British geography. *Area*, **20**, 121–6

Potter, R.B. and Unwin, T. (eds) (1992) *Teaching the Geography of Developing Areas.* Monograph 7, Developing Areas Research Group of the Institute of British Geographers

Potter, R.B. and Unwin, T. (1995) Urban–rural interaction: physical form and political process in the Third World. *Cities*, **12**, 67–73

Potter, R.B. and Welch, B. (1996) Indigenization and development in the Caribbean. *Caribbean Week*, **8**, 13–4

Potts, D. (1995) Shall we go home? Increasing urban poverty in African cities and migration processes. *Geographical Journal*, **161**(3), 245–64

Prebisch, R. (1950) *The Economic Development of Latin America.* New York: United Nations

Pred, A. (1977) *City-Systems in Advanced Economies.* London: Hutchinson

Pred, A.R. (1973) The growth and development of systems of cities in advanced economies, in Pred, A. and Törnqvist, G. (eds) *Systems of Cities and Information Flows: Two Essays.* Lund, 9–82

Preston, D. (1987) Population mobility and the creation of new landscapes, in Preston, D. (ed) *Latin American development: geographical perspectives.* London: Longman, 229–59

Preston, P.W. (1985): *New Trends in Development Theory.* London: Routledge

Preston, P.W. (1987) *Making Sense of Development: An Introduction to Classical and Contemporary Theories of Development and their Application to Southeast Asia.* London: Routledge

Preston, P.W. (1996) *Development Theory: An Introduction.* Oxford: Blackwell

Prothero, R.M. (1959) *Migrant labour from Sokoto Province, Northern Nigeria.* Kaduna: Government Printer, 46

Prothero, R.M. (1994) Forced movements of population and health hazards in tropical Africa. *International Journal of Epidemiology*, **23**(4), 657–64

Prothero, R.M. (1996) Migration and AIDS in West Africa. *Geography*, **81**(4), 374–7

Pryer, J. (1987) Production and reproduction of malnutrition in an urban slum in Khulna, Bangladesh, in Momsen, J.H. and Townsend, J. (eds) *Geography of gender in the Third World.* London: Hutchinson, 131–49

Pugh, C. (ed) (1996) *Sustainability, the Environment and Urbanization.* London: Earthscan

Pye-Smith, C. and Feyerarbend, G.B. (1995) What next? in Kirkby, J., O'Keefe, P. and Timberlake, L. (eds) *The Earthscan reader in sustainable development.* London: Earthscan, 303–9

Raath, J. (1997) Mugabe wants aid to seize white land. *The Times*, 20 October

Rakodi, C. (1995) Poverty lines or household strategies? *Habitat International*, **19**(4), 407–26

Reading, A.J., Thompson, R.D. and Millington, A.C. (1995) *Humid Tropical Environments.* Oxford: Blackwell

Redclift, M. (1987) *Sustainable Development: Exploring the Contradictions.* London: Methuen

Redclift, M. (1997) Sustainable development: needs, values and rights, in Owen, L. and Unwin, T. (eds) *Environmental Management: Readings and Case Studies.* Oxford: Blackwell, 438–50

Reed, D. (ed) (1996) *Structural adjustment: the environment and sustainable development.* London: Earthscan

Rees, J. (1990) *Natural resources: allocation, economics and policy*, 2nd edn. London: Methuen

Renaud, B. (1981) *National Urbanization Policy in Developing Countries.* Oxford: Oxford University Press for the World Bank

Rich, B. (1994) *Mortgaging the earth: the World Bank, environmental impoverishment and the crisis of development.* London: Earthscan

Richards, P. (1985) *Indigenous Agricultural Revolution.* London: Hutchinson

Richards, P. (1986) *Coping with Hunger: Hazard and experiment in an African rice-farming system.* London: George Allen & Unwin

Richards, P. and Thomson, A. (1984) *Basic Needs and the Urban Poor.* London: Croom Helm

Richardson, H.W. (1973) *The Economics of Urban Size.* Farnborough: Saxon House

Richardson, H.W. (1976) The argument for very large cities reconsidered: a comment. *Urban Studies*, **13**, 307–10

Richardson, H.W. (1977) City size and national spatial strategies in developing countries. World Bank Staff Working Paper 252

Richardson, H.W. (1980) Polarization reversal in developing countries. *Papers of the Regional Science Association*, **45**, 67–85

Richardson, H.W. (1981) National urban development strategies in developing countries. *Urban Studies*, **18**, 267–83

Riddell, J.B. (1970) *The Spatial Dynamics of Modernization in Sierra Leone: Structure, Diffusion and Response.* Evanston IL: Northwestern University Press

Riddell, J.B. (1978) The migration to the cities of West Africa: some policy considerations. *Journal of Modern African Studies*, **16**(2), 241–60

Rigg, J. (1996) *Southeast Asia.* London: Routledge

Righter, R. (1995) *Utopia Lost: the United Nations and the World Order.* New York: Twentieth Century Fund Press

Riley, S. (1988) Structural adjustment and the new urban poor: the case of Freetown. Paper presented at the Workshop on the New Urban Poor in Africa, School of Oriental and African Studies, London, May 1988

Robins, K. (1989) Global times, *Marxism Today*, December 1989, 20–27

Robins, K. (1995) The new spaces of global media, in Knox, P.C. and Taylor, P.J. (eds) *World Cities in a World-System.* Cambridge: Cambridge University Press, 248–62

Robson, B.T. (1973) *Urban Growth: An Approach.* London: Methuen

Robson, E. (1996) Working girls and boys: children's contributions to household survival in West Africa. *Geography*, **81**(4), 43–7

Rojas, E. (1989) Human settlements of the Eastern Caribbean: development problems and policy options. *Cities*, **6**, 243–58

Rojas, E. (1995) Commentary: government–market interactions in urban development policy. *Cities*, **12**, 399–400

Rostow, W.W. (1960) *The Stages of Economic Growth: a non-communist manifesto.* Cambridge: Cambridge University Press

Rowley, C. (1978) *The Destruction of Aboriginal Society.* Ringwood: Penguin

Ruthenberg, H. (1976) *Farming systems in the tropics*, 2nd edn. Oxford: Clarendon Press

Sachs, W. (1992) *The Development Dictionary.* London: Zed Books

Safier, M. (1969) Towards the definition of patterns in the distribution of economic development over East Africa. *East African Geographical Review*, **7**, 1–13

Sahnoun, M. (1994) Flashlights over Mogadishu. *New Internationalist*, **262**, 9–11

Said, E. (1979) *Orientalism.* New York: Village Books

Said, E. (1993) *Culture and Imperialism.* London: Chatto

Sandford, S. (1983) *Management of pastoral development in the Third World.* Chichester: John Wiley

Santos, M. (1979) *The Shared Space: the Two Circuits of the Urban Economy in Underdeveloped Countries.* London: Methuen

Satterthwaite, D. (1997) Sustainable cities or cities that contribute to sustainable development. *Urban Studies*, **35**

SCF (Save the Children Fund) (1995) *Towards a children's agenda: new challenges for social development.* London: SCF

Schneider, F. and Frey, B. (1985) Economic and political determinants of foreign direct investment. *World Development*, **13**(2), 167–75

Schultz, T.W. (1953) *The Economic Organization of Agriculture.* New York: McGraw-Hill

Schumacher, E.F. (1974) *Small is beautiful.* London: Abacus

Schumpeter, J.A. (1911) *Die Theorie des Wirtschaftlichen Entwicklung.* Leipzig

Schumpeter, J.A. (1934) *The Theory of Economic Development.* Cambridge, MA: Harvard University Press

Schuurman, F. (ed) (1993) *Beyond the Impasse: New Directions in Development Theory.* London: Zed

Scoones, I. (1995) Policies for pastoralists: new directions for pastoral development in Africa, in Binns, T. (ed) *People and environment in Africa.* Chichester: John Wiley, 23–30

Scoones, I. and Thompson, J. (1994) *Beyond farmer first.* London: Intermediate Technology Productions

Scott, J.C. (1976) *The moral economy of the peasant: rebellion and subsistence in south-east Asia.* New Haven CT: Yale University Press

Secrett, C. (1986) The environmental impact of transmigration. *The Ecologist*, **16**(2/3), 77–89

Seers, D. (1969) The meaning of development. *International Development Review*, **11**(4), 2–6

Seers, D. (1972) What are we trying to measure? *Journal of Development Studies*, **8**(3), 21–36

Seers, D. (1979) The new meaning of development, in Lehmann, D. (ed) *Development Theory: Four Critical Studies.* London: Frank Cass, 25–30

Shah, P. (1994) Participatory watershed management in India: the experience of the Aga Khan Rural Support Programme, in Scoones, I. and Thompson, J. (eds) *Beyond Farmer First: rural people's knowledge, agricultural research and extension practice.* London: Intermediate Technology Publications

Shankland, A. (1991) The devil's design. *New Internationalist*, **219**, 11–13

Shariff, I. (1987) Agricultural development and land tenure in India. *Land Use Policy*

Sharp, R. (1992) Organising for change: people-power and the role of institutions, in Holmberg, J. (ed) *Policies for a small planet*. London: Earthscan/IIED, 39–65

Shaw, J. and Clay, E. (eds) (1993) *World food aid: experiences of recipients and donors*. London: James Currey

Shibusawa, M., Ahmad, Z.H. and Bridges, B. (1992) *Pacific Asia in the 1990s*. London: Routledge

Shipton, P. and Goteen, M. (1992) Understanding African land-holding: power, wealth and meaning. *Africa*, **62**(3), 307–25

Sidaway, J.D. (1990) Post-Fordism, post-modernity and the Third World. *Area*, **22**, 301–3

Siddle, D. and Swindell, K. (1990) *Rural change in tropical Africa*. Oxford: Blackwell

Silvers, J. (1995) Death of a slave. *Sunday Times*, 10 October, 36–41

Simon, D. (1992a) *Cities, Capital and Development: African Cities in the World Economy*. London: Belhaven

Simon, D. (1992b) Conceptualizing small towns in African development, in Baker, J. and Pedersen, P.O. (eds) *The Rural–Urban Interface in Africa*. Uppsala: Nordic Institute for African Studies, 29–50

Simon, D. (1993) The world city hypothesis: reflections from the periphery. CEDAR Research Paper 7, Royal Holloway College, University of London

Simon, J.L. (1981) *The ultimate resource*. London: Martin Robertson

Singer, H. (1980) The Brandt Report: a north-western point of view. *Third World Quarterly*, **2**(4), 694–700

Slater, D. (1992a) On the borders of social theory: learning from other regions. *Environment and Planning D*, **10**, 307–27

Slater, D. (1992b) Theories of development and politics of the post-modern: exploring a border zone. *Development and Change*, **23**, 283–319

Slater, D. (1993) The geopolitical imagination and the enframing of development theory. *Transactions of the Institute of British Geographers, New Series*, **18**, 419–37

Smith, D.W. (1994) On professional responsibility to distant others. *Area*, **26**, 359–67

Smith, D.W. (1998) Urban food systems and the poor in developing countries, *Transactions of the Institute of British Geographers*, **23**, 207–19

So, Chin-Hung (1997) Economic development, state control and labour migration of women in China. Unpublished PhD thesis, University of Sussex, Brighton

Soja, E.W. (1968) *The Geography of Modernization in Kenya: A Spatial Analysis of Social, Economic and Political Change*. Syracuse NY: Syracuse University Press

Soja, E.W. (1974) The geography of modernization: paths, patterns, and processes of spatial change in developing countries, in Bruner, R. and Brewer, G. (eds) *A Policy Approach to the Study of Political Development and Change*. New York: Free Press

Soja, E.W. (1989) *Postmodern Geographies: The Reassertion of Space in Critical Social Theory*. London: Verso

Soliman, A.M. (1996) Legitimising informal housing: accommodating low income groups in Alexandria, Egypt. *Environment and Urbanisation*, **1**, 183–194

Soussan, J. (1988) *Primary resources in the third world*. London: Routledge

Sparr, P. (1994) *Mortgaging women's lives: feminist critiques of structural adjustment*. London: Zed Books

Stewart, C. (1995) One more river to cross. *New Internationalist*, **273**, 16–17

Stock, R. (1995) *Africa south of the Sahara: a geographical interpretation*. New York: Guildford Press

Stocking, M. (1987) Measuring land degradation, in Blaikie, P. and Brookfield, H. (eds) *Land Degradation and Society*. London: Methuen, 49–64

Stocking, M. (1995) Soil erosion and land degradation, in O'Riordan, T. (ed) *Environmental science for environmental management*. London: Longman, 223–43

Stöhr, W.B. (1981) Development from below: the bottom-up and periphery-inward development paradigm, in Stöhr, W.B. and Taylor, D.R.F. (eds) *Development from Above or Below?* Chichester: John Wiley, 39–72

Stöhr, W.B. and Taylor, D.R.F. (1981) *Development from Above or Below? The Dialectics of Regional Planning in Developing Countries*. Chichester: John Wiley

Streeten, P. (1995) *Thinking About Development*. Cambridge: Cambridge University Press

Stren, R., White, R. and Whitney, J. (eds) (1992) *Sustainable Cities: Urbanization and the Environment in International Perspective.* Oxford: Westview Press

Stycos, J.M. (1971) Family planning and American goals, in Chaplin, D. (ed) *Population policies and growth in Latin America.* Lexington KY: Heath, 111–31

Taaffe, E.J., Morrill, R.L. and Gould, P.R. (1963) Transport expansion in underdeveloped countries: a comparative analysis. *Geographical Review,* **53,** 503–29

Tasker, R. (1995) Tee Masters, *Far Eastern Economic Review,* 5 January

Tata, R. and Schultz, R. (1988) World variations in human welfare: a new index of development status. *Annals of the Association of American Geographers,* **78**(4), 580–92

Taylor, D.R. and Mackenzie, F. (eds) (1992) *Development from within: survival in rural Africa.* London: Routledge

Taylor, P. (1985) *Political Geography.* London: Longman

Taylor, P.J. (1986) The world-systems project, in Johnston, R.J. and Taylor, P.J. (eds) *A World in Crisis? Geographical Perspectives.* Oxford: Basil Blackwell, 333–54

Teo, P. and Ooi, G.L. (1996) Ethnic differences and public policy in Singapore, in Dwyer, D.J. and Drakakis-Smith, D. (eds) *Ethnicity and Geography.* London: John Wiley, 249–70

Thomas, A. (1992) Non-governmental organisations and the limits to empowerment, in Wuyts, M., Mackintosh, M. and Hewitt, T. (eds) *Development policy and public action.* Oxford: Oxford University Press, 117–46

Thomas, C.Y. (1989) *The Poor and the Powerless: Economic Policy and Change in the Caribbean.* London: Latin American Bureau

Thomas, D.H.L. (1996) Fisheries, tenure and mobility in a West African floodplain. *Geography,* **81**(4), 35–40

Thomas, D.S.G. (1993) Storm in a teacup? Understanding desertification. *Geographical Journal,* **159**(3), 318–31

Thomas, G.A. (1991) The gentrification of paradise: St John's, Antigua. *Urban Geography,* **12,** 469–87

Thompson, M. and Warburton, M. (1985) Uncertainty on a Himalayan scale. *Mountain Research and Development,* **5,** 115–35

Thompson, R.D. (1992) The changing atmosphere and its impact on Planet Earth, in Mannion, A.M. and Bowlby, S.R. (eds) *Environmental issues in the 1990s.* London: John Wiley, 61–78

Thrift, N. and Forbes, D. (1986) *The Price of War: Urbanisation in Vietnam 1954–1986,* London: Allen & Unwin

Tiffen, M. and Mortimore, M. (1990) *Theory and practice in plantation agriculture: an economic review.* London: Overseas Development Institute

Tiffen, M., Mortimore, M.J. and Gichuki, F. (1994) *More People, Less Erosion: Environmental Recovery in Kenya.* Chichester: John Wiley

Todaro, M. (1994) *Economic Development.* Harlow: Longman

Todd, H. (ed) (1996) *Cloning Grameen Bank: replicating a poverty reduction model in India, Nepal and Vietnam.* London: Intermediate Technology Publications

Toffler, A. (1970) *Future Shock.* London: Bodley Head

Tordoff, W. (1992) The impact of ideology or development in the Third World. *Journal of International Development,* **4**(1), 41–53

Toye, J. (1987) *Dilemmas of Development.* Oxford: Blackwell

Traisawasdichai, M. (1995) Chasing the little white ball. *New Internationalist,* **263,** 16–17

ul-Haq, M. (1994) The new deal. *New Internationalist,* **262,** 20–23

UNDP (United Nations Development Programme) (1991) *Cities, People and Poverty.* New York: UNDP

UNDP (United Nations Development Programme) (1993) *Human Development Report, 1993.* Oxford: Oxford University Press

UNDP (United Nations Development Programme) (1994) *Human Development Report, 1994.* Oxford: Oxford University Press

UNDP (United Nations Development Programme) (1996) *Human Development Report, 1996.* Oxford: Oxford University Press

UNDP (United Nations Development Programme) (1997) *Human Development Report 1997.* Oxford: Oxford University Press

UNEP (United Nations Environment Programme) (1995) *Global biodiversity assessment.* Cambridge: Cambridge University Press

UNEP (United Nations Environment Programme) (1997) *World Atlas of Desertification,* 2nd edn. London: John Wiley

UNEP/WHO (United Nations Environment Programme/World Health Organisation) (1993) *City air quality trends*, Vol. 2, Nairobi: UNEP

UNICEF (United National Children's Fund) (1997) *The state of the world's children*. Oxford: Oxford University Press

United Nations (1988) *World Population Trends and Policies, 1987: Monitoring Report*. New York: United Nations Department of International Economic and Social Affairs

United Nations (1989) *Prospects for World Urbanization 1988*. New York: United Nations

United Nations (1993) *The Global Partnership for Environment and Development: A Guide to Agenda 21*. New York: United Nations

United Nations Centre for Human Settlements (Habitat) (1996) *An Urbanizing World: Global Report on Human Settlements, 1996*. Oxford: Oxford University Press

United States Department of Energy (1994) *Energy use and carbon emissions: some international comparisons*. Washington DC: Energy Information Administration

UNRISD (United Nations Research Institute for Social Development) (1995) *States of disarray: the social effects of globalisation*. Geneva: UNRISD

Unwin, T. and Potter, R.B. (1992) Undergraduate and postgraduate teaching on the geography of the Third World. *Area*, **24**, 56–62

Urban Foundation (1993) *Managing Urban Poverty*. Johannesburg: Urban Foundation

Urry, J. (1990) *The Tourist Gaze*. London: Sage

US Bureau of the Census (1994) *Trends and patterns of HIV/AIDS infection in selected developing countries*. Country Profiles, Research Note 15, Health Studies Branch. Washington DC: Center for International Research

Vance, J.E. (1970) *The Merchant's World: The Geography of Wholesaling*. Englewood Cliffs NJ: Prentice Hall

van der Gaag, N. (1997) Gene dream. *New Internationalist*, **293**, 7–10

Vapnarsky, C.A. (1969) On rank-size distributions of cities: an ecological approach. *Economic Development and Cultural Change*, **17**, 584–95

Vesiland, P.J. (1993) Water: the Middle East's critical resource. *National Geographic*, **183**(5), 38–71

Vivian, J.M. (1992) Foundations for sustainable development: participation, empowerment and local resource management, in Ghai, D. and Vivian, J.M. (eds) *Grassroots environmental action: peoples' participation in sustainable development*. London: Routledge, 50–80

von Moltke (1994) The World Trade Organisation: its implications for sustainable development. *Journal of Environment and Development*, **3**(1), 43–57

Wallerstein, I. (1974) *The Modern World System I*. New York: Academic Press

Wallerstein, I. (1979) *The Capitalist World Economy*. Cambridge: Cambridge University Press

Wallerstein, I. (1980) *The Modern World System II*. New York: Academic Press

Ward, P. and Macoloo, C. (1992) Articulation theory and self-help housing practice in the 1990s. *International Journal of Urban and Regional Research*, **16**, 60–80

Watkins, K. (1995) *The Oxfam Poverty Report*. Oxford: Oxfam

Watts, M. (1984) The demise of the moral economy: food and famine in a Sudano-Sahelian region in historical perspective, in Scott, E. (ed) *Life before the drought*. Boston MA: Allen & Unwin, 124–48

Watts, M. (1996a) Development III: the global agro-food system and late twentieth century development. *Progress in Human Geography*, **20**(2), 230–45

Watts, M. (1996b) Development in the global agrofood system and late twentieth-century development (or Kautsky reduxe). *Progress in Human Geography*, **20**(2), 230–45

Watts, M. and McCarthy, J. (1997) Nature as artifice, nature as artefact: development, environment and modernity in the late twentieth century. Paper presented in the Lectures in Human Geography Series, University of St Andrews

WCED (World Commission on Environment and Development) (1987) *Our common future*. Oxford: Oxford University Press

Weidelt, H.J. (1993) Agroforestry systems in the tropics – recent developments and results of research. *Applied Geography and Development*, **41**, 39–50

Wellard, K. and Copestake, J. (1993) *Non-governmental Organizations and the State in Africa*. London: Routledge

Werksman, J. (1995) Greening Bretton Woods, in Kirkby, J., O'Keefe, P. and Timberlake, L. (eds) *The Earthscan reader in sustainable development*. London: Earthscan, 274–87

Werksman, J. (ed) (1996) *Greening international institutions*. London: Earthscan

Wheat, S. (1993) Playing around with nature. *Geographical Magazine*, **LXV**(8), 10–14

WHO (World Health Organisation) (1994) *The current global situation of the HIV/AIDS pandemic*. Geneva: WHO

Williams, I. (1995) *The UN for beginners*. London: Writers and Readers

Williams, M. (1994) Making golf greener. *Far Eastern Economic Review*, May, 40–41

Williamson, J.G. (1965) Regional inequality and the process of national development: a description of the patterns. *Economic Development and Cultural Change*, **13**, 3–45

Wilson, F. (1994) Reflections on the present predicament of the Mexican garment industry, in Pedersen, P. (ed) *Flexible Specialization*. London: IT Publications, 147–58

Wolfe-Phillips, L. (1987) Why Third World – origins, definitions and usage. *Third World Quarterly*, **9**(4), 1311–9

Wolpe, H. (1975) The theory of internal colonialism, in Oxaal, J., Barnett, T. and Booth, D. (eds) *Beyond the Sociology of Development*. London: Routledge & Kegan Paul, 229–52

Women and Geography Study Group (1997) *Feminist geographies: explorations in diversity and difference*. London: Longman

World Bank (1988) *World Development Report*. Washington DC: World Bank

World Bank (1989) *Sub-Saharan Africa: from crisis to sustainable growth: a long term perspective study*. Washington DC: World Bank

World Bank (1991) *Urban Policy and Economic Development: An Agenda for the 1990s*. Washington DC: World Bank

World Bank (1992) *World Development Report*. Washington DC: World Bank

World Bank (1993) *East Asian Miracle*. Washington DC: World Bank

World Bank (1994a) *World Bank and the Environment: Fiscal 1993*. Washington DC: World Bank

World Bank (1994b) *World Development Report, 1994*. Oxford: Oxford University Press

World Bank (1995) *World Development Report, 1995*. Oxford: Oxford University Press

World Bank (1996a) *A Review of World Bank Experience in Irrigation*. Washington DC: World Bank Operations Evaluation Department

World Bank (1996b) *World Development Report, 1996*. Oxford: Oxford University Press

World Bank (1997) *World Development Report: the state in a changing world*. Oxford: Oxford University Press

World Resources Institute (1990) *World Resources 1990–1991*. Oxford: Oxford University Press

World Resources Institute (1992) *World Resources 1992–1993*. Oxford: Oxford University Press

World Resources Institute (1994) *World Resources 1994–1995*. Oxford: Oxford University Press

World Resources Institute (1996) *World Resources 1996–1997*. Oxford: Oxford University Press

Worsley, P. (1964) *The Third World*. London: Weidenfeld & Nicolson

Worsley, P. (1979) How many worlds? *Third World Quarterly*, **1**(2), 100–108

Wratten, E. (1995) Conceptualizing urban poverty. *Environment and Urbanization*, **7**(1), 11–37

Wuyts, M., Mackintosh, M. and Hewitt, T. (eds) (1992) *Development policy and public action*. Oxford: Oxford University Press

Yeh, A.G.O. and Wu, F.L. (1995) Internal structure of Chinese cities in the midst of economic reform. *Urban Geography*, **16**(6), 521–54

Yeung, Y.-M. (1995) Commentary: urbanization and the NPE: an Asia-Pacific perspective. *Cities*, **12**, 409–11

Young, E.M. (1996) *World Hunger*. London: Routledge

Zimmerman, E.W. (1951) *World resources and industries*. New York: Harper & Row

Index